"THEY SHALL BEAT THEIR SWORDS INTO PLOWSHARES AND THEIR SPEARS INTO PRUNING-HOOKS"

THE LEOPARD SHALL LIE DOWN WITH THE KID

THE FARM AND THE FIRESIDE

THE WOLF ALSO AND THE LAMB SHALL DWELL TOGETHER

THE LION SHALL EAT STRAW LIKE THE OX.

HERRICK

JOCELYN SC

TO IMPROVE THE SOIL AND THE MIND.

THE FARM AND THE FIRESIDE;

OR

THE ROMANCE OF AGRICULTURE.

BEING

HALF HOUR SKETCHES

OF

LIFE IN THE COUNTRY.

BY THE REV. JOHN L. BLAKE, D. D.,

AUTHOR OF FAMILY ENCYCLOPEDIA; GENERAL BIOGRAPHICAL DICTIONARY; FARMER'S EVERY-DAY BOOK; AGRICULTURE FOR SCHOOLS; AND THE FARMER AT HOME.

AUBURN:
ALDEN, BEARDSLEY & COMPANY.
ROCHESTER:
WANZER, BEARDSLEY & COMPANY.

1852.

J. P. JONES & Co., STEREOTYPERS,
183 WILLIAM-STREET.

INTRODUCTION.

WHY did the gold of California remain untouched towards six thousand years, unless it is admitted that King Solomon knew of it, as the gold of Ophir, when building his memorable temple? It is simply because all the world, the Yankees as well as others, did not know it was there. Supposing some visionary scholar had written a book, stating that probably in that region this precious metal existed in profusion. Would such a book have put the whole creation agog in search of it? But, as soon as a few straggling soldiers and sailors from the Mexican war discovered and gathered up by handfuls the shining dust and the solid masses, then there was from every part of the earth, not excepting the Empire of the Sun, an incessant scramble to share in it. This is all consistent, because the subject now became a matter of fact. Instead of theories and moonshine dreams, here was exhibited to the human senses—vision and feeling—the very thing in all ages and in all countries so much desired. Ordinarily, sensible men must know a thing exists and can be had, before they embark in a stampede for its attainment, leaving every thing else in neglect.

Why was it left to Sir Richard Arkwright to revolutionize the labors of the fireside, by applying machinery to the production of those fabrics in human apparel and domestic economy, hitherto made by hand, thus resolving large portions of our race into new combinations for all coming time, and leaving his heirs rich like princes? Why was this thing not done before? Why were those tedious processes continued down through the days of our grandmothers? We have had no new races of men in the world. We all have a common pedigree up to creation's dawn, with the same natural capabilities that were possessed thousands of years ago. Or, why was it left to John Jacob Astor, in our own time, as it were, to roll himself in the furs of the Rocky Mountains, like the silk-worm in its cocoon, and then die with his twenty millions of dollars? Why did men not do the same thing before him? Furs were as plenty in the days of the Plymouth Pilgrims and the Jamestown colony—nay, more—than at any subsequent period. The reason is simply this. Age after age was dreaming about something else. There was either a deathlike slumber of the human intellect, or all aspirations were for renown in the mitre or the crown—in feudalism or the crusades—or their equally ignoble concomitants. The thing was never thought of till these men proved it could be done—till they placed the boon in the bright light of heaven, and before a gazing, wondering world.

And why was it left to Professor Morse and his co-laborers to convert the lightning's flash into one of the most beneficial and useful agents of nature? annihilating space; regarding mountains and rivers as fictions; becoming a messenger of joy or grief from one side of the continent to the other with the rapidity of thought; at one moment standing by the bedside of the dying or in view of the hearse at New Orleans, and the next moment at Boston, paralyzing the relatives with the agonizing intelligence; or, on the other hand, at one moment in Charleston, near the banquet table, or some other occasion for joy, and at the next moment at Cincinnati or St. Louis, or Chicago, revealing to kindred spirits the sounds of exhileration and conviviality! Why was it left to him or his co-laborers to do this? Why was it left to Professor Morse in our own day to become such a benefactor to mankind, and at the same time, it may be, to enrich himself thereby, annually a million of dollars? The reason is the same as in the other cases; because no one thought it could be done. But when it was done, how the way is crowded by competitors! How many now seek the golden prizes which glitter in their imagination!

Let us have the moral from the above prominent cases. It is believed there is an important moral from them that can be applied to agriculture and rural economy. The opinion is very general among farmers, that agriculture is a degrading occupation, and that money cannot be made from it. The exceptions are so few and are connected with circumstances so dissimilar to the general circumstances of the laboring yeomanry, that these exceptions are comparatively without influence. These masses have been led, and in many cases taught to believe that such is the fact. They have never realized that there is tangible evidence to the contrary. They have never witnessed ocular demonstration that agriculture is not a degrading occupation, and that persons generally from it become rich, or what is better, derive as much pecuniary interest from it, as from other labors. Demonstrations to this effect, to be availing, must be manifest to the human senses, as in the cases to which we have adverted. There is to be nothing equivocal about them. They are to be susceptible of no evasion. They are to be invulnerable to argument, or wit, or sarcasm.

Now we hazard the assertion, that if an array of facts in agriculture and rural economy can be well spread before those engaged in these pursuits, every way analogous to those mentioned of California gold, of manufacturing by machinery, of the fur trade, and of magnetic communication, the result will be much the same with the common sense of the farmer that it has been in relation to the latter on the community at large. The question then arises, are such facts in existence? If in existence, can they be found? If found, can they be so arrayed as to make the desired impression? We know they exist; we think enough of them can be found; and we have hope they can be so arrayed, as to be effective to the end in view. Our present effort is to do this, and if we could labor with our pen, as we can imagine how it can be done, or if we wielded a pen as apt and as potent as that of some of our brethren, we know it would be done.

We have many excellent books on agricultural chemistry and the kindred branches of rural knowledge; no matter if as many again, if they would be read and studied. We have annually a score or two of annual addresses on these subjects; some of which should place their authors in the highest grade of mental accomplishment; they are learned and eloquent. But is it not a fact, that upon the agricultural community at large, these books and these addresses exert about as much influence as did the speculations respecting a western continent on the masses of mankind, before the discovery of Columbus? Common people do not want to go through a course of study before engaging in any new labor; they want something that can be seen by their eyes, and moved about by their hands. That they understand without effort. That commends a thing to their consideration. Thus an hundred boys will learn to swim, by plunging into the water one after another, the untaught watching the movements of the skilful, while a single one will learn to do it by studying the rules for swimming by Franklin and others, in books on the subject. It is much so with many parts of good farming. One practical farmer, in a town, from having been penniless, and has become rich, by the regular profits of agriculture; one that actually makes a hundred dollars annually from his poultry—or, from half a dozen cows—or from the yield of his garden—or in manure from his piggery—which everybody can see, and which all his neighbors do see and know to be a fact—this farmer will do more towards stimulating all his rural brethren of the town to efforts that will end in the same results, than thousands of didactic treatises on the subject. Books of this kind are excellent in their places; but ordinarily, farmers will not study them till they become stimulated by such examples as are here indicated. It is not in our power to teach all the world by examples of the sort under their own eye. We hence collect a few of them to be read in the following work; and we trust they will not be without influence in leading all our readers to believe that all may make agriculture productive, and as not being a degrading occupation.

All instances of a man's becoming rich from farming alone, and of his becoming eminent for intelligence when laboring on his farm with his own hands; all instances of improvement in stock so as to be the admiration of a whole county; all instances of neatness and elegance in and about the premises upon a farm; all instances of restoring worn out lands to great fertility; all instances of extraordinary crops on account of judicious tillage; all instances of great profit from poultry, the dairy, or fruit; all instances of well ordered households, combining personal accomplishments and well chosen and well regulated furniture from the profits of the farm; all instances of associated taste and frugality, whether in fences, court-yards, shrubbery or flowers; all instances of pleasing natural scenery upon the farm, or in view of the mansion; and, above all, every labor-saving implement for the work on the farm, in the kitchen, or in the dairy department, may be termed the gems of rural life, and being described and written in books are the gems of rural literature. If they do not deserve that name what does deserve it? Compared with them, what are the gems that glitter and dazzle the eye, in the ball-room? Compared with them, what are the gems that decorate the crowns of royalty? These may be gems scarcely noticed by the masses;

neglected because they have received no polish, as precious stones in their native beds are unnoticed and unadmired, till passing through the hands of the lapidary. Facts of the character named, according to Webster, may be called the "Romance of Agriculture." We hope it will be our fortune to be successful in thus rendering them worthy of occupying attention at the fireside of the farm mansion.

We believe, the fireside of the farmer, with his family about him, is the proper place for contemplating the elements and productions of the farm. Here, with the aid of experience, and agricultural books, and agricultural papers, the members of the family may literally every winter evening make an agricultural club; here they may amuse and enlighten each other on every matter relating to the staple interests to which they are allied, and on which they subsist. Here too, they may make themselves familiar, not only with the literature of rural economy, but all current literature—with history—with biography—with poetry—with music—and with whatever else may come within the range of their taste and fancy.

It is known and considered, that farmers have or should have wives like other men; that they may have sons rising to manhood and daughters budding and swelling into womanhood. Neither of them is supposed to be made up of cold abstractions, but all have social affinities, requiring a mental element corresponding to the development and nourishment of these affinities. Far be it from us to stifle or starve these affinities. They should live under a genial sun, be fed with refreshing showers, and then receive the care of a wise social culturist. A farm and a farm house without matrimonial influences would be cheerless like an Arabian desert. Without the mental and physical tendencies that lead to wedlock, youth would be less interesting than they now are. Accordingly, in preparing a work for the farmer's fireside, for the young as well as the old—for one sex as well as for the other—for each supposable juncture in the family organization, we have collected some gems of the upheavings of the young bosom, as well as of the soil; of the flowers that spring up in the heart, as well as those that cluster in the garden and about the door-side. If these social germs are permitted to shoot up in their native glades, they will resemble the daisy and the rose, that develope not half their inherent elements of beauty and fragrance, till transplanted to the flower garden; and, the asparagus and other rich esculents, that, till recieving the skilful supervision of the gardener, were small austere plants, unfit for use, instead of being the luscious components of a savory dinner, as they now are. It is the especial duty of the mother in her own house, and of all guardians of mental culture to watch over these uprising elements of the human bósom. Here is a garden to be enriched, that may yield flowers and balsams for the adornment and vigor of life, and, even for the glory of the world.

It is left for the reader to judge how far the author has judiciously adapted his present labor to a social exigency, for which no adequate provision had been made. If any merit belong to him, it is in the discovery of the exigency, rather than in the adequacy of his ability to administer to it. And although this labor may have been impaired by the limited time allowed for it, and by an unusual concurrent pressure of other duties, it is, nevertheless, hoped, that it has some little claim, at least, to the favorable consideration of those, for whom it is designed.

"View them near
At home, where all their worth and power is placed;
And there their hospitable fires burn clear,
And there the lowest farm-house hearth is graced
With manly hearts in piety sincere,
Faithful in love, in honor stern and chaste,
In friendship warm and true, in danger brave,
Beloved in life and sainted in the grave."

AGRICULTURAL PARADOX.

It is verily a great paradox, that agriculture should ever be held in low estimation. Such, nevertheless, is the fact. It is no uncommon thing, that we hear that even those who are engaged in it, and are dependant on it for a living, express a great abhorrence of it. Did we not witness the ridiculous absurdity of such conduct, we could not believe it true; for it is in opposition to the clearest evidence on which any hypothesis can be predicated. Who may not see with his own eyes, that on the products of agriculture, the entire life of the animal world—of man and beast—is sustained; and, that these products are the very elements of nearly every kind of business in the whole range of society? Were it not for them the animal kingdom would be blotted out of existence; and, the world itself would become one wide field of solitude and desolation. Yet, it often happens, that persons in other departments of labor, though dependant therein on agriculture, speak of it in derision and with assumed contempt; and, that farmers themselves seem to feel ashamed that they are farmers.

Let us look into the subject a little further, both in order to find the origin of such false conceptions, and to make their absurdity appear so palpable and ridiculous, that they may be discarded. Who does not know, that when the harvest is abundant, manufactures increase, and the whole country is prosperous; that, from one extremity of the land to the other there is an increasing hum of business, and on every countenance a glow of animation and joy. But, if Heaven, for a single season, frown upon the earth, with drawing its rain or its sunshine, or sending mildew and blight, and all this round of prosperity is stopped—machinery becomes motionless, vessels are laid up on their moorings, the efforts of genius are paralyzed, and the distortions of want and despair fill the places of departed plenty and gladness. It is but a few years since we saw famine, disease, and death spread over, with frightful visage, one entire realm, only in consequence of the failure of a single esculent. Nor were the effects resulting from the failure of the potato crop confined to one country. Heavy bankruptcies were occasioned by it in other countries; and, no small portion of the Christian world was brought to a solemn pause in its career of enterprise and business, and to engage in ministrations of charity. So important to society are the products of agriculture! So interwoven are they with all the interests of life!

It might be supposed, that a thing so indispensable to the existence as well as to the enjoyment of mankind, as the cultivation of the earth, would have received, in all time, the highest honors and the highest place in the affections of the people, as well as all possible attention in rendering it perfect. Such has not been the case. In looking on the works of man, it is seen, that agriculture has been strangely neglected; and, that this neglect has been most apparent in those most interested in its results. Great and successful efforts have been made in devising ingenious implements for working the soil, and machinery for appropriating its products; but little among the large body of the farmers of this country, to improve the kind and quantity of these products has been done; nor can this neglect be ascribed to any deficiency in the development of science. Science has shone forth with peculiar lustre upon the pathway of the farmer, but too often has it been wholly unheeded. A prejudice, propagated and handed down from one to another, and to which he has adhered with as much tenacity as to a choice relic of a distinguished sire, has bound him hand and foot. Till within a short period, rarely has there been found in the farmer's house a book on scientific agriculture; and, even at the present time, where it may be found in one house, in fifty it will be wanting.

It would be difficult to account satisfactorily for this general apathy on a subject of so much importance. At best, our inquiries will reach no clear solution. This is about all we know of it. The idea has been entertained, that any one can be a farmer; that a farmer is a mere spontaneous production; and that from instinct alone, and without the aid of science, he can perform all that is necessary in that employment; and, that success depends, not upon his skill, but entirely on the amount of physical labor he bestows. Hence it has been the practice, that when an individual of ingenuity and fond of research, or a youth of promise and fond of distinction, has appeared in the ranks of farmers, his attention has been immediately turned from the field of agriculture to some other, and, as has been erroneously supposed, more favorable department for the exercise of his faculties. The operation of such a policy is to deprive a rural community of its best talents; and, in doing this, to prevent elevation of character, as well as success in the development of its appropriate resources. This is inevitable. No other inference in regard to it can be drawn. We need no other evidence of this conclusion, than a hasty glance at the leading features of the process.

Now let us suppose, that a farmer discovers in one of his sons a taste for knowledge, and an inclination for reading and study. Does he give this boy an education to render him peculiarly useful on the farm, not only by applying to it scientific agriculture, but also by enlightening his father and brothers in this and other useful branches of learning? He does no such thing. He forever exscinds him from the homestead, in giving him an education for one of the learned professions. If he has another son of superior address and enterprise, he is sent to the city to become a clerk with a merchant. And, if he has one that evinces unusual genius in the construction of curious things, he is fitted to become an artisan. The remaining one, two, or more, for a supposed want of talents, are doomed without education, saving knowing a little of arithmetic, and how to read and write, to work on the farm. This course necessarily induces the favored boys to despise the occupation of their father, and to feel that it is an employment unworthy of their talents, while those who are destined to it not only feel themselves degraded, but are taught thereby to believe that nothing but muscular strength is needed in the performance of their own duties.

It may be well to note the career of these neglected boys, who are pronounced destitute of genius and competent only to cultivate the ground. Is it possible they should not feel degraded? Is it possible that whatever of talents are possessed by them should not become paralyzed and stagnated. They have been told, in a way that cannot be mistaken, that they are inferior to their brothers, and that they must spend their life in a service requiring no more thought than that of the ox who is to toil in company with them. Nor does the mischief end on the premises where this takes place. The tale is told to others; it spreads through the neighborhood; it is known throughout the town or the county; it becomes an element of public opinion, that agriculture is a servile employment, requiring only the lowest grade of talent. With such dogmas in the community, so far as they have weight, is it strange that the mass of agriculturists should place a low estimate on their vocation as well as on themselves? How could it be otherwise? Not to suppose it would indicate an entire ignorance of human sympathy and metaphysical science. Human opinions are ordinarily the result of some conventional influence; in this matter, especially, and in all others, to a great extent.

Under the influence of such prejudices and erroneous opinions, the farmer begins to cultivate the ground. His aspirations rise not above a comfortable subsistence. He dreams not of acquiring reputation in society like that of men in other spheres of life; or, of acquiring property, unless by the most intense personal application to toil, and by self deprivation. He views his own career to be as monotonous as that of the traveler who traverses a South American pampas, or one of our own western prairies, having no diversification of scenery, and the immeasurable plain in every direction uniting with the concave sky; and, in unmitigated and unvaried physical effort like that of the culprit doomed to spend his days on a tread-mill. Apprehending no change, he soon becomes attached to this very monotony, and to whatever on his labors have been associated with it. And after a while, although he may look upon it with a kind of sullen abhorrence, he would cleave to it, if from no other motive, because he would imagine himself not at home elsewhere.

Those who commence an agricultural course of labor in the manner we have supposed, ere long become firmly wedded to their own forms and usages, and guarding them, as is natural, with a jealous eye, delight in being able to look back and say that they have not departed from the ways of their fathers. A certain degree of reverence for antiquity, and the habits of a past generation, is seemingly an instinct of man, and is not to be treated with severity. Indeed, it might perhaps be well, if there were more of it in our country than we usually witness. It is almost exclusively in rural life that we are enabled to gaze upon the evidences of this lovely attribute of human sympathy. While in the city there is an universal impatience for change and novelty, in the breast of the farmer is an unfailing depository for reverence to things that come from and are associated with those who sleep in dust. In this trait of rural character we can readily find an apology for some of the evil incident to it. Nevertheless they should not, as they frequently do, give it scope to the hindrance of increasing knowledge. But we well know that having formed an adherence to the time-honored customs of their progenitors in tilling the ground sufficient to give them a comfortable subsistence, they recognize no means of increase other than an increase of toil; and a labor-saving machine for a long time is treated by them with suspicion, as being merely an instrument for the encouragement of idleness.

The respectability and the amount of profits in agriculture depend entirely upon the policy with which farmers pursue their vocation. If they desire to see it duly honored in public estimation, they must not dishonor it themselves. If they desire to see those of other occupations engage in it, they must on no account evince an aversion to it, or a desire to forsake it. And if they would have their sons place a just appreciation upon agriculture, give them a good education, and then it may be presumed the vocation will be rendered additionally lucrative and honorable. Let youth grow up with as much ambition to excel in farming, and to make it profitable, as is requisite with those of a corresponding age in mechanic trades, mercantile pursuits, and the learned professions, and we shall witness a new era in rural economy, Let these directions be observed, and we shall hear no more complaints among farmers, that they are not properly remunerated for their labor, or that their vocation is less desirable or less honorable than other branches of industry and enterprise. Fathers and sons will both be satisfied with it, and they will be among the most honorable and useful members of the community.

It is an undeniable truth, that if apprentices and journeymen in mechanic trades, or clerks and accountants in the mercantile departments of society, or students and young men in the professions, were as destitute of ambition and enterprise, and as slovenly in their habits of attention to business, as boys and young men usually are on the farm, not one in a thousand would find success, and seven-eighths of them would die paupers. With all these there must be an unwearied vigilance to discover every new avenue to patronage, to the acquisition of wealth, and to honorable fame, or sad indeed will be the memoir of their life. Without this vigilance who have ever reached a proud eminence? Without it, who have ever become the pride of their country? Without it, who have ever established a family name that became a cherished legacy to a succeeding generation? Rarely one may have done it; rarely something like irrepressible destiny may have led to such a result; but it was only an exception to a law that is nearly universal. No one can calculate on this. No one should presume it within his own reach.

But, where is a corresponding ambition or vigilance with young farmers? Allusion is not made here to severe manual labor; to an intense application to toil, ten or twelve, or fourteen hours each day; to great feats of physical power or endurance! Something besides this is needful. Where, then, is their ambition to improve the general appearance of the farm?

to cause the family mansion, the barns, the stables, the outbuildings, the fences, the courtyards, the gardens, to present an aspect of neatness, durability, and well-defined beauty? Where is their vigilence to avail themselves of every implement for the work of the farm that will save enough each season to pay for itself? Where is their vigilance in collecting fertilizing agents and improving their mode of tillage and their breeds of stock so as to double the annual profits of their labor! This is what they ought to do. This is what can be done. When not occupied in manual labor, let them cultivate and improve their minds in reference to higher attainments in agriculture, and they will soon find that their heads are of more value than their hands—that the usually unoccupied season of winter and stormy weather can be made of more avail than that which is devoted to the most severe and unremitting physical toil.

Agriculture is not only the means of supporting life, but it is to be venerated for its antiquity. Its origin has priority over all other arts. This fact alone should give it a deep place in our affections. It might seem, therefore, that the individual which casts reproach upon it is incapable of just appreciation and of logical deduction; and, that he is a stranger to refined moral perception, as well as guilty of a species of impiety. It is an attribute of our nature, and a dictate of revealed religion that we reverence the institutions of Heaven. Is not agriculture one of these institutions? Is it not the first of them? Did not man receive his commission to till the ground from the Deity himself? Was it not, too, on the very completion of the material creation, as if to constitute man his associate in a ministration of beneficence, that God placed him in the garden of Paradise, to dress it and keep it? And, as if to make this labor of man a sacred adjunct to the labor of Heaven while imparting life and joy to God's rational creatures in all coming life, was not the commission for it bestowed the very day of nature's grand jubilee, when the morning stars shouted and sang in a loud anthem of praise? Was it not granted beneath the delightful bowers of Eden, where fragrant odors and spicy aromas floated on every breeze!

To our apprehension, the circumstances attendant on the institution of agriculture, should give it the same pre-eminence in physical economy that the Christian ministry has in the moral world; a pre-eminence that should shield it from reproach and desecration of every kind. These circumstances have an impressive sanctity which cannot be resisted by the well trained mind. In order to see an object in the full splendor of its own beauty, we are often constrained to place it in company with other objects. Thus, how much more beautiful appears each hue of the rainbow when placed in juxtaposition with the others, than though it were seen alone!

And thus many of the institutions of life derive much of their overpowering suasion from the array of influences in which they had birth. Is it not much the case in the institution and solemnization of matrimony? If the parties to this holy allegiance make their vows in private, how little is there to cause deep impressions on themselves or others? But, when these vows are made at the altar, before the congregation of the church, amidst the ardent excitement of parents and friends, with all the responsibilities and solicitudes of the future to overwhelm them, how does the occasion gather pathos and undying sanctity, for a seal to their plighted faith? And, if we would contemplate American Independence in all its inbred sublimity, we must, in imagination, carry ourselves to the Hall where the fathers of the Republic were inscribing, amidst the ringing of bells and the shouts of freemen, their names upon that chart of human liberty. So our awe and reverence, in a multitude of cases, arise rather from a magnificent display of attending circumstances, than from any abstract convictions of a particular truth. Hence, as much as we may love agriculture for its power to administer to human wants, we cannot be unmindful of the moral grandeur with which man was commissioned to be its minister. In the one view, we experience a rational conviction. In the other, we involuntarily yield ourselves up to a social impulse as sacred as it is powerful.

THE POULTRY YARD.

Among the novelties of the age is the excitement that has been manifested within the last few years, particularly in some of the New England States, on the subject of improved breeds of poultry. The extravagances that have grown out of it have afforded the lovers of fun not a few occasions for jest and merriment; for, not a few of our notable savans in business and professional fame became as much absorbed with this branch of research, as they would have been previously, in matters out of which fortunes were to be made. Indeed we have seen these gentlemen as much galvanized with sleepless zeal to ascertain whether a particular variety of fowls should have four toes or five, as in collecting and adjusting the newly found bones of the notable sea-serpent, or of a mammoth, in a new locality. Positively it was ungenerous to laugh at them for this new kind of mental effervescence. It exhibits no new type of human character. Thousands as notable as they are have evinced, in relation to other matters, similar gushing impulses. Rarely does a year roll round and pass away, without leaving on its tomb-stone some corresponding inscription of a new fledged zeal that marked its authors for unenviable notoriety.

The motive which led these gentlemen into such perils to their reputation was excellent. The result to the community will be good, without doubt. The chaff from their harvest will be blown away or burnt up, but there will be left a residue for use equal to the best wheat. Improvement in the breeds of farm animals is, undoubtedly, one of the most rational topics that has claim on the attention of the farmer. Individuals who have distinguished themselves in it, and there are several in Great Britain, have achieved a reputation for themselves as undying and far more honorable than that of the greatest generals the world ever had. The feasibility of such improvement has been demonstrated to an extent that places it in the first class of objects on which successful enterprise, in rural economy, can be promoted. The principles on which such improvement is predicated are pretty well defined ; but the philosophy of these principles is among the unfathomable mysteries of nature. In this matter, as in numerous other ones, human science may advance to certain points ; the facts discovered in the progress may be as prominent and incontrovertible as mathematical problems ; may stand forth like pillars in a magnificent temple, firm as the foundations of a gigantic mountain, and transparent like the clear light of noon ; but beyond these points human science stands appalled—not a step onward can be made—not a gleam of light dawns up on the untrodden path : in the perspective all is dark and incomprehensible.

SHANGHÆ FOWLS.

All this is literally and emphatically true in relation to animal and vegetable physiology. We know that the different races of men, according to common received theories, have been occasioned in a long succession of generations by meteorological influences. But who can tell why these influences in the human species should have led to the difference in organization, complexion, and mental endowment, obviously characterized in the native American, the Asiatic, the African, the Malay, and European races? No one can tell. Conjectures may be raised; hypothetical explanations may be propounded; but the real truth lies deeply hidden from human investigation. And who can tell, in the feathered tribes, why there is such an infinite diversity in the plumage, for instance? We mean not different species, but simply different varieties in a single species. Why is there such an assemblage of varying hues, and such a silky lustre in the vesture of the proud and exulting peacock, of the delicate and matchless bird of paradise, or even of the beautiful little humming bird, which seeks nourishment like the honey bee from the flower garden. Man can no more explain this than he can explain, why the same vegetable element whitens in the lily and reddens in the rose; or why in one plant it becomes sweet, in another bitter, and in another acid. These things and all similar ones are among the unrevealed canons of infinite wisdom. In relation to them, the Author of them may and does say to us, as He says to the waves of the sea—hitherto shalt thou come, but no farther!

Let us look at the tenants of the poultry yard, and much indeed will be presented to our view as worthy the consideration of the philosopher as the rural economist. The latter may easily estimate the pecuniary value of this branch of his investment and care; but, can the farmer as easily tell us why there is an almost infinite diversification in the development of the charming birds that enliven the mansion and the surrounding enclosures on the well disposed farm—diversification of form, of color, of voice, and of social attribute? Here is a countless number of mysteries in the animal kingdom, which a profound philosopher can no more explain than the most unlettered peasant. These things are beyond human comprehension. We can no more tell why there is such a diversification and commingling of the colors in the plumage of the poultry yard, and why there are such deteriorations in the muscular organization and development arising from successive reproductions, than we can tell by what strange process a portion of the human family have become the pigmies called Aztecs, now attracting so much attention among the curious and the philosophical. We may indeed say it is from the operation of the law of nature; but, of the operative principle of this law, we are as ignorant as we are of the law of gravitation that causes the magnetic needle to point to the poles of the earth. We know it is so—we know it must be so from the tens of thousands of cases in which it has been demonstrated; but this is all we know, or all we can know on the subject.

With persons of taste, it is no matter of surprise that there has been of late such an effort to improve the poultry yard. For a much longer period there has been a systematic effort to improve other farm animals, particularly cattle, horses, sheep, and swine. In this case there has been an evident increased pecuniary profit from what has already been accomplished. When these improvements are more generally spread over the country, that profit will be far greater. It is not irrational to suppose that improvement in the breeds of the poultry yard tenants, will be attended with corresponding pecuniary profit. It is believed moreover, that in this case the profit will be greater than in the other case; that is, a greater percentage on the money invested in the stock and fixtures. What may be denominated the profits of the poultry yard have generally, by farmers, been overlooked. With a vast majority—perhaps with a thousand to one—it has never been asked if there can be any profit from it. But with the increase of better defined principles of political economy, and particularly of that portion of them applying immediately to agriculture and rural life, the world is not to move on in this sluggish manner. If labor is now to be applied to any object, it must be understood, as a preliminary, that the process is to be a remunerating one. If money is to be expended, it is now to be soberly estimated whether or not it is again to return to its owner with a fair amount of accumulation. The intelligent farmer now wishes to know beforehand, whether there will be a net profit from cows or sheep, or swine, before he attempts to rear them.

The same calculation is appropriate to the poultry yard. It is therefore rational and commendable that, in this department of rural interest, there should be the same enlightened policy that has been extended to other branches of enterprise on the farm. If poultry may be made profitable, let those varieties be selected that are most profitable. In introducing improved breeds, there have indeed been occasional absurd extravagances. The hen fever has now and then risen so high, and been so violent, as to threaten the extinction of rational pulsations. The sufferers have been the legitimate subjects of ridicule and sarcasm. But after all, is it not as absurd to pay five hundred dollars for a hog, or fifteen hundred for a sheep or one of the ox tribe, as to pay one hundred and fifty for a pair of Asiatic hens? Let the reader brush up his mnemonics, and he will perceive that all this has been done. Such extravagances are the mere incidents of ardent minds in new enterprises.

Connected with every farm establishment there should be a poultry yard. Without it the farm is as incomplete, as it would be without a piggery. Indeed, every family in the country, although not devoted to agriculture, should have one. To the mechanic it is important, so is it to the professional man and to the merchant. No direction or rule can be given, as to the size of it; whether it shall contain ten hens, fifty, or an hundred. If it is partly designed to supply materials for a market, it may of course be proportioned to the demand there is for its products. If these products are wanted for home consumption only, the size of the family should regulate the size of the poultry house and the number of its tenants. And in each case, it is apparent, that the amount of feed produced on the premises for the use of the fowls, and the local conveniences which can be appropriated to their accommodation, should have an influence in deciding how many should be kept. These are matters which all can decide for themselves. What might be expedient for one family would be inappropriate to others. Some, too, are excessively fond of eggs; others care less about them. The same is true in regard to the flesh of poultry. This also will have its influence.

Nor can any general advice be given as to the expediency of keeping other kinds of poultry. One may be disposed to keep turkies, another geese, another guinea hens, and another ducks; and perhaps rarely the same family will keep them all. This is mainly to be determined by taste and fancy, provided there are adequate conveniences for any or all the different kinds. Rarely will the web-footed birds do well without ample supplies of water, and running water is best because it is fresh and clean. If there is a river or a pond close by, most families would do well to keep geese and ducks. Here, in the season of warm weather, they would collect most of their food, and would keep themselves clean and glossy. Sometimes both geese and ducks do well without this natural supply of water.

The rearing and keeping of poultry has become an important branch of rural economy. Books are becoming numerous on the subject. Bird fanciers are devoting their whole time to it; studying the habits and the profit of each kind; and, also ascertaining the best modes of treatment for them. Their labors may prove of great value to the community. The subject, in this country, till very recently has attracted little or no attention. It may at first be viewed, as too insignificant to merit serious consideration. This is natural. Little things are frequently treated with contempt, although in the aggregate they assume magnitude surpassing credibility. This is literally so with poultry. Because a fair stock of hens can be bought for two dollars or so, they are regarded as beneath the rank that entitles them even to kind treatment, especially if viewed in connection with expected remuneration. But, although the winter stock of hens on a common farm may be estimated at two dollars only, the fair valuation of all the hens in the country, gives them a commercial importance ranging with some of our best products. It is not the design of this article to anticipate or supersede books exclusively on the culture of poultry. All didactics on that branch of rural economy are by us purposely omitted. Our readers are referred to them for instruction. The present labor is to recommend the subject to all persons living in the country. This we shall do, by setting forth, as well as we can, the pecuniary profits of the poultry yard; and, also, the benignant influences that may arise from it, to the various residents of the contiguous mansion, both old and young—male and female, particularly the youthful and female members of the family.

Our agricultural journals, it is well known, every now and then, contain some detached account of the profits of hens; but rarely has there been any general collection of poultry statistics, that would convey an adequate idea of the aggregate amount of these profits in the country, or of the value of this branch of rural investment. In the absence of general statistics, we must take isolated ones, and from them, draw general conclusions. Such conclusions, if not perfectly accurate, in this case will be as likely to fall below the reality as to rise above it. It is believed that by this hypothetical process we shall be able to satisfy the reader that the culture of poultry is of much more importance than generally imagined; and, that consequently it should become one of the first objects of attention with every family in the country. Let it therefore be supposed, that there are in this country three millions of families that possess all the conveniences for keeping poultry more or less. This calculation cannot possibly be too high. The number is doubtless greater. Then, let it be supposed, that to each of these families, belong ten hens; surely a moderate allowance; but this will make thirty millions for the entire country, which, at thirty cents each, constitutes a permanent investment of nine millions of dollars. Four chickens to each old hen is probably raised for the table; that is, one hundred and twenty millions of chickens raised every year for the table, which at the same price, will yield thirty-six millions of dollars; or forty millions of dollars for both. Again, if each of the stock hens lays only sixteen dozens of eggs in the year; less than one dozen in three weeks; there will be a product of eggs in the entire country of four hundred and eighty millions of dozens. These eggs are worth at least, two dollars for each hen. But allowing one half to go for feeding them, there will be left a net profit from the eggs of thirty millions of dollars annually; that is, a net profit of sixty millions of dollars annually in the country from hens' eggs and chickens raised.

Let this result be placed with some of the leading staples of the country. The value of the flour of the country in 1847, has been set down at $140,000,000. If one half of this is deducted for cost of production, and that is not enough, the value of the poultry is worth more to the country than our wheat crop. And taking similar data for comparison, it is worth double of our oat crop, double of our potato crop, double of our cotton crop, and is equal to our crop of hay. Indeed, taking the statistics of our agricultural productions that year as a guide, there is but one of them that yielded, according to most favorable calculation, so large a net profit as the poultry. Or, if the poultry did not yield as much as supposed, it is because the poultry yard is unduly neglected, and its products are under-estimated. The value of the Indian corn crop was but a little below four hundred millions of dollars, deducting nothing for production and sending it to market. It is affirmed that with the exception of prime cows, there is not on the farm a single article of produce, whether animal or vegetable, that according to the value of the original investment, and of the expense and labor of production, that yields as much clear profit as will come from the poultry yard, if properly regulated. The assertion is made with confidence, because it is sustained by our own experience, and by a careful examination of the subject. What better occupation ; rather, what better amusement can the young members of a family have than to feed and watch over the poultry of a farm ! In this way they may clothe themselves, and pay for their books, without interfering with the school exercises, or any reasonable labors expected from them in other things ! If it should be thought advisable let this be one of their standing perquisites.

However, it is not designed to treat the reader solely with suppositions, or hypothetic calculations. We have a few solid facts for a basis to our theory. The census of 1840 fixed the value of the poultry of the country ; that is, as it is to be presumed, the stock or brood poultry, at between twelve and thirteen millions of dollars, a quarter higher than we have placed it. And this was a dozen years ago. The whole amout of eggs yielded in a year we have supposed to be 480,000,000 dozens—that is, nineteen dozens for each individual in the country to be used in a year, or a fraction more than four eggs a week for each person, or in a family of six persons, an average allowance of two dozen a week. This is a moderate allowance; for in France the annual consumption of eggs is 8,000,000,000, being about twenty dozen to each person; and in Paris alone the annual consumption is 140,000,000. We have no method of ascertaining the quantity of eggs used in Boston, New-York, or the other large cities of this country; but as it is well understood that the inhabitants are fond of good fare, and will have

COCHIN CHINA HENS.

it when the means are at command. In evidence of it, statistics show that in Boston the annual consumption, with a population of less than 150,000, is to the amount of one million of dollars for poultry alone, and that in New-York and its dependencies three times the amount is expended for the same article. The egg trade in Cincinnati a few years since was put down at

25,000,000, in a single year, which since, without doubt, has much increased.

We have estimated the profits of the hen at one dollar each in the year, in addition to paying for her food. But she must have good accommodations, suitable food, and enough of it; then our estimate is a low one, provided, also, that she is of any good common native breed. To show this, we will give a few statistics on that point. A correspondent in the "New England Farmer," under date of March 3d, 1851, says that one of his neighbors the year previous kept 54 hens, three geese, and nine turkeys, which he valued at $32 50. In the year they ate 90 bushels of corn, which cost $58 58. The money received from eggs and the carcases sold was $174 59; or a clear profit of $83 51. A correspondent of the "Genesee Farmer," who kept twenty-five hens, says the profit on them, in a year, after paying all expenses, was $25 92—a trifle above our own estimate. Col. M. Thayer has stated that he can make more profit from one hundred good hens than from his farm of two hundred acres. His farm is called a good one, and he has been accustomed to poultry for fifty years. J. H. Austin, of Canton, Conn., has stated the net profits for one year on fifteen hens to have been twenty dollars. Mr. Crocker, of Sunderland, N. H., had a net profit of $16 97 on seventeen hens, in a year. The above are taken almost at random from the different agricultural journals at command, excluding such as gave a much higher rate of profit, preferring those in the range of our own estimate. Of this class we might extend the number to almost any indefinite extent.

To show how much a few hens will contribute to the subsistence of a family in case of necessity, the following affecting anecdote is given, which shows their value, and the ingenuity and tact of the individual alluded to. A few days since we read the account of a widow having six children to support. She was, at the time in question, entirely without means to procure them food save eight hens which laid daily. She did not feel willing to make known her necessities, or to solicit any assistance.

She resolved to try the experiment of sustaining herself and children on these eight eggs. To eat them would do but little beyond maintaining life. To exchange them, therefore, for other food was the plan adopted. Accordingly she daily exchanged six of her eggs for beans, and two of them for a small piece of coarse meat, both of which were cooked together for soup, which was, under those circumstances, quite palatable, and in amount sufficient to satisfy hunger. In this way they were all sustained, till more ample means were at control: an instance of management rarely or never excelled.

A poultry establishment, connected with every country residence, has another claim to favorable consideration. As a branch of rural productive investment it has been shown to deserve a place on every farm. It is also to be prized for the facilities afforded by it, in case of sickness or any unexpected occasion for a delicacy or luxury upon the table, when the butcher's cart is not at hand. With fat chickens and fresh eggs in abundance the cook is never at a loss in furnishing something acceptable to the invalid or epicure. When in our life occasionally making a sudden call on a country friend, in some secluded locality, within thirty minutes of time we have usually been refreshed with as nice a meal as ordinarily expected at a good hotel. It is the pride and delight of the well bred wife and daughters of the farmer to have at command such means for entertaining a city relative and a distant friend honoring their mansion with a visit. The treasures of a good country larder, with what can be produced on a farm, seem to the stranger almost inexhaustible; and the skill and tact ever ready to deck them off in a style of professional gastronomy seem incredible to those who have had no opportunity for sharing in them. But without the poultry-yard for such emergencies, there would be an irreparable deficiency. Where could be found a fair substitute for the roasted or boiled capon at dinner, or a broiled one for an early breakfast? Or where could be found an equally acceptable equivalent for the omelet, or the boiled, or fried, or poached egg, or for the ever-welcome custard? Without these requisites for good living, epicures would nearly pine away, and become as it were out of humor with life.

And there is yet another consideration to be offered in favor of the poultry-yard. Is there nothing in the feathered tribes that dwell there to gratify the eye or the ear of those who watch over and nourish them? Cannot the lover of natural beauty see something for admiration in the well-rounded breast and the gradually tapering and gracefully curved necks of these well-chosen and well-fed birds? Is there no beauty in their infinitely variegated plumage? Can human art successfully imitate the silky fineness and lustre of their feathers? There is in the human bosom an apprehension of what is comely and adapted to a well-designed purpose, denoting wisdom and beneficence on the part of its author; and the human senses are modulated so as to convey to the mind the images of objects thus administering to its gratification. Where, it may be asked, is there in the broad creation aught that contains so much to delight the eye as in the poultry-yard filled with a choice collection of beautiful fowls? Burke, the well-known writer on natural beauty, selected the neck and breast of certain birds, as well as the neck and the fully developed breast of the youthful human figure, as an illustration of his subject. And is it possible, with all the dye stuffs on the land or in the sea, for human skill to produce such an assemblage of delicate and brilliant colors, combined and commingled in ten thousand aspects, as are every day displaying themselves, in the poultry-yard, to the gaze of contemplation? Ca: they be seen among the rich fabrics of Turkey or Persia? Can the artists of France, or Germany, or England equal them? If the mansion of the farmer is not ornamented like the drawing-room and boudoir of the rich merchant, with costly drapery and tapestry, he may have a poultry-yard exhibiting specimens of beauty that may be the envy of princes.

It is admitted, that in the want of artistic decorations, and sometimes the want of polished society, the country, to those not familiar with it, may seem cheerless and dull. Yet it seems to us there is a sufficient counterpart for this, in the rich verdure of the meadow, the waving foliage of the forest, the buoyancy of farm-yard animals, the singing of birds, and especially in the gaiety and sprightliness of the feathered corps that surround the neat cottage or cluster in the poultry yard, displaying to the eye their gaudy plumage, and to the ear the melody of their music. It might be thought, that to be dull and cheerless amidst such constant exhibitions of animated nature, is equivalent to being destitute of the amiable and lovely susceptibilities of the human heart Even in the stillness of a thick darkness that lulls to rest earth's wide realm, do we experience no pleasure at the unfailing notes of chanticleer announcing the hour of midnight or of dawning day ? We never hear those notes without an emotion of joy. They tell us there is to be a resurrection to the duties and the hopes of life. Does the rattle of the watchman or the rumbling wheels, at that hour, on the city pavement, create an emotion like this ? We think not. The one denotes the reign of order throughout earth's domain. The other may and often does denote the discovery of burglars or assassins, or the constant liability to their depredations.

And with the rising sun what an uninterrupted concert opens among these feathered groups ! What a jubilee begins to greet the new born energies of the world ! With an hundred hens constantly leaving their nests, there is an uninterrupted succession of their joyful notes. Their mouths do not appear to be large enough to emit all the boisterous emotion that animates them. In the midst of this ceaseless cackling, every now and then their lordly mates cause the surrounding forests to echo with their shrill crowing—as overpowering as the diapason of the cathedral ; and rising above this at measured intervals, is heard the pompous shout of the gobbler, almost causing the ground, like the discharge of cannon, to tremble beneath him. And, if the Guinea fowl belong to this community, as if to increase this vocal jargon, or to make burlesque upon it, his harsh voice, not unlike the filing in a machine shop, is heard for half a mile. If there be not music in all this, there is life in it—there is animation in it. Wherever witnessed there can be no stupid languor ; no lugubrious dullness for want of objects to inspire a feeling of interest ; no painful sensation of solitude and loneliness. The human being that gives no responsive emotion to the sounds of this scene must have a heart as impenetrable as adamant, affections frigid like the ice of the poles, and is indeed an outcast from Nature's temple.

GUINEA FOWL.

Yet far more important in a social or sympathetic view is the pleasure experienced in feeding poultry, and thus making them, as it were, companions. Animal nature is full of social impulses, and these impulses are not confined in their operation to the particular species in which they severally originate. These impulses belong to other animals as well as to man. If primarily exercised on the members of their own species, they may be taught to have scope on others. The little chicks separated from their mother, and the calf or the lamb from its dam, suffers as does the unweaned child when separated from the object that ministers to the preservation of its life. Each will then seek companionship with those which belong to other species of animals, wherever kindness and nourishment are to be had. The wolf may again, as once in olden time, become the nurse of the human infant, and there grow up between them a mutual and strong friendship, unnatural as this may seem. And frequently does the half-grown pullet, in want of society and protection among its kindred, learn to seek for it at the kitchen hearth, in the mistress who kindly feeds it. Under similar circumstances of loneliness and destitution, the calf and the colt will herd together; so will the dog and the goose; and, even the cat and its legitimate victim of prey have been known to do the same. Instances of analogous facts have been numerous.

If man were in solitary confinement like Baron Trenck, or like La Fayette in the dungeons of Olmutz; or cast away on an uninhabited island, like Robinson Crusoe on Juan Fernandez, how dear to him would become the companionship of a dog, a goat, a bird of any kind, or even a mouse! In all such cases there would be mutual attachment; and if the parties were to be separated, the consequence would be mutually painful. It is well known that the constant supervision of most farm animals leads to affectionate familiarity, mutually cherished. These animals well know their kind guardian from all others; when fed, they express their gratitude in grateful looks, and wanting feed, their attitude, their motions, their beaming eyes, are so many beseeching manifestations of hunger. It is not indeed human speech; it is not a written language; but their keeper understands them perfectly, and if he is a man of kindness and Christian feeling, he fails not to administer to their wants and to cultivate the exercise of their sympathies. Let it be asked, if he experiences no pleasure in his intercourse with them, and especially in his ministrations to their necessities? Does their seeming fondness for him inspire no corresponding emotion in his own bosom? Can he look upon them with the heartless indifference that he surveys the stones beneath his feet? Does no reciprocity of feeling spring up between them?

It is not possible. If it were, he would not deserve the name of a man.

The greatest social pleasure felt from an intercourse with, and a supervision of, dumb animals, is that which arises from a supervision of the poultry yard. This is to be expected. The intercourse is more constant than with any other farm animals. In the time of rearing the young, it is seemingly every hour in the day. The feebleness, and the recklessness of young birds, render this indispensable. In all cases, the strength of mutual attachments is proportioned to the degree and constancy of intercourse kept up. This is true in human society. It is equally true in the brute creation. It is also true, where they exist between a human being and dumb animals. Let a young lady illustrate the idea. We will suppose a young lady recently removed from the city to the country, a young lady having little or no society of her own species, age, and sex, and cut off from the means of amusement to which she had been accustomed, what a relief would she experience in becoming the guardian and nurse of the poultry yard! Such an one, although not having constantly before her the graceful flitting of female youthful forms, although not animated for days and for months with their cheerful conversation, and the buoyant melody of their joyful hearts, yet, she will soon find objects claiming her most ardent sympathies, and cheering her with their own companionship and glee. Does she now feel lonely? Do the paralising vapors of solitude rise up before her and depress her spirits? Are there no animated forms before her, that in a measure supply the place of human forms? Does she hear no accents of joyous or affectionate breathing, to supply the place of human speech and human melody?

No one can affirm to the contrary. To her there will be no dreariness, no tedious monotony, no impatience for change of scenery or companions. The broad earth is spread out before her in all its native loveliness; the verdure and the flowers fill the air with fagrance; she inhales the rich odors; the sky is over her like a magnificent canopy; and her ear and her heart are open to the tones, whether of supplication or gratitude of the dependant creatures that live on her ministrations, and rejoice in her smiles. In the midst of such influences, the human bosom, and especially the female bosom never fails to receive an impressive culture; the sterner passions are mollified; the amiable affections become deeper rooted, and throw off more broadly and higher, the aromas of kindness; and above all, the heart, the sanctuary of the divinity that presides over the young sparrow and the lillies of the field, becomes more and more expansive with those sublimated breathings that connect mortals with the throne of Heaven.

Less than fifteen years ago, in one of the most memorable mercantile revulsions that ever cast its paralizing wand over our beloved country, hundreds of families, some reared in affluence and others in elegance, were reduced to poverty. We select one to show the benignant rural influence on human sensibilities. Their pecuniary means were reduced to the lowest point. There is a mildew that falls upon the human form as well as on vegetation, that paralizes all vitality. Two of this family in the two first succeeding years, from the effects of this mildew, were consigned to the tomb. The others, in order to live cheaper, and especially to avoid the daily mortification of meeting their former associates, sought residence in the country. Here was selected a cottage, humble in size and in external attraction; and inwardly equally destitute of superfluities. Indeed, the furniture was rigidly in accordance with the simplicity of olden times; nearly every article that had enriched their spacious mansion in the city, and would command cash, had been sent to the pawnbroker or auction room, to furnish means to supply the table with food. This is the common waning shadow of departing competence in a commercial emporium; and, its reflections form the rising shadow of future destitution and broken hearts. In such a cottage, midway between perfect solitude, at the mountain's base, and the full glare of the country village, were they in poverty to meditate upon the vanity of this world's glory. They had no taste for society; but, having a good conscience, they did not shun it, knowing that no personal disgrace had fallen upon any of their number. And, the remains of a former well filled wardrobe—true, much worn and faded—enabled them with constancy to attend church. This is as persons of sense, and Christians, always should do in similar circumstances. If persons go to church to display new or gay clothing; or, stay away from it because their clothing is cheap, much worn, or old fashioned, the conclusion is irresistible that they have a religion of no value in this world or in the world to come.

For many a month their names were not known; but, their discreet appearance every Sunday soon attracted notice. The reminiscence exhibited by them of better days, created sympathy; and, their good sense, and the cheerful acquiescence in their destiny, showed that this sympathy was well deserved. In this quiet retreat they eat and slept, mingling their congratulations for what remained to them, and their sorrows for the departed ones that once had place with them; and, in it they were not forgetful of the daily morning and evening oblation to that Being whose Providence always provides for the motherless and the destitute. What a comfort it is that there is One to whom persons in such straits can always resort for help; that there is one deep fountain always full and always open for the supply of their necessities! In this poverty-stricken and excellent family was a young lady of an age to feel most severely the blight that had come over them. She had just seen the world at a distance: she had not entered society, as it is called; from her the shackles of the boarding school had not wholly fallen off; she had seen the prairie of life covered with gaudy flowers, but had not been near enough to them to inhale their odors; she had simply heard the far away sounds of fashionable merriment, but had never approached so near as to feel, in her own bosom, the witchery of its delusion. The curtain dropped even before she had opportunity to test its reality. Perhaps she was in the neighborhood of fifteen years; and had as lovely a spirit as we ever knew. The remembrance of her, to this day, cheers and depresses us with a strangely pleasing sadness.

In such an exigency there are two considerations that become paramount to all others: the first is, how the table is to be supplied: and the second is, what occupation can be had to prevent a feeling of ennui. In past times a check was to be drawn on the bank, and money became as plenty as water. Now all, old and young, were to be fed, like the young ravens, from the rich stores of their Heavenly Father. Our dear young friend had read something of the productiveness of the poultry yard, and of the agreeable associations with it. She resolved to procure a few dozen of hens. She did so—perhaps by the sale of an article from the wardrobe—and, in a short season, these prolific creatures superseded the table necessity for bank checks. There seemed to be no end to the eggs, and in a few months there was a fine new grown race of the parent stock. What a luxury did they impart to a hitherto scanty repast! Once more the hour of midday was hailed with joy! Even the scent of dinner gave additional zest to life. The experiment was one from which thousands, there is no doubt, will learn wisdom, when in conflict with adversity.

It was revealed to us, two thousand years since, that mortals cannot "live on bread alone;" the mind as well as the body is to be nourished; there are social instincts to be fed as well as the gnawings of hunger. This was shown in the case of which we are speaking. Ellen, the name of this self-taught guardian of the poultry yard, here soon found a delightful panacea for the want of society. These beautiful birds became to her like one's own children. She would spend hour after hour with them, feeding them and caressing them. Indeed their society was never unwelcome to her. She would sometimes sit down on her feet upon the grass, and they would all gather round her: apparently striving each to get nearest to her: all looking her fully in the face with their peculiarly expressive eyes: hopping up in her lap, eating out of her hand, and occasionally flying upon her shoulders, reaching out their graceful necks as if the better to catch the full glance of her own eye. Oh, had they been gifted with human speech, what enrapturing sounds of affection would they have uttered—making her very soul burn within her, as if touched by a living coal from friendship's holy altar! Did not Ellen love these beautiful and prolific creatures? She did love them, and she nourished them better than their own mother had nourished them!

But, alas, Ellen's ministrations to the poultry yard, like all her joys and her sorrows of earth, were hastening to a premature termination. That terrible mildew which had been so destructive in her family, had left on her own delicate form some indelible scars. They had been impregnating her system with the virus of their poison till it reached her vitals, and spreading upon the surface, like the leprosy, till the whole was

SPANGLED POLAND HENS.

covered. After a few years she became unable to visit in their accustomed haunts the objects that had yielded to the family so much sustenance, and to herself so much pleasure. But while confined to her room, if placed on a chair by her window where the poultry could see her, they would gather round it, reaching up their fair necks and greeting her with their bright eyes as if to ask for food, to inspire her with hope of renewed health. Such hope had become impossible ; for she already felt the seal of her own destiny : she already felt that the light of her own life was burning dimly in its socket, and would soon flicker away and become extinct. All that she could do was to throw from the window the few crumbs with which she had been furnished. These, indeed, would be gathered up by the affectionate group—but, they seemed to evince a premonition of the coming doom of their kind mistress—as soon as eaten they would quietly retire, one by one, in seeming sadness, as neighbors and friends at a funeral retire from the cold grave that contains the object of their love. Here are two of them !

Where disease had made such irrevocable havoc upon the functions of an animal form, there could have been but few opportunities, even at the window, for similar demonstrations of mutual attachment. Long before such a crisis was imagined, except it may have been in secret by herself, she suddenly swooned away, and never again opened her eyes upon a world that had raised hopes in her, only to be disappointed, and that had opened upon her a dazzling effulgence so soon to be extinguished by the mantle of midnight darkness. If there is in this vale of vicissitude and sorrow, any one thing that, above all others, paralises our firmest energies, and causes the heart to melt in disappointment and woe, it is the death of a young lady, beautiful in person, amiable in temper, accomplished in manners, and adorned with Christian graces, before she is called to fulfil the most important duties of her earthly mission. Why is such an one so called away? Why are not those called first that have accomplished all the good they can do? Is she no longer needed by agonized relatives? Has the world no more occasion for her kind offices? Are there yet to be on earth, no more hungry to be fed; no more naked to be clothed ; no more tears to be wiped away ; no more broken hearts to be bound up, by such a lovely personification of Christian excellence? Infinite wisdom alone can tell why it is so!

Delightful to us would be the thought that this sketch is a work of fiction. Such is not the fact. We knew her ministrations to the poultry yard; and, when these ministrations became impracticable, we oft saw her emaciated form, and her long tiny fingers, colorless and almost transparent like a skeleton hand which had been dipped in white wax. We oft saw her sunken eye and her faded cheek. We oft heard her impaired voice, and witnessed her utmost efforts to appear quiet amidst pain, and cheerful amidst the withering joys of life. And, also, we saw her lifeless remains, without a father or a mother to weep for her, carried to the cold grave! There she still sleeps; and, as she died in the faith of the gospel, her few surviving relatives, rejoice in the hope, that in the opening of another life, she will receive the crown of unfading glory.

A WIDOW WORTH HAVING,

OR A SHORT CHAPTER ON FEMALE REPUTATION AND THE RIGHTS OF WOMEN.

The recent discussion respecting the relative rights and talents of men and women has partaken more of comedy than of tragedy. Some of the leaders in this discussion have suffered not a little from incipient fever, but we are not aware that the termination has in any case been fatal. There have been two or three conventions composed mainly of the feminine gender, who, if they had been fortunate enough to have husbands, left them at home to rock the cradles and dry-nurse the babies. This seemed to us rather an awkward and unnatural transposition in social economy; and, not being furnished with statistics on the subject, we are unable to say which succeeded best in their respective new occupations, pantaloons in the management of Russia diaper, or petticoats in making speeches and passing fiery resolutions for the recognization of the community. It would have been unreasonable to expect, as long as the doctrine of innate ideas, since the days of John Locke has been on the wane, that either party could in their first essays acquire enviable success or reputation. The last of these demonstrations was adjourned, it is understood without day and without year, to speak in parliamentary technicality; that is, not to meet again; and, the only known remaining spasm of it is the occasional repetition of three august lectures—one on manhood—one on womanhood—and the other on something else.

It was thought in olden times, that much learning might make one mad. In our own time, the more probable presumption is, that the want of competent learning may make persons appear like fools. No one acquainted with our views on this subject, or the labors we have directed in relation to it, can suspect we do not duly appreciate the female character, and the mission woman is ordained by Providence to accomplish. Far from it! Let her agency become paralised, and the moral world degenerates into a waste place; sterility would spread out before us her broad and uncomely bosom; the most fertile plants of earthly growth would wither and die; and, there would seemingly be left, nothing but a bleak desolation. Let the fountain of woman's pity be dried up; let the overflowing of her kindness cease; let her forever buoyant assiduities and her heaven-inspired smiles fail in their ministrations to the unfortunate, the sick, and the broken hearted, and where—oh, where, will the lords of creation find aught to render existence desirable? Where she is not, there may be drunkenness and revelry and crime; but, who in such a place, would seek for the sunshine of the soul! Without her presence, the fireside would radiate no gleams of joy; and without it, even the sanctuary of religion would more resemble the sculptor's studio than a temple of devout and sincere worshippers. Man seems to be the mere bone and muscle of his species; woman is comparatively the soul of it—the genius which gives it life and beauty.

When we look at the stations respectively filled by the two sexes, and the labors of them respectively performed, in the best conditions of human society; the one devoted to the sterner and more arduous occupations of life, tilling the soil, buffeting the ocean storm, wielding the sword, or in the councils of state, where everything is exciting and exhausting, while the other is devoted to the gentler and more placid avocations of the domestic sanctuary; both alike necessary—both alike honorable; we feel no disposition to break in upon, or to disturb the present order of society on the subject. We admire the wisdom and beauty of this allotment. Man thus seems so necessary to woman, and woman so necessary to man, we cannot doubt it is in accordance with the divine wisdom and benevolence. Hence, we think it as unnecessary and inappropriate, and we feel no more inclined to inquire whether man is intellectually superior to woman, or woman is superior to man, than to inquire if a horse can be made capable of teaching algebra—a labor now performed by our college professors. It is possible, that by a long course of discipline, woman may become large and muscular, and athletic, and brawny, as the farmer and the blacksmith now are, and that her feelings may become so correspondingly changed, that she may handle the scapel as well as the surgeon, and the meat axe as well as the butcher; while, on the other hand, men may become as familiar with mending old stockings and the care of children, nursing only excepted, as women now are; yet, it appears to us, that a long time would be required for it. Moreover, if this can be done, we are at a loss to perceive wherein will be the gain; and in what respect the world will be made better. Perhaps some of our petticoat lecturers could tell. It is beyond our comprehension.

It is well known to those conversant with the laws that govern in animal physiology; the mind as well as the muscular organization; that development is the result of circumstances. The brute creation is supplied with all necessary instincts; but, in the human species there is reason instead of instinct, and this reasoning power is made available by culture. We know little or nothing, unless we learn it. Ordinarily, we are learning something the most of our life. We are taught by others about us; also, by the force of our own necessities. If left in destitute, or cast under perilous circumstances, we are thereby straitened to make efforts for our own relief. This is a progressive work. Each step taken prepares the way for an additional step. Each object reached stimulates us to press forward to an additional one. In this manner persons rise from poverty to affluence; from obscurity to fame; from ignorance to eminence in knowledge. Those who have not studied the laws of mind are perplexed to know how individuals do sometimes thus become distinguished. The inference is legitimate, from these cases, that all might accomplish much more than they do; and, that there is no physical or mental necessity for the stationary or non-progressive conditions of most persons. The serfs of Russia and the peasants of all feudal countries, if removed from the influences that now depress them, might rise, there is reason to believe, to rank with the best of their fellow men.

It is admitted that circumstances, we think not an evil, restrain most women to what may be deemed a monotonous course of life. There is little, and we think ordinarily there should be nothing, to entice or drive them from the routine of domestic avocations. The good of society demands this. Their own ease and happiness are promoted by it. If it be any loss to them not to be in a condition to enter into competition with men for wealth or fame, this loss is more than counterbalanced by the consequent exemption they enjoy from the hardships and perils of men. Now and then a woman, most frequently a widow, is left in such destitution and with such responsibilities for the support of herself or children resting upon her, as to break through the timidity of most women, and to make efforts in some kind or other of business, till she becomes as much distinguished as the best of men for enterprise and success. Now and then under those circumstances a woman will become rich by her own unaided efforts.

Under our own observation, cases of the kind have occurred. Were we to take time for brushing up our mnemonics, we could name dozens of them. Without effort we can name a few. Among those in cities who were obliged to resort to the business of keeping boarders, some by good management became rich, and in shrewdness would stand creditably by the side of the best merchants. Of the number is Mrs. Ballard, of New-York. She is indeed too wise to tell every one how much she has made; but, if she were to draw a check on her banker for ten thousand dollars, hundreds would pronounce it good. The same might be said of many others in the same business. Take also the case of Mrs. Beman, nearly as well known in a large circle as William B. Astor, doing an extensive business in the City of New-York. Only a few years since, she was, we believe, a poor seamstress; it may be, like thousands of poor females in that city, earning in making shirts for the slop-shops, only eighteen pence a day. Look at her now with a large establishment for furnishing gentlemen with costly under garments, and giving employment and subsistence, to how many of her own sex, no one but herself can tell. From the apparent extent of her operations, one might fairly conclude, that the City Park cannot conveniently accommodate with fresh air all that she must necessarily employ. And, if she were desirous of purchasing ten thousand dollars worth, on credit, of Irish linens, muslin, and flannels, from a large jobbing house, her custom would be much desired, or we are much mistaken in regard to her means and known responsibility.

Take, too, the case of Mrs. Farnham, who, a few years since, went to California. The newspapers are frequently giving accounts of her progress. Her first intention was to induce a few hundred reputable single women to accompany her to that land of gold: it might be to become the wives of the first lonely emigrants. This shows her enterprise and energy; and there was nothing indelicate in the idea. The same thing has been done before on a small scale. If our foreign missionaries lose their wives, there has been no scruple with the most pure in mind, to send out to them accomplished ladies, although the parties to be united had, perhaps, never heard of each other before. And it is no unusual thing that those of royal families are brought into matrimonial alliances, who had never seen each other. In such dilemmas matrimony is a matter of propriety, or of business, or of taste. In such exigences, without any freaks of romantic love, men, and women too, are glad to take such companions as are at hand and have been deemed fit for them. Had father Adam declined to receive and take to his bosom mother Eve, because he did not fancy the color of her hair and eyes, it is very problematical whether there would be one like ourself, in the nineteenth century, to write for the amusement and the edification of the thousands who will read the Farm and the Fireside.

Mrs. Farnham, however, either because she could not procure an assorted cargo of petticoats equal to her estimate, or because some other project dawned upon her fertile imagination, abandoned this enterprise. Nevertheless, ere long she was in the land which had filled her mind with bright visions. Shortly after her arrival, it was reported by a correspondent that she was seen boarding and shingling the roof of a house that was to be her own home. At short intervals we receive through the newspapers fresh intelligence of the magnitude of her agricultural achievements; and we have had before us, when at our writing table, hanging upon the wall, a few ears of the wheat she raised in 1851. We obtained and retained them, as a curiosity of female enterprise, and as a sample of better grain than we had before seen. Should Mrs. Farnham continue her present operations ten years longer with undiminished energy, Roswell R. Colt, Esq., of Patterson, N. J., Col. J. M. Sherwood, of Auburn, N. Y., and the Hon. John Delafield, the President of the New-York State Agricultural Society, will be thrown into the shade by her. And not less deserving of notice is the case of a poor woman, mentioned in the papers not a month since, who went to California, and had baked ten thousand dollars' worth of pies for the gold diggers in her vicinity ; the greater part of it, without even a cooking stove, in a single open iron vessel, and collecting with her own hands on the mountain side, the wood used in the process. This is a fine illustration of Woman's Rights.

It may be well to inquire, before we leave these instances of female enterprise and success, by what development of human destiny, or by what unprecedented concatenation of circumstances, these achievements were perfected ? Did any unusual astrological concomitants attend their birth ? Were there seen in them during infancy and childhood, any precocious indications of extraordinary advancement ? Did they receive an impulse from the stringent resolutions of some female convention for the elevation of the female character ? Did they become galvanized from hearing lectures on manhood and womanhood ? Or have they been stimulated onward in their novel career by the magical and the overpowering influence of Bloomer unmentionables ? If such is the fact, at present it is not generally known. If such is the fact, history will doubtless embody it on her pages, and transmit it to coming generations. For one, we do not believe it. Our own impression is that they have done what thousands of other women might do, if disposed and straitened by poverty as these were. It is doubtful if ever the members of one of those conventions, or a baptized disciple of a half-yard long petticoat ever will do it ; it may rather be expected that they would prefer to live pensioners on their more fortunate relatives, or even to die in an almshouse.

Some time last year, or in the year previous, there appeared first in a Greenfield paper, Massachusetts, then in a Troy paper, N. Y., an article headed—" A Widow Worth Having." As usual with such choice bits, it passed the circuit of the newspaper press, and became a fragment of periodical literature. On its first appearance, the editor sanctioned it with the seal of authenticity, and furnished it with the above cognomen, indicative of the merits of the writer. The article was in the form of a letter addressed to him, and signed by the widow aforesaid—Marinda Hines. Extracts from the letter are as follow:—"I have five cows, and have sold, the past season, thirteen hundred pounds of butter, besides milk, cream, and butter for family use. Our family will average eight. I raised seven calves last Spring, and some late; two of them I got from my neighbors. I have fatted seven hundred and sixty-eight pounds of pork, mostly on sour milk. Now, let us leave one of the cows for family use, and set the credit to the other four: say thirteen hundred, divided by four, makes three hundred and twenty-five pounds, at sixteen and a half cents per pound—the price at which it sold—fifty-three dollars and sixty-two cents, to each cow. The calves were worth, say thirty dollars; twenty-five to four cows; and five hundred of the pork, at six dollars, will be thirty; and the twenty-five for the calves makes fifty-five dollars; and divided by four, leaves thirteen dollars and seventy-five cents to each cow; and this, added to what was received for the butter, will make sixty-seven dollars and thirty-seven cents of receipts for the produce of each cow; or two hundred and sixty-nine dollars and fifty cents for the produce of the four."

Mrs. Hines afterwards stated that her cows were of the native breed, and of medium size; that they have good pasture in summer, and good hay in winter; and no other feed, except a few apples, small potatoes, apple parings, and the like, in the winter, to Old Brindle, to make her milk hold out till some of the others came in. Thirty dollars is a large allowance for the feed of each of her cows. Now, suppose she were ambitious, and disposed to keep fifty cows instead of her present number, and there would be a clear profit of that number, of $1,868 75 cents annually. Who can say there is no profit in agriculture, under proper management? Such a widow would indeed make a wife, for a farmer, worth having. The dairy may, and should, properly, be under the direction of the female head of the family. Let all wives of farmers imitate the example of this widow, and there will be a new aspect in rural communities. All may do it, if they will. Mrs. Hines may be taken as a model agriculturalist. She has connected the improvement of the mind with the improvement of the farm. She has been accustomed to read the periodicals. The editor of the paper which introduced her to the public, said she had taken, and promptly paid him for, his journal, twenty successive years.

Of such instances of female talent our agricultural societies should take cognisance. The authors of them deserve publicity and tokens of public appreciation. The community at large would be stimulated by them. Hence we make record of them wherever found ; and of the benefit that may be derived from that of Mrs. Hines we think so highly, we would cheerfully render her a more substantial tribute of regard. One thought we had was to procure her portrait for an illustration in the present volume The difficulty of procuring it seemed to present an obstacle. Then it occurred to us, that instead of a kindness to the worthy widow, it might prove an evil to her. It can scarcely be supposed that a woman who can take care of herself and her farm as well as she can, needs a husband. And, if she were to try the experiment, and obtain one less competent for these duties than herself, she might naturally despise him. If so, the consequences would be the most inauspicious. Besides, it occurred to us, that if she were a handsome woman, she might be greatly annoyed from the crowds of rural widowers and old bachelors that would be thronging her from every part of the country where our book is read. And, on the other hand, it occurred to us, that possibly, notwithstanding her skill in the matters aforesaid, instead of being handsome, her features are coarse and repulsive. If so, she would by no means indulge a feeling of complacency in being paraded before the public. We happen to know that two of our most distinguished literary ladies have absolutely prohibited their publishers from prefixing their portraits to their published works, simply because they are notoriously ugly in their looks. We know not why a dairy woman may not have some pride in such a matter as well as a writer of prose or poetry.

The conclusion, therefore, is that, if any of the feminine gender are not satisfied with in-door avocations, and the customary apparel for women, they go immediately to the tailor and obtain a full equipment of clothing now customary for the male sex. Then let them abandon their needles and their scissors, their wash tubs and their scrubbing boards, their cook book and their saucepans,—one taking in the place thereof the brick mason's trôwel or the stone mason's hammer, another the carpenter's jack-plane or hand saw, another the farmer's spade or dung fork, another the blacksmith's bellows or sledge hammer, —each taking whatever occupation is most agreeable. Do this without the delay of attending a convention for women's rights; all that is unnecessary; and the regeneration of society will be nicely under way. True, it may be, the use of the broad axe and the sledge hammer will make your back ache and your arms ache; true, standing in the muck and water when digging ditches, you may be assailed with chills and fever; true, when stepping astride of the nether leg of an unruly horse to repair his understanding he may most ungallantly kick you over; but no matter for all that; all that is not to be thought of in comparison with the degradation of wearing petticoats and chemisettes, and performing the effeminate labors of domestic economy. Very well! we yield the point! In matters of taste there can be no argument; all must have their own way.

But, say our fine muslin, silk, and lace female reformers, we have no idea of any such servile occupation; our aim is to take the place of dandies cutting antics with ratans and cigarettes; or of gentlemen going to our Congress, or on national foreign missions, or, at the most, giving directions in the counting-room to male clerks, and drawing checks on the bank. This does, indeed, look better. Well, try this! The equipment of the dandy is quickly and cheaply provided; and the elevation to which he aspires is quickly reached, brains or no brains. If you prefer political distinction devote yourselves to the reading of political papers, visit porter houses in the city and taverns in the country, proclaiming yourselves as candidates for office, and soliciting, for months in succession, the support of loafers and rowdies; and, prior to elections, traverse the State, and make blackguard stump speeches. Possibly you will obtain a majority of the public suffrage. If not, you can try again, as do politicians of the other gender. Or, if mercantile attainments are the boon you seek, become apprentices, for five or seven years, like young men, to learn the mysteries of trade and commerce; and, for aught we know, you will stand one chance in a hundred, like them, after a life of anxiety and fatigue, to become rich.

Nevertheless, it seems to us, that if the pleasure and the honor of being the chief objects of admiration in the domestic and the social circle; of being good wives and good mothers; of preparing the other sex as well as their own for eminent usefulness in the world; of making the fireside the most endearing and lovely place on earth—with its quietude and sanctified affections, a type of heaven itself; is not sufficient to satisfy their aspirations, there are other means within their reach more congenial to the taste and physical organization of women, than the noise, and toil, and commotion in the routine of business or politics. An enviable and imperishable fame is by no means confined to those hitherto masculine pursuits. There are gems more precious than gold. There are laurels less prone to fade than those of the statesman or the warrior. Genius may weave a crown that will be brilliant when the crowns of kings and queens shall have become worthless. Are not the names of Mrs. Montague, Mrs. Chapone, Mrs. Hemans, and Hannah More, destined to live in freshness for centures to come? What more honorable distinction can a woman desire? And is there not room for others in the same pathway to reach the same distinction.

Nor is this the only pathway to honorable distinction, congenial to female instincts. If the female heart is not warm and impulsive far beyond that of the other sex in philanthropy, as yet we have never understood the female character. In ministrations of mercy, she shines with undying effulgence. In these miinstrations she is surrounded with a halo surpassing the beams of the mid-day sun. Did not Mrs. Fry, in mitigating the miseries of the female convicts of an hundred prisons experience a moral luxury, attained from no earthly pleasure? If those wretched creatures under her auspices became Christian penitents, will not the anthems of heaven have a never ending increased melody and pathos? And will her name ever cease to be resplendent in the annals of virtue? Surely not! Let our female reformers give up their quixotic crusade and follow her example. The world is full of wretchedness. Broken hearts too are more agonising than broken limbs. Bind up those wounds that medical science was never designed to heal. Probe and mollify those depths of sorrow the surgical instrument was never able to reach. Here should be a woman's glory.

There is now one living monument in our own country, of the fame and moral exaltation to which woman can aspire, without derogation to the gentleness of her nature; it is in restoring to its original symmetry and beauty the reins of the human mind. Miss Dix, of Boston, has spent most of her time for twenty years, in angel visits to ameliorate the physical suffering and to soften the hearts in our criminal penitentiaries; and especially in founding asylums for the comfort and the restoration of lunatics. In hundreds, and it may be, in thousands of raving maniacs, she has been instrumental in causing, about their own firesides, the irradiations of reason again to impart pleasure to all that behold them. Let the gifted lecturer, to whom allusion has been made, turn her attention to some kindred work of Christian charity. Here she might labor to the end of life; and when dead, thousands would rise up and call her blessed. In such an enterprise her deeds would follow her—or, to the end of time would survive her, in the plenitude of their freshness. There is still opportunity to do it. The deep recesses of human woe have not yet all been explored.

We close this chapter with the following anecdote of Miss Dix. When at Greenville, S. C., a few years since, a lady said to her: "Are you not afraid to travel all over the country alone, and have you not encountered dangers and been in perilous situations?" "I am naturally timid," said Miss Dix, "and diffident, like all my sex; but in order to carry out my purposes, I know, it is necessary to make sacrifices and encounter dangers. It is true I have been in my travels through the different States, in perilous situations, I will mention one which occurred in the State of Michigan. I had hired a carriage and driver to convey me some distance through an uninhabited country. In starting, I discovered that the driver, a young lad, had a pair of pistols with him. Inquiring what he was doing with arms, he said he carried them to protect us, as he had heard that robberies had been committed on our road. I said to him, give me the pistols, I will take care of them. He did so reluctantly.

"In pursuing our journey through a dismal looking forest, a man rushed into the road, caught his horse by the bridle, and demanded my purse. I said to him with as much self-possession as I could command, 'Are you not ashamed to rob a woman? I have but little money, and that I want to defray my expenses in visiting prisons and poor-houses, and occasionly in giving to objects of charity. If you have been unfortunate, are in distress, and in want of money, I will give you some.'—Whilst thus speaking to him, I discovered his countenance changing, and he became deadly pale. 'My God,' he exclaimed, 'that voice!' and immediately told me that he had been in the Philadelphia Penitentiary, and had heard me lecturing some of the prisoners in an adjoining cell, and that he now recognized my voice. He then desired me to pass on, and expressed deep sorrow at the outrage he had committed. But I drew out my purse, and said to him, 'I will give you something to support you until you can get into honest employment.' He declined at first taking any thing, until I insisted on his doing so, for fear he might be tempted to rob some one else before he could get into honest employment."

Had not Miss Dix taken possession of the pistols, in all probability they would have been used by her driver, and perhaps both of them been murdered. "That voice" was more powerful, in subduing the heart of a robber than the sight of a brace of pistols. Rarely is a human heart so hard, as to be unmoved by accents of kindness. They will cause a listening ear, and will secure a chastened homage when all other agents are powerless.

GOOD HUSBANDRY IN MASSACHUSETTS.

In the fall of 1850, the Hon. F. Holbrook, of Brattleborough, Vermont, visited some of the best farms in Massachusetts; among others, that of George Pearce, Esq., in West Cambridge, a few miles west of Boston. The year previous to this visit, the farm attracted particular notice from the large amount of its net profits, as reported by a committee of the Middlesex Society of Husbandmen and Manufacturers. These profits were stated to be two thousand and thirty-seven dollars and seventy-nine cents in that year; a sum showing that according to the capital invested, rarely can more money be made than from agriculture; for, if one person can do it, others may also. With such facts, and many of them may be collected, it is difficult to account for the general incredulity on the subject.

The committee awarded the first premium on farms to that of Mr. Pearce. In their report they say, this farm consists of about forty acres. The soil, except about four acres of swamp, is of a sandy loam. Two acres of this swamp has been cultivated for the first time the present season—the remainder is thickly overgrown with white birch, sprung from the stumps of a previous growth. The farm is cultivated for the express purpose of supplying vegetables and fruits for the Boston market. There are two peach orchards on the farm, one of which contains about eight hundred trees on three acres and a half. These were all procured from New-Jersey, and were set out in the spring of 1846. Many of them have borne fruit. The spaces between the rows are planted with melons, beans, and cauliflowers. The beans had been gathered in before the examination of the committee, and the vines thrown around the roots of the peach trees. The soil of this orchard was in a high state of cultivation before the putting in of the trees in 1846, but has since received no manure, except a shovelful placed in each of the melon-hills, twelve feet apart, at the time of planting. Some acres produced tomatoes,—on others are now growing celery, cauliflowers, cabbages, spinach, corn, potatoes, and, what the committee have nowhere else seen cultivated for the market, dandelions. In the judgment of the committee, there were about three acres covered with this latter vegetable, (which generally passes for worthless weeds,) and which affords a rich return for the labor and expense of cultivation. The following statement of the expenses and value of the produce was given to the committee, viz :—

Hands employed from April to October, at an average of $16 per month, ...	$672 00
Labor paid by the day.	80 00
Board of men at $10 per month, . . .	420 00
Night soil from ten vaults, ...	30 00
Manure from one stable in Boston, . . .	400 00
Teaming the same, ...	30 00
Manure from one stable in Charlestown, the produce of 44 horses, at $10 per horse, delivered at the farm, ...	440 00
Manure from Porter's stable in Cambridge, 30 cords at $5 50 per cord, ...	165 00
	$2,507 00
Proceeds of sales from March 23d, as rendered by the market-men, of which a daily account is kept, . -	$4,514 79
	$2,037 00

Leaving a balance of $2,037 79 in favor of the farm, exclusive of all the crops now on the land, probably worth as much more. Mr. Pierce generally manages so as to have his manure from the stables brought at a time when it can be spread and ploughed in. He makes no compost, but what is manufactured by two hogs. His domestic animals are one or two cows, three horses, used for ploughing and transporting his fruit and vegetables to market.

A leading principle of culture on this farm is to get all kinds of vegetables into market at the earliest possible period ; for any article, appearing there a week or two before its usual time, commands a very high price, which richly rewards any extra labor or pains. The proprietor has extensive hot beds for forwarding his various productions for an early market. He has 250 sashes, or some 1400 surface feet of glass, under which all sorts of vegetables are started. As a novelty in gardening, in this way he forces the growth of dandelions for greens. In March and April of 1850, the receipts from this source alone, amounted to three dollars per sash, or one shilling per surface foot of ground. At one picking of tomatoes in the same season he gathered thirty-two bushels, which, from being very early, sold for one dollar and seventy-five cents per bushel. Pole beans are produced early by digging large deep holes for the hills and filling them partly with fresh hot horse manure ; over that a suitable covering of earth is placed, and the beans are planted. Early potatoes are first started either on manure heaps undergoing fermentation, or in hot beds ; and when the weather will admit, and the sprouts are six to eight inches long, they are transplanted in drills upon open flats. This process forwards the crop from fifteen to twenty days. On one quarter of an acre, the same season, eighty-one bushels of marketable potatoes were dug, which sold at one dollar and seventy-five cents per bushel, or at the rate of $567 per acre.

ANECDOTES OF THE GOOSE.

WILD GEESE.

THE value and usefulness of geese is scarcely calculable. Had it not been for the quill of this bird much of the ignorance of the dark ages would still overshadow the world. Is it not to the agency of it that we are indebted for ninety-nine hundredths of all the books in existence? Were not the 500,000 volumes in the Bodleian library of Oxford; the 600,000 volumes in the library of Lyons; the 400,000 volumes in the royal library of Paris; the 300,000 volumes in the imperial library of St. Petersburg; the 300,000 volumes in the university library of Gottingen; and the 100,000 volumes in the library of Harvard college, all, or nearly all written, by the quills of geese? Let these, and their kindred illuminations, become extinct, and where would be our literature and our science, as needful in the realms of mind as the material sun in the material creation? True, the present century, abounding in novelties, is now furnishing us with pens of steel, and silver, and gold, in no less profusion than were the locusts of Egypt; but, still, give me in preference to all such trickery, the quill of an old white gander; in the use of such an instrument, as firm as it is elastic, there is some pleasure in writing. To such an one the reader is now indebted for this tribute of respect to the much abused goose. Before our hand had become so clumsy from more than fifty years of toil, it seems we might have written a volume with one such, whereas it would require a peck measure of these little stubborn, brittle, yellow scratchers, to enable us to do it.

And the goose has other claims on our affectionate consideration. Ask the gouty old man, and fidgety old woman, which is the chief alleviation to their aching bones, and they will tell you, it is an old fashioned bed filled with the feathers and the down of geese. How many millions of persons consider this one of the greatest of human luxuries! Who can tell how many midnight watchings it has lulled into the sweetest slumbers? Who can tell how many agonizing pains it has supplanted by its bewitching sorcery? Who can tell how many fretful crying infants it has soothed into the sweet languor of a quiet repose? If there were not untold charms in the feather bed, would it have so many votaries? Would such multitudes be clinging to it twelve hours out of twenty-four, when two-thirds of that time are sufficient for rest, were there not a magical witchery about it found no where else? Could the feather bed write its own annals how many reveries of its own creation would it unfold to the world? There we should see the solitary female beauty in all her loveliness; the ennui of young manhood in the enjoyment of a rich inheritance; and not less, the new made disciples of Hymen's altar, all—all dreaming away their mortal existence, as if it was to have no end.

Recently, however, there has been gotten up a crusade against feather beds as well as against goose quills; and, the whole object seems to originate in a conspiracy to drive from the pale of public favor the bird which produces them. It is said that feather beds produce effeminacy; that curled hair made into mattresses, only is compatible with health and vigor. It is possible that there is some philosophy in this. So far as there is, we hesitate not to make the admission. We are no enemies to rational progress. We rejoice in all valuable improvements; and we frankly admit that human health and vigor are frequently impaired by a want of physical culture; that long and habitual exemption from atmospherical influences and muscular action will lead to feebleness of body and mind. We see this exemplified in thousands of cases. The city is crowded with them. The mansions of the rich are frequently made desolate by them. Nevertheless, it is laughable, to hear the lugubrious homilies against feather beds from persons who sleep in rooms heated by furnaces to a degree to bake custards, or to produce, in luxuriance, the most tender of tropical plants. If it were true, which it probably is, that such persons might advantageously sleep on a mattress, or even the soft side of a pine board, farmers or the families of farmers are in no danger of effeminancy from sleeping on feather beds. Till their ploughing is done by steam, and till their planting, and hoeing, and mowing, and reaping, and thrashing is done by steam or horse power, we recommend to them the softest feather bed they can procure. We will be responsible for all injury from it.

The main profit, however, from the rearing of geese, is in their flesh. So highly prized is the meat of this bird that a Mr. Bagshaw, of Norwich, England, annually fattens thousands for the London market. It is stated that occasionally he has as many as two thousand, and on his extensive premises at the same time. They are collected from all parts wherever to be obtained. Some are raised in England, more are from Holland, but the greater part procured even from Prussia. It would be a curious labor to collect statistics of the goose tribe, particularly the number in our large cities, where they are most sought, dressed for the table. For ourself we prefer the flesh of the goose to that of the turkey. Others may have a different preference. In matters of taste there can be no conclusive argument. All have a right to their own fancy.

It has been supposed that geese can be raised with greater profit, according to the cost of feeding them and the labor of watching over them, than most other poultry. All the fears and anxieties to rear the turkey and prepare it for making a proper appearance at table, are with them unnecessary ; grass by day, a dry shelter by night, and a tolerably assiduous mother, being all that is required. They will thrive well, too, on grass that is too short for any other animal. It is known that sheep will do well on pasturage that has been cropped by cattle and horses as closely as they are able to do it. Yet, after sheep have gone over it, and with their sharp teeth have given it another trimming, a flock of geese will find upon it luxuriant fare. The little grass that is found upon the road side will afford them decent sustenance, although most other animals would not notice it. A writer on the subject supposes that any common village green, with the adjacent highway, would support fifty brood geese. If from each one ten goslings were raised, which is not a great estimate, there would be a produce to the inhabitants of the village of five hundred geese, at a mere trifle of expense, save what would otherwise be of no use to any one. How many village greens, and how many verdant acres of highway sides and public lands are there in our country, that would sustain these valuable birds ? It is believed that thousands of farmers would derive as much profits from such incidental appendages to agriculture as they do from their staple crops. At any rate, they would furnish the family mansion, as it were, without material cost, with the materials for good living. It is indeed true that geese will thrive best where they can have free access to the shores of a pond or the margin of a river. Here they will feast themselves on worms, and whatever is found by them in the mud. It is amusing to see what havoc they will make on the little fish, and no one can tell what it is that attracts their notice. They dive and seize their prey, the young goslings as well as the old ones, with the precision that a Cape-Coder will harpoon the whale. Woe be to the finny and vermiculous tribes, if an old white gander and his fifty associates can have a fair chance at them.

It is a common opinion that geese are silly and stupid. Hence, if a biped in pantaloons or petticoats is peculiarly deficient in common sense, he or she is dubbed with the cognomen of goose-head. According to the prevalent apprehensions of gooseology, such designations, in numerous cases, would be very appropriate; but, oftentimes, when applied, we think the feathered gentry, of the two races, is the more slandered. "We once had," said the author of British Husbandry, "four geese for our own use, which were constantly turned upon the neighboring common, at the distance of a quarter of a mile, to which they regularly walked, through the village, headed by the gander, calling at every house, from which fragments of bread were sometimes thrown, and returning at the usual hour in the evening. Though also accused of dirty habits, because they grope for food in any pond or ditch to which they may have access, yet no bird is more careful of its plumage, or fonder of a clean and dry bed. They are very capable of attachment to those who treat them with kindness; extremely quiet in their habits, and peaceable towards all the other poultry, but courageous in defence of their young; and the hiss of the gander in their protection is full of menace." "When riding, sometime ago," says the same gentleman, "rather sharply across the common, our mare unheedingly trod upon a gosling, when, instantaneously, both geese and gander flew upon her with such violence that it was difficult to either restrain her from running away, or to beat them off; and for a long time afterwards the mare, upon seeing a flock of geese, snorted and started off with affright."

One of the most pleasing anecdotes illustrating the character of the goose is given by the Rev. Mr. Ottway. He says "that at the flour mills near Clonmel, in Ireland, while in the possession of a Mrs. Newbold, there was a goose which, by some accident, was left solitary, without mate or offspring, gander or gosling. Now, it happened, as is common, that the miller's wife had set a number of duck eggs under a hen, which in due course of time incubated; and, of course, the ducklings, as soon as they came forth, ran with natural instinct to the water, and the hen was in a sad pucker—her maternity urging her to follow the brood, and her selfishness disposing her to keep on dry land. In the meanwhile up sailed the goose, and with a noisy gabble, which certainly, being interpreted, meant, 'leave them to my care,' she swam up and down with the ducklings; and, when they were tired with their acquatic excursion, she consigned them to the care of the hen. The next morning, down came again the ducklings to the pond, and there was the goose waiting for them, and there stood the hen in her great flustration. We are not at all sure that the goose invited the hen, observing her maternal trouble, but it is a fact, she being near the shore, the hen jumped on her back, and there sat—the ducklings swimming, and the goose and the hen after them, up and down the pond. And this was not a solitary event. Day after day, the hen was seen on board the goose, attending the ducklings up and down, in perfect contentedness and good humor, numbers of people coming to witness the circumstance, which continued until the ducklings, coming to the days of discretion, required no longer the joint guardianship of the goose and hen."

Our own domestic goose is too common not to be familiarly known to the reader. Among other varieties that have been domesticated, and are the best known, are the African goose, the Indian goose, the Poland goose, the Toulouse goose, the Canada goose, the Chinese goose, and the Bremen goose. Some beautiful specimens of the Chinese goose are on the farm of the Hon. Daniel Webster, having been obtained by his son, Fletcher Webster, Esq., a few years ago, when on a public mission to the celestial empire. Col. Jaques, of Ten Hills Farm, has also fine samples of them, and also of the Bremen geese. Of the latter, Mr. James Sisson, of Warren, R. I., has a good stock. In a description of them, he says—"In the fall of 1826, I imported from Bremen three full blooded perfectly white geese. I have sold their progeny, for three successive seasons—the first year at $15 per pair, the two successive years at $12. Their properties are peculiar; they lay in February, sit and hatch with more certainty than common barn-yard geese; will weigh nearly, in some instances quite, twice the weight; have double the quantity of feathers; never fly, and are all of a beautiful snowy whiteness. I have sold them all over the interior of New-York, two or three pairs in Virginia, as many in Baltimore, North Carolina, and Connecticut, and in several towns in the vicinity of Boston. I have one flock, half blooded, that weigh, on an average, when fatted, thirteen pounds; the full blooded weigh twenty." The African or Guinea goose, is said to be the largest of the varieties named.

A writer in the Farmer's Gazette makes the following calculations on the profits of keeping geese. He says, let us estimate the profit of ten old geese, kept in the manner usual. Purchased in the spring of the year, before they commence laying, and they will cost, at most, but a dollar each. Two of the ten should be ganders; eight geese, that will have, on an average, ten goslings each. But allowing one half for paper calculation, it will leave at the end of the season a flock of fifty, old and young, worth, when dressed for the market, we will suppose, only fifty cents a piece, or twenty-five dollars. In addition to this, every old goose will yield one pound of feathers, and every young one three-fourths of a pound, making in all forty pounds, which, added to the twenty-five dollars will give us forty dollars. This may be called nett profit, for there is not one goose keeper in ten that feeds his geese, either old or young, after the grass has started in the spring, until fattening time in the fall; and then the large quills will more than pay for the extra food. This is the common way farmers keep geese; but if they were to provide good warm houses for their accommodation, and attached to a good pasture with a small stream or pond of water, the prospect of gain would be still better.

On the advantages of keeping geese, a writer in the Maine Farmer says—" I once knew a couple of industrious sisters, who lived near a never-failing brook or stream in Massachusetts, and they generally kept through the winter thirty geese, male and female. They had erected some suitable but not costly sheds, in which were apartments for them to lay, set and hatch. Their food in the winter was meal of various kinds, to some extent, but principally apples and roots. In the summer they had a pasture enclosed with a stone wall or board fence which embraced the water. Their wings were clipped so that they could not fly over the fence. The owners well knew what we all know, that live geese feathers are a cash article at a fair price. They picked off the feathers three times in the season. Those thirty geese wintered would raise from seventy-five to an hundred goslings or young geese, and of course they had that number to dispose of every fall, or in the beginning of winter, when they were sent to market; from these they obtained feathers twice—from those wintered, four times. It would be easy to be certified that there must be a handsome profit in such a case. How many families there are, in all our towns, so situated that they may make the raising of geese an advantageous business? There is no mystery about it. The process is a very simple one. It is scarcely possible under any circumstances that it should be attended with loss, unless the stock is killed by the foxes, a nuisance not very common now a days.

Having known something of the American Wild Goose from childhood, and having, perhaps, seen tens of thousands of them, we are tempted to give a few particulars respecting this variety of the goose family. The first in order is from Col. M. Thayer, of Braintree, Massachusetts, dated August 18th, 1848. "A few years since," says he, "a neighbor of mine shot at a flock of wild geese while passing to the south, wounded one in the wing, took him alive, and very soon domesticated him. He soon became very tame, and went with the other geese. I bought him, and kept him three years, and then mated him with an old native goose. The wild goose does not breed till four years old. They had several broods of young ones, and the old goose became very feeble; so much so, that she could not sit long enough to hatch out her eggs. Accordingly I put them under another goose where they did very well. In the fall of the year I gave her away, and mated the wild gander with another. In the spring following, about six months after, I heard that the old goose had got better, and was in good health. She was brought home and put in the poultry yard. The wild gander and his new mate were at a distance of about eighty rods, in another pasture, on a high hill. As soon as the old goose was put into the yard, she made a loud noise, which the wild gander heard. He immediately left his new mate, and came down to the yard, recognised his old mate, entered into close conversation, and appeared extremely happy in seeing her again. His other mate followed him, and wished to join the party, but he appeared much offended, treated her with the greatest indifference, and drove her from him.

"The old goose soon began to lay; and as soon as she set, I put under her, besides her own eggs, three laid by another goose. They all hatched out, and the goslings all looked precisely alike —so no one could tell the difference; but as soon as the wild gander saw them, he appeared to take particular notice of three of them, and looked at them for several minutes. He then began to peck and push them away. I thought nothing of this, and left them. In the course of the day, I looked at them again. He was then pecking and trying to kill them. I took a stick, and struck him several times on his wings, and drove him away. The next morning I went to see them, and found him still pecking them, and had almost killed them. I then whipped him more severely, when he soon left them, shaking his head several times, a signal the wild goose always gives previous to migrating. I then left them till the next morning, when I found he had gone and could not be found.

"About ten days afterwards, I heard that a wild goose had been taken about two miles from my farm, while swimming down the bay. I sent a man after him, and it proved to be my gander. He was brought home, put with the old goose and goslings, but took no notice of them, and would not go near them, keeping at a distance of four or five rods from them, thus continuing for about three months. I then killed the three goslings, immediately after which he went to his old mate and goslings, appeared to converse with them for several minutes, made all up, and continued a faithful and affectionate husband and father, and remained with his mate till he was accidentally killed."

The following account of the wild goose is taken from Wilson. "I have never visited any part of the country," says that gentleman, " where the inhabitants are not familiarly acquainted with the passing and repassing of the wild goose. The general opinion, in Canada and the northern part of the States, is, that they are on their way to the lakes to breed ; but the inhabitants on the confines of the great lakes are equally ignorant with ourselves of the particular breeding places of these birds. Their north is there but commencing, and how far it extends it is impossible for us at present to ascertain. They were seen by Hearne in large flocks within the arctic circle, and were then pursuing their way farther north. They have also been seen on the dreary coast of Spitsbergen, feeding on the water's edge. It is highly probable that they extend their migrations under the very pole itself, amid the silent desolation of unknown countries, shut out from the eye of man by everlasting barriers of ice. That such places abound with suitable food, we cannot for a moment doubt."

The flight of the wild goose is heavy and laborious, generally in a straight line or in two lines approximating a point like the letter V. In both cases the van is led by an old gander, who every now and then pipes his well known *honk*, as if to ask how they come on ; and the *honk* of all's well is generally returned by some of the party. When bewildered in foggy weather, they appear sometimes to be in great distress, flying about in an irregular manner, making great clamor. On these occasions, should they alight on the ground, as they sometimes do, they are liable to speedy death and destruction. The autumnal flight south lasts from the middle of August to the middle of November ; and the vernal flight north from the middle of April to the middle of May.

The wild goose breeds only at the north. Some of their favorite haunts are the shores of Labrador and Hudson's Bay. We have been informed by fishermen and others who have visited those regions, that they are found on the banks of creeks and streams, during the breeding season, in immense numbers. The fishing vessels sometimes send boats, and procure their eggs in large quantities.

In becoming domesticated, the wild goose retains for a long time, as already intimated, many of its original habits, and is much disposed to return to its natural state. "An instance was related to us," says a writer in the Albany Cultivator, "of a wild gander having been wing-broken by a shot, and caught. His wound healed, but he was so much disabled he could never fly. He was kept with the domestic geese of the farm, with one of which he finally mated. He remained on the farm many years, and became very tame; yet as the migratory season returned, he always manifested the greatest uneasiness. Impelled by an instinctive desire to fly away, he would wander about, often stretching his wings and pluming his feathers; refusing food, and even leaving unnoticed his domestic mate, to which he seemed generally much attached. If a flock of his former associates chanced to fly over, salutes were quickly interchanged; but the thrilling note of the free ones in the air, in the language of our informant, seemed to set the poor creature crazy for a time, and for a day or two after he would be mute and melancholy. Except during the flying season he was very contented."

The English at Hudson's Bay depend greatly on geese, and in favorable seasons kill three or four thousand, and barrel them up for use. They send out their servants, as well as Indians, to kill them on their passage. They mimic the cackle of the geese so well, that many of them are allured to the spot where they are concealed, and are thus easily shot. The gunners sometimes take one or two of the domesticated wild geese with them to those places over which the wild ones are accustomed to fly, and concealing themselves, wait for a flight, which is no sooner perceived by the decoy geese, than they begin calling aloud, until the flock approaches so near, that the gunners are enabled to make great havoc among them with their musket shot.

The author of the Naturalist's Library mentions that Mr. Platt, a respectable farmer on Long Island, being out, several years ago, in one of the bays which, in that part of the country, abound with waterfowl, wounded a wild goose. Being unable to fly, he caught it and brought it home alive. It proved to be a female, and turning it into his yard with a flock of tame geese, it soon became quite familiar, and in a little time its wounded wing became entirely healed. In the following spring, when the wild geese migrate to the northward, a flock passed over Mr. Platt's barn yard, and just at that moment, ther leader happening to sound the bugle note, our goose, in whom its new habits and enjoyments had not quite extinguished the love of liberty, and remembering the well-known sound, spread its wings, mounted into the air, joined the travellers, and soon disappeared. In the succeeding autumn, the wild geese, as usual, returned from the northward in great numbers, to pass the winter in warmer climes. Mr. Platt happened to be standing in his yard, when the flock passed directly over his barn. At that instant he observed three geese detach themselves from the rest, and after wheeling several times, alight in the middle of the yard. Imagine his surprise and pleasure, when, by certain well remembered signs, he recognised in one of the three his long lost fugitive. It was she indeed! She had travelled many hundred miles to the lakes; had there hatched and reared her offspring; and had now returned with her little family, to share with them the sweets of civilized life.

In Willoughby's Ornithology, among others we find the annexed curious anecdote. "It is well known," says he, "that Canada geese are not fond of a poultry yard, but are of a rambling disposition. One of these birds, however, was observed to attach itself in the strongest and most affectionate manner to the house dog, and would never quit the kennel, except for the purpose of feeding, when it would return again immediately. It always sat by the dog, but never presumed to go into the kennel, except in rainy weather. Whenever the dog barked, the goose would cackle and run at the person she supposed the dog barked at, and try to bite him by the heels. Sometimes she would attempt to feed with the dog, but this the dog, who treated his faithful companion rather with indifference, would not permit. This bird would not go to roost with the others at night, unless driven by main force; and when in the morning she was turned into the field: she would never stir from the yard gate, but sit there the whole day in sight of the dog. At last orders were given that she should be no longer molested, but suffered to accompany it as she liked. Being thus left to herself, she ran about the yard with him all the night; and, what is particularly extraordinary, and can be well authenticated, whenever the dog went out of the yard and ran into the village, the goose always accompanied him, contriving to keep with him by the assistance of her wings; and in this way of running and flying, followed him all over the parish.

This extraordinary affection of the goose towards the dog, which continued to his death, two years after it was first observed, is supposed to have originated from his having accidentally saved her from a fox in the very moment of distress. While the dog was ill, the goose never quitted him, day or night, not even to feed; and it was apprehended she would have been starved to death had not order been given for a pan of corn to be set every day close to the kennel. At this time the goose generally sat in the kennel, and would not suffer any one to approach, except the person who brought the dog's or her own food. The end of this faithful bird was melancholy: for when the dog died, she would still keep possession of the kennel, and a new house dog being introduced, which in size and color resembled the one lately lost, the poor goose was unhappily deceived, and going into the kennel as usual, the new inhabitant siezed her by the throat and killed her.

The goose has been supposed as long lived as the swan, and numberless instances have been recorded of their having outlived the age of man. Moubray indeed, mentions as an established fact that there was, in 1824, a goose living in the possession of a Mr. Hewson, in Lincolnshire, England, which was then upwards of a century old. It has been throughout that term, in the constant possession of Mr. Hewson's forefathers and himself; and, on quitting his farm, he with a feeling which does him credit, would not suffer it to be sold with the other stock, but made a present of it to the incoming tenant—" that the venerable fowl might terminate its career on the spot where its useful life has been spent such a length of days."

The goose has for many ages been celebrated on account of its vigilance. The story of the saving of Rome by the alarm they gave, when the Gauls were attempting the Capitol, is well known, and was probably the first time of their watchfulness being recorded, and, on that account, they were afterwards held in the highest estimation by the Roman people. The story is simply this; that while the watch dogs and sentinels of the city, at the hour of midnight, were either asleep, or were remiss in duty, a few of the enemy scaled the precipitous and perilous heights which led to the capitol. The geese, however, heard the noise, commenced cackling, and awakened the consul, Marcus Manlius. The alarm was immediately given, many of the garrison were collected, and consequently the assailants were repulsed with a fearful slaughter. And it is certain to all acquainted with the habits of this bird, that nothing can stir at night, nor the least or most distant noise can be made, but the whole flock are roused, and begin to hold their cackling converse; and on the nearer approach of apprehended danger, they set up their more shrill and clamorous cries. Even the cunning and stealthy fox is almost always foiled in his schemes for a late supper on goose. He is too wily to fail in making a sudden retreat whenever the old gander wakes up the dog or the wearied farmer; for he seems to have an unconquerable aversion to the smell of gunpowder, or the sharp white teeth of that quadruped. On account of this trait in the character of the goose, it is by many persons esteemed the best of all country protections against the depredations of horse thieves and burglars.

HISTORICAL SKETCH OF THE TURKEY.

THE domestic turkey, in some important respects, is the most valuable bird that has place in the farmer's poultry yard.

It is large, comely in appearance, and its flesh furnishes to the epicure one of our richest dainties. To the careful observer its habits are interesting, although somewhat eccentric ; and, what is greatly in its favor, the more we study these habits, the more we are pleased with it. There is one trait in the male that is

never unobserved. His shouts of exultation when surrounded by his female companions, and when calling together their broods of young, may sometimes be heard nearly a mile. It is wonderful to observe, how the little progeny will respond to his voice, if at a distance of twenty or thirty rods in the rear, as led by him in their daily explorations for food, and especially at the close of the day, when returning for repose at their usual place of spending the night. It cannot be denied, however, that in this latter respect, turkies are deficient in punctuality, and not unfrequently are overtaken by night before reaching home. If so they make an encampment wherever they may happen to be. But this is not the result of indifference to home, as in the case of the tipler and the gambler, so much as to a defect in the science of geometry, not remembering how far they have wandered from it, or to a deficiency of astronomical observation, not having observed how rapidly time had sped.

The well fed male turkey, especially if rendered social by a numerous family of female attendants, is a very important character about the homestead. No one attracts more notice than his lordship. No one is more tenacious of his rights, or more complacent in the enjoyment of them. He is an original character truly ; but has numerous imitators. The incessant pompous display of his plumage has ever been deemed an appropriate counterpart of the human being which struts and seeks, by ostentatious exhibitions of exterior embellishment, to attract attention beyond any claims founded on intrinsic merit. We cannot fail to be amused on seeing either of these animals of the masculine gender thus struggling for the ascendancy ; but we cherish less respect for the one in broadcloth than his prototype in feathers. Indeed, the latter, although not celebrated for his mental endowments, possesses more intelligence than is usually attributed to him ; and, moreover, as the representative of his family, occupies no inferior rank in respectability or the elements of being useful. He is led by instinct, if not by reason, to be a pattern of devotion to the safety of the community of which he is the legitimate head. He watches over the turkey chicks with the assiduity of the most faithful shepherd when guarding his flocks. He will never leave them ; and is apparently unmindful of his own wants, so long as they require his watchful care. On one occasion a flock of forty odd, more than half grown young turkies, with the old ones were overtaken by night before reaching home. The consequence was they roosted on the fence adjoining our corn field. In the night, eight of the young brood were killed, by we know not what, and dropped on the ground. For hours in the morning the living ones remained on the spot around those that had been killed, the gobbler and his mates making the most piteous lamentation, till we were thus drawn thither. For a long period afterwards, they were not seen to go near the place of this calamity ; but, daily went in an opposite direction, which previously they had not done.

It is frequently said that turkeys are very stupid. We formerly thought so ; but on being more acquainted with them have become somewhat sceptical in regard to such an opinion. If they possess naught of what is usually termed reason, they have a kind of cunning much resembling it. The hen turkies are noted for stealing away their nests; and, if they do it, no little difficulty is experienced in finding the place of concealment. If you follow them, the probability is, should they perceive your intentions, they will lead you in a wrong direction, or will wander about for hours, till you become wearied out and leave them, when they will immediately go and deposit their eggs. On one occasion, it became apparent, that a favorite hen of ours, daily left the yard by flying over the fence, to visit her nest. It was usually about eleven o'clock in the morning, and after being absent an hour or two, would return and join the flock. Her direction was through an adjacent wheat field in a line apparently as straight as could be drawn by a land surveyor. This we noticed for several days in succession. Her course was always in the same beaten track. Every now and then she would stop, reach upwards her head, and look round, to see if she was observed. At length we concluded to follow her, at a distance of thirty rods or so, keeping behind the apple trees; but after a while she caught a glimpse of us, and although at such a distance, and the wheat was more than two feet in height, she turned about and came back nearly in the same path, and without enabling us to be the wiser for our labor. This we did several times with similar results, and at last gave up the attempt. Those whose metaphysics will not allow them to credit this turkey with the possession of reason, will do us a favor by telling us what we should call it.

We have made similar efforts in other cases, and usually with no better success. In some instances, however, we attained our object, yet, in others, the pursued, on discovering us, would turn off in a right angular direction, or go to a distance evidently far beyond the object of our search, in order to deceive us. What was that but reason? In man it would be called so. And, when they are setting, though in concealment, unless at a great distance from home, they will daily, or once in two days, leave

4

their nests, and return to the one accustomed to feed them, if to be found, asking for their accustomed allowance, as plainly as close and continued pursuit and beseeching looks will enable them to do so. When they receive it, their departure will be speedy ; but, if you follow them, especially in the early period of their incubation, they will be likely to beguile you as they had previously done. And, we have noticed another fact, allied to the above, very curious indeed. We say fact, because it has the appearance of fact. When the hens are laying in privacy, the gobbler, every time he has a chance, will approach us to the distance of a few feet only, reach out towards us his head, as if he wanted to tell us something, making a peculiar noise, and standing before us for some minutes. Perhaps we have seen it hundreds of times. If it is not to make such a revelation, the object is unknown to us. The same thing is not done at other times.

Turkies, too, have a language known and understood among themselves, as well as their owners understand written language. It may not be Hebrew, or Greek, or Dutch, or Esquimaux, but it answers their own purpose. It cannot be as extensive as that of the Chinese, the whole of which cannot be learned in the lifetime of man ; but, by it, young and old of a turkey community will learn to communicate their thoughts to each other in far less time than small children in an infant school will learn the elements of our vernacular dialect. By a particular word —to them it is a word—or whistle, or sound uttered by one of the elder members of the flock, particularly if by the gobbler, should there be fifty of them, all will stop feeding and look up to learn what is wanted. This will be invariably done, if the old gentleman or one of the old ladies, should discover a hawk a mile distant in the air. How do turkies know the character of those Goths and Vandals of the feathered races, till they shall have experienced their ravages ? They apparently do know it from instinct. Persons wiser than ourselves may answer the question. We cannot.

We feed our poultry mostly on Indian corn. Two years

ago, some of our seed corn for soiling purposes, which had been soaked, and was left in the barrel, became a little mouldy. One morning we took a peck measure partly filled, and commenced scattering it on the ground among the turkies, as we had frequently done before with dry corn. All as usual seemed delighted with their breakfast. There was a complete scramble among them, old and young, apparently to see who would get most. All at once the gobbler did not like the taste of it. He suddenly suspended operations, first twisting his head half way round to take a more careful view of it with his right eye, then the other way to scrutinize the suspected grain with the left eye. This was repeated several times with as much naivete as the school boy will look through a piece of smoked glass to observe the shadow of an approaching eclipse on the sun's disc. Quickly his gobblership became satisfied that something was wrong about the corn ; this mould might be poison from the apothecary shop for aught he knew ; it might have been prepared to kill the villainous rats, and by accident given to the honest turkies—possibly thought he ; at any rate he deemed caution a primary canon in his code, and resolved to give the alarm. He did so by one of his peculiar sounds, or words, which have not been translated into the English language, holding up his head, and looking as wise as a bronze statue of Confucius. Anon the old ladies, like other old ladies, rarely second and ordinarily in advance of their lords temporal in espying mischief, stopped eating, held up their heads, and repeated the same signal of danger. All then, old and young, stopped eating, and held up their heads, apparently asking what was the matter. Gobbler now again turned his head this way and that way, first to inspect the corn with one eye and then with the other, for them to see. They all did the same, with as much accuracy as a battalion of soldiers will go through Steuben's Manual Exercise, under the command of their colonel. In five or ten minutes they all disappeared, without eating another atom. Lucky would it be for parents, although unlucky for doctors and grave-diggers, if our own children would as readily give heed to our cautions, when we tell them not to eat plum cake and sweetmeats.

A few incidents may be related illustrative of the fidelity of the gobbler in watching over the young brood. Our own practice is the common one, in a few days after the process of hatching is completed, to put the hens into a large coop or pen, of a rod or two in extent, with one side at least like pale fence, to admit egress and ingress of the chicks while their mothers, naturally great gossips, are restrained from long peregrinations too toilsome and hazardous for the tender offspring. On the outside and in the immediate vicinity of the pen, the gobbler spends his days in becoming assiduities to the infant family, and his nights in roosting upon it or close by it. Quickly does it happen, that they become more fond of his society, than of their pent up mothers. As soon as his supremacy over them is well understood, and their strength admits of it, he will abduct them, no one can guess how far, as stealthily as the unprincipled swain runs off with his improvident lass for a clandestine marriage. Search for the missing ones is usually as unsuccessful in one case as in the other. However, gobbler is a far better protector of his treasure than the speculating lover, who steals a wife from her fond parents. The former never abandons his charge; whereas, the latter frequently does, leaving his deceived fair one, after being robbed of all she possessed, to return in disgrace and poverty to her broken-hearted family. It is amusing to see how faithfully the male turkey, when thus the sole guardian of his children, will seek to provide them with food, and to protect them from injury. In the night and in stormy weather he spreads over them his broad wings; and if a hawk is seen, the same is done to shelter them from his marauding descent upon them; if they have become too large to be thus sheltered, they collect around him as close as possible, while his gorgeous crest rises above them, not more captivating and alluring to an enemy than the expressive banner which floats in the breeze over the well-mounted and strongly manned fort.

We annex the following anecdote of a gobbler in Rhode

Island, of recent occurrence, not very dissimilar in character to what we have often seen ourselves. The Providence Post is voucher for the facts. It appears that a male turkey kept in 1851 on the farm of Mr. Paris Mathewson, in Johnston, resolved on a revolution in turkeyism. Accordingly, he drove from the nest one of his better halves, where there were 21 eggs, and performed the duties of incubation himself. The duties were so well performed, that 18 young turkies duly made their appearance. Nor was this all. He became so pleased with the female cares of domestic life, that he spurned all interference from the gentler sex. When his own brood was fairly out of the shell, and finding that others of the household had been occupied in the same labor, so that there were in all 67 young turkies to be taken care of, he determined to have undivided dominion in the domestic realms of turkeydom. This he did by turning the entire female fraternity out of doors, and taking the entire care of the nursery upon himself. The Post did not inform the reader whether this was a mere freak of oddity or eccentricity in his worship, or was designed to test the philosophy of the expurgated members of the family, who might have had a convention in regard to female rights, or had otherwise become tinctured with Bloomerism. If the latter be true, the incident may be of use to all turkey hens, whether in feathers or chemisettes.

To describe the domestic turkey is superfluous; the voice of the male, the changing colors of the skin of the head and neck, his proud strut with expanded tail and lowered wings jarring on the ground; his irascibility, which is readily excited by red or scarlet colors, are points with which all are conversant. Turkey cocks are pugnacious and vindictive, and often ill-treat the hens; and they are known occasionally to attack children. If there are two gobblers to the same family of hens, provided they are equally matched, their contests will be incessant. One season from this cause, the effect was the same to our own turkey profits as though we had had no gobbler; instead of a young progeny our eggs were all rotten. Between the turkey gobbler and chanticleer, particularly the game-cock, there is a most invincible antipathy, and consequently they are in almost perpetual collision. The pomposity of the one and the provoking insolence of the other, furnish ample occasion for testing their gladiatorial skill. One has the advantage in size, but the dashing activity of the other, aided as he is by a ready and efficient application of his spurs, often oblige his bulky antagonist to relinquish the contest, second best, if not fairly subdued. Of course, when a little rested, it requires no great provocation on either part again to encounter the perils of war.

We do not imagine, that as a general thing, the raising of turkeys is attended with great profit. Such profit would depend on what might be termed good luck, and against this it cannot be denied there are numerous contingencies, to obviate which no foresight or caution can always be made efficacious. If one happens to have such luck, the profit is most satisfactory. However, other considerations may be taken into the account. They are among the most highly prized luxuries, within the reach of the farmer, and which he should not deny himself. Very soon after their introduction into England, so essential were they deemed, that Archbishop Cranmer recognised them as a constituent of the Christmas festival. From that time to this, these birds have seemed almost canonically to constitute the fare of the farmer, especially on that joyous anniversary. Our forefathers assigned to them a corresponding dignity at the New England Thanksgiving dinner-table. We raise turkies, therefore, not simply for profit, but for a kind of conventional decency, as we polish our boots; or as we use sugar in our tea and coffee, because thereby they become more palatable; or, as we use an hundred other things, simply to gratify the eye or the ear, or because we relish them. If no attention were paid to motives of this kind, our course would be retrograde—to a state of barbarism.

All domestic turkies had their origin in the Wild Turkey of America. This is a noble bird, far exceeding its domestic relative both in size and beauty. Crosses, however, in our own country occasionally take place between the wild and the tame race, and are highly valued, both for external qualities and for the table. Indeed, in districts where the wild turkey is common, such crosses are frequent, the wild male driving away the domesticated rival, and usurping the sultanship of the seraglio. Eggs of the wild turkey have been frequently taken from their nests, and hatched under the tame hen. The young preserve a portion of their uncivilized nature, and exhibit some knowledge of the difference between themselves and their foster mother, roosting apart from the tame ones, and in other respects showing the force of hereditary disposition. There have also been instances where wild turkies have been caught when so small, as to be domesticated; but, they always retained more or less of their original character, roosting by themselves, and on the highest branches of trees, above the tame birds.

The native country of the wild turkey extends from the north western territory of the United States to the Isthmus of Panama, south of which it is not found, notwithstanding the statements of some authors to the contrary. In Canada and the now densely populated portions of our country, wild turkies were formerly very abundant, but like the American Bison, they have been compelled to yield to the ingenuity of the settlers, and seek refuge in the remotest parts of the wilderness. The wooded parts of Arkansas, Louisiana, Tennessee, and Alabama, the unsettled portions of Kentucky, Indiana, Illinois, the vast expanse of territory north west of the States, on the Mississippi and Missouri, as far as the forests extends, are more supplied than any other parts of the Union with this valuable game, which forms an important part of the subsistence of the hunter and traveller in the wilderness.

GREAT PROFITS FROM A SMALL FARM.

One of the greatest mistakes made by most farmers is the disposition to have large farms. This is mostly a matter of pride, or it may be misapprehension in relation to the principles of tillage. The fact is well proved, that a small farm, well cultivated, will yield more than one double in size, if badly cultivated. Rich farmers naturally have large ones—as large as they are able to keep in good condition ; for if there is profit on small ones there will be corresponding profits on large ones, provided the culture be the same ; but poor farmers, or those of limited pecuniary means, will always remain poor, and will have to toil hard, if in this respect they undertake to follow the example of such as have unlimited means. If the products of the soil cannot be increased to any indefinite extent by progressively high culture, it is known, because it has been demonstrated in thousands of cases, that they may be increased far beyond what most ruralists think possible. The common mode of impressing them with this fact is a reference to the difference between the products on half an acre of land under ordinary garden culture, and the products on the same quantity under ordinary farm culture. This difference must be entirely the result of difference in the degree of manuring and in the quality of the tillage.

Allusion is also made for the same purpose to Flemish agriculture. In Flanders a family of six persons can be supported on the products of about four acres ; that is, two-thirds of an acre to each person ; and in the same proportion, whether the family be larger or smaller. Upon this principle of apportionment the land is arranged in farms ; and we have cases in our own country which will show that in many localities the same can be done here. Even more than this has been accomplished. In 1849, the editor of the Maine Cultivator published the products of a single acre, in that State, which were sufficient to support the family occupying the little farm or garden, whichever it be denominated. If one family can obtain a good living from one acre, another family can do the same. If it can be done in the State of Maine, it can be done in the State of New Jersey or Connecticut. So it may be done elsewhere. On one-third of this acre were raised thirty bushels of sound Indian corn, besides the refuse. This was sufficient for the use of the family, and to fatten the pork. From the same ground, or in connection with the corn, there were raised between two and three hundred pumpkins, and a family supply of dry beans. From a bed of six rods square, sixty bushels of onions were obtained. These were sold at one dollar per bushel, and the proceeds converted into flour. Thus from one-third of an acre and the onion bed the breadstuffs were furnished. The rest of the ground was used for vegetables; potatoes, cabbage, parsnips, beets, sweet corn, peas, beans, cucumbers, squashes, melons, with some fifty or sixty bushels of sugar beets and carrots for the cow. Besides, there were on the acre strawberries, raspberries, currants, and gooseberries, in great abundance, and a few choice apple, pear, plum, cherry, peach and quince trees, and even a flower garden. Would not a farmer become rich on a hundred—or fifty—twenty-five acres, cultivated in the same way?

There was an anecdote told, a few years since, in one of the foreign magazines, of an English farmer, illustrating in a larger way the same principle; that is, of the advantage of small farms. The individual in question, from inheritance probably, had one thousand acres of land, and a small cash capital, but was just able to live comfortably, having no surplus income. He had three daughters. and on the marriage of the eldest he gave her two hundred and fifty acres of his land cultivating the rest as previously. Not long after, he gave two hundred and fifty acres more to the second daughter on her marriage, thus having left but five hundred acres for cultivation. Next the third daughter married, and according to arrangement received one-half of the five hundred acres, so that the father had left for his own use but two hundred and fifty acres To him the prospects appeared disheartening. However, he adopted a higher mode of tillage, increased his own diligence, and applied all his cash capital to these two hundred and fifty acres. Greatly to his surprise he soon found that he was able to raise double now to what he had before raised on a thousand acres. The consequence was, he was annually able to lay up in cash more than half of his income; and at his death he was found to have become a rich man. In the above facts there was every appearance of authenticity. Moreover, they are so much in accordance with those known to us, that we have no doubt of their truth.

The following statement comes under the sanction of the Hon. James Tallmadge, President of the American Institute. Connected with the Bloomingdale Asylum, within the northern limits of the city of New-York, is a farm of forty acres; ten of it in woodland, and the other thirty occupied as a farm and garden, and for the buildings of the establishment. It appears that General Tallmadge, in 1848, had occasion to visit the premises, with which he was much pleased; and he accordingly highly commended the prudent and judicious management which had led to the results annexed. It may accordingly be relied upon that the following statement is a faithful account of the expenditures on that farm, and the income from it in the year named, showing also the net profits over and above the expenditure; being the pretty little sum of two thousand one hundred and forty-two dollars, and forty-six cents, from thirty acres of land, in a single year.

Income for the Garden.

370	bushels of	Potatoes,	at 50 cents,	. . .	$185 00
260	"	Sugar Beets,	at 37½ cents,	.	97 50
95	"	Blood Beets,	" 50 "	. .	47 50
730	"	Turnips	" 31¼ "	. .	228 12
120	"	Parsnips,	" 50 "	. .	60 00
30	"	Carrots,	" 50 "	.	15 00
50	"	Onions,	" 75 "	. .	37 50
60	"	Rhubarb,	" 200 "	.	120 00
45	"	Asparagus,	" 300 "	. .	135 00
125	"	Radishes,	" 100 "	.	125 00
150	"	Tomatoes,	" 50 "	. .	75 00
96	"	Cucumbers,	" 75 "	.	72 00
150	"	Green Corn,	" 37½ "	. .	56 25
30	"	Egg Plants,	" 50 "	.	15 00

130 bushels of Beans,	at 50 cents,	.	$65 00
65 " Peas,	" 75 "	. .	48 75
210 " Spinach,	" 755 "	.	157 50
125 " Squashes,	" 37½ "	. .	46 87
90 " Pumpkins,	" 37½ "	.	33 75
1 " Nasturtions,	" 200 "	. .	2 00
5 " Peppers,	" 75 "	.	3 75
80 Citron Melons,	" 12½ " each,	.	10 00
3000 heads of Celery,	" 3 " "		90 00
4000 " Cabbage,	" 4 " "	.	160 00
4000 " Lettuce,	" 2 " "		80 00
2500 " Salsify,	" 1 " "	.	25 00
1000 " Leeks,	" ½ " "		5 00
			$1996 49

Incomé from the Farm.

40 tons of Hay at 1250 cents,		. . .	$500 00
14 " Oats, in milk, at 1000 cents,		. .	140 00
60 bushels ripe Oats,	" 37½ "	.	22 50
1155 bundles of Straw,	" 4 "	. .	46 20
2730 pounds of Pork,	" 6 "	.	163 80
601 " Butter,	" 22 "	. .	132 22
4409 gallons of milk,	" 18¾ "	.	826 62
322 dozens of Eggs,	" 14 "	. .	45 08
165 pounds of Poultry,	" 8 "	.	13 20
			$1889 62

Income from the Orchard.

75 bushels of Apples,	at 150 cents,	.	$112 00
50 " Pears,	" 100 "	. .	50 00
150 " Cherries,	" 100 "	.	150 00
25 " Currants,	" 100 "	. .	25 00
10 " Peaches,	" 100 "	.	10 00
2 " Strawberries,	" 600 "	. .	12 00
			$359 50

Recapitulation of Entire Income.

Income from the Garden, ...	$1996 49
" " Farm, ...	1889 62
" " Orchard, ...	359 50
" " Sale of Live Stock, ...	97 50
	$4,343 11

Expenses of the Bloomingdale Asylum Fund—1848.

Farmer's Wages, including one man the whole year, at $25 per month; one teamster, who does the ploughing and is employed the whole year, at $12 per month; and four hands, employed about half the year, at $10 per month. A woman is also employed to take charge of the milk, and to assist in other labors of the establishment. . . .	$773 90
Board of persons thus employed, ...	520 00
Expenses upon Farm Implements, ...	12 80
Grain fed out by the Cows, ...	120 00
Hay and vegetables consumed by Cows, ...	388 00
Manure purchased, in addition to what was made on premises, ...	308 76
Live Stock purchased, ...	78 09
	$2200 65

Thus it will be seen that when the expenses are taken from the income, there will be remaining a net balance of two thousand one hundred and forty-two dollars, and forty-six cents, for the profits of these thirty acres of land; or at the rate of seventy-one dollars and forty-one cents per acre. Here is a case of the profits of agriculture about which there can be no mistake. True, the farm is close to a market; but with our present rail road facilities it matters little whether the farm is ten or thirty miles from market, with most products. Besides, crops may be regulated according to distance from market, and consequent conveniences for reaching it. Indeed, many articles formerly raised of necessity within a few miles of the market, are now raised fifty or sixty miles from it ; and other articles, yielding as good a profit, may be produced, contiguous to a rail road, one hundred miles from market.

THE FARMER'S BOY.

As the richest ore is sometimes hidden in the coarse rock, so is genius often to be found beneath a rough and unpolished exterior. Robert Bloomfield, the author of that celebrated poem called "The Farmer's Boy," was born at Honington, a small village in England, on the 3d of December, 1766. His parents were very respectable people, although in humble circumstances. His father, who worked as a tailor, died during Robert's infancy; and the widow, finding herself obliged to maintain a family of six children, opened a small school. Under her instruction he learned to read, but, at the age of six years, he was sent for some two or three months to a writing master. From this school, he was removed, at the time of his mother's second marriage, probably because the new husband did not feel inclined to educate the offspring of his predecessor.

The accounts of Robert's childhood, are scant and meagre, for we next hear of him as having been taken into the employ of his uncle, a respectable farmer. This was when he was in his twelfth year. It does not appear that any favor was shown him on the ground of relationship, and he worked industriously in the field like his cousins and the hired laborers. He, however, received good food and comfortable lodging, but all his clothing was furnished by his mother. For a year or two he remained on the farm, and, at the expiration of that time, his frame was pronounced too diminutive, and his constitution quite too delicate, for so arduous an occupation. This circumstance, together with the difficulty which his mother experienced in providing him with clothes, induced an offer from his brothers who lived in London, to bring him up, provided he would come there and learn the trade of a shoemaker.

In 1781, when he was in his fifteenth year, his mother brought him to London, and left him with many strict injunctions to follow a pious and upright course through life. The elder brother, George, says, " She charged me, as I valued a mother's blessing, to watch over him, to set good examples for him, and never to forget that he had lost his father." This good brother is worthy of all praise for the strict fidelity with which he executed his trust.

Robert was immediately taken into a shoemaker's garret, where George and four others were at work. They appear to have had a strong love of knowledge, as they were subscribers to several works—such as a Geography, a History of England, and the British Traveller, which were published weekly in sixpenny numbers. They also took a periodical by the name of "The London Magazine," and the boy who came regularly from the public-house to receive orders for porter, was in the habit of bringing with him a newspaper of the previous day. Before Robert came, the men were accustomed to take turns in reading aloud, but, his time being considered least valuable, he was at once appointed the reader. He seemed little pleased with the duty, and performed it more as a task than as a relief from labor on the bench. He also acted as errand boy, and daily brought the men's dinners from the cook's shop. He was at all times kindly treated, and the men were ever ready to assist, or instruct him about his work.

In his situation of reader, Robert found not a little difficulty in pronouncing words properly, as well as in understanding their meanings. Accident, one evening, led him to a dissenting meeting-house, where he was so much pleased with the eloquence and choice language of the preacher, that he attended the service regularly on every succeeding Sabbath. He also derived much benefit from an occasional visit to the room of a debating-society and to the Covent Garden Theatre. His brother purchased, at a book-stall, an old dictionary, for which, in consequence of its being small and ill-used, he paid but four pence. This proved a great treasure to Robert, and, by studying it carefully, he soon became able to comprehend the parliamentary debates, so that he read them with interest. He, however, took most pleasure in the "Poet's Corner" of the Magazine, and its notices of new publications and the movements of literary men.

George was one day much surprised to hear Robert repeating a little song, that he had composed to an old tune, and suggested its being offered to the editor of the Magazine. Under the title of "The Village Girl" it was sent, and duly accepted. A second one, called "The Sailor's Return," met with like favor, and the youthful poet was induced to make several other contributions. We can well imagine his gratification at thus finding himself in print, and that too in the very "Corner" which he had regarded with such interest.

A very unfortunate occurrence for Robert, was his brother's removal to another house, where one of the inmates had a good many books. Among the number were Thompson's "Seasons" and Milton's "Paradise Lost." The first of these two pleased Robert exceedingly, and, perhaps, gave him the idea of a long poem descriptive of rural employments. This appears probable from other circumstances than that of "The Farmer's Boy" being divided into four books—corresponding with the four seasons of the year. But Bloomfield owes little or nothing more to Thompson, for, while the latter contemplates nearly every phenomenon of nature, he first restricts his muse to the humbler circuit of the farm.

In the year 1786, some trouble was occasioned by the journeymen shoemakers of London about those workmen who had not served a regular apprenticeship, and Robert, being of a quiet and peaceable disposition, thought it prudent to leave the bench for a season, and go to his uncle's house in the country. With the idea of a great poem still uppermost in his mind, he was pleased to have this opportunity of reviewing former scenes, and reviving the recollections of his youthful labors. His stay was not of long continuance, for George's landlord, a benevolent and kind-hearted man, offered to take him as an apprentice, and at the same time to allow him to work with his brother as heretofore. He, therefore, returned to London, and, before the expiration of two years for which he was bound by the indentures, had become an expert at his trade.

When twenty-four years old, he was married to a person in his own humble rank of life, but the two had little more to depend upon, than good health and a deep reliance on the protection of Providence. He sold his violin, on which he had become a very respectable performer, and devoted himself even yet more assiduously to his business. He had not forgotten his literary project, and in all the noise of a shoemaker's garret, where six or seven persons were at work, he actually composed near six hundred lines—about half the whole poem,—before one was committed to paper. The feat of stamping them upon the memory, sinks into utter insignificance beside the labor of composition under such extraordinary circumstances.

"The Farmer's Boy" is an attempt to depict the charms which hang around rural life, by a connected account of the varied labors of the husbandman. The theme is very humble, and is not disfigured by any pretensions to ornament. The style is smooth and flowing, while the versification is remarkably correct, but throughout the whole there is an evident lack of strength and passion. Simplicity and harmony of numbers are its chief merits as a work of art, while its apparent truthfulness and fidelity to nature give it the impress of genius. Had it been the production of one more favored by birth or education, its popularity might have been ephemeral; but, for its author's sake, if on no other consideration, is it entitled to a prominent position among the works of British poets.

The labor of composition proceeded slowly, and, when completed, the manuscript was shown to several leading publishers of the metropolis. No one was disposed to undertake its publication, although the poet's hope of pecuniary advantage was small enough to remove any objection on that score. In some places, where common politeness should have dictated a kind word to the humble aspirant for fame, he was turned away with ridicule. By some means—we had almost said, accidentally,—a certain

Mr. Capel Lofft was mentioned as a proper person to decide upon the character of the poem, and it was, therefore, sent to him, in company with a modest note from George Bloomfield. Mr. Lofft possessed a taste for intellectual pursuits, and being a resident of the poet's native district, was induced to examine what others had thrust aside with contempt. So much was he pleased with the beauties of the work, that he ventured to put it in the hands of an intelligent friend, upon whose opinion he placed reliance. The decision of this friend was equally favorable, and then Mr. Lofft undertook to prepare the work for the press. His alterations of the text were very few, as he was willing to collect little more than the bad orthography and occasional grammatical errors. He added some notes, and prefixed a short prefatory notice of the author.

Mr. Lofft's kindness did not stop here, but he procured publishers, and read the proofs as they were taken. The book appeared in the year 1800, and was almost immediately received into favor. The critics were astonished at the beauty of the poem, and the attention of the wisest and most eminent of the land was drawn to the genius which ennobled this indigent shoemaker. The preface caused a rapid and extensive sale. Seven editions, twenty-six thousand copies in all, were disposed of in three years, and the publishers generously added £200, together with an interest in the copyright, to the sum of £50, which they had at first given. A subscription for the author's benefit, was commenced in the neighborhood of Hadleigh ; the Duke of York made him a liberal present ; while the Duke of Grafton gave him a small annuity, and a situation in the seal office.

This was the season of Bloomfield's greatest prosperity. His work was translated into French and Italian, and the first part was rendered into Latin. He was the centre of general remark and observation, while the sums paid him by the publishers afforded him a very comfortable maintenance. This was so much more than he had dared to anticipate, that he declared " his good fortune appeared to him like a dream." But this unexpected success did not make him proud, nor forgetful of his humble origin. He scorned to neglect those with whom he had been associated, and felt happy to recognise and extend a helping hand to his relations and old friends. The only pride he seemed to experience was in giving a copy of his book to his dearly-beloved mother. All this must have been peculiarly gratifying to that good brother, by whom his young intellect had been trained and nurtured; as well as to that prince of friends, Mr. Capel Lofft, by whom his genius had been discovered and made public.

His career of prosperity was, however, of short continuance. Ill health forced him to resign the situation in the seal office, the duties of which had, perhaps, been all along irksome and disagreeable. He was induced to engage in the book trade, but he soon became bankrupt. He then returned to the work-bench, and amused himself in leisure moments by the manufacture of Æolian harps, that he sold to friends. His subsequent literary efforts—a collection of rural tales, Hazlewood Hall, &c., did not add to his reputation, although they were all well received. In the latter part of his life, he was greatly troubled by headaches and the gradual loss of eyesight. Being of a nervous temperament, he was in an almost constant state of excitement, so that it was necessary for him to give up all mechanical labor. This state of health continued so long that fears were, with good cause, entertained of his becoming insane. Several warm-hearted friends exerted themselves at this juncture to smooth the downward path of life. In the hope of being benefited by a removal to those fields from which he had been taken in early years, he left the metropolis, and went to reside in a small country town. The change was not attended with the desired effect; his disease soon assumed a hopeless form, and terminated by death on the 19th of August, 1823.

Bloomfield's character as a poet is respectable, but by no means brilliant. For a person in his situation, having had few advantages, and forced to work diligently for daily bread, the production of such a work as "The Farmer's Boy" is extraordinary. A valuable lesson may be drawn from his success.

He was pious, and throughout life experienced the advantages of early religious training. His character as a man was eminently worthy of praise and imitation. In all the social relations no one could be more exemplary. He was a good neighbor, a devoted husband, and a kind, indulgent parent. We have here an illustration of the old truth, that neither genius nor virtue is peculiar to a single class of society, and that neither is to be concealed by the untoward circumstances of humble birth and defective education.

It is not easy to imagine a more interesting and impressive scene than that of the farmer's son, at the age of fourteen or fifteen, fired with an impressible ardor for some imperfectly conceived idea of future greatness, forever leaving the place of his birth. For months this idea, notwithstanding all remonstrance, has pressed upon him like the nightmare. At length, he collects together his entire substance, a few dollars in silver money, saved, it may be, from the avails of poultry perquisites, and a few articles of cheap apparel, manufactured from the raw material by his mother and an elder sister. These are all compressed into a package not larger than a peck measure. The never-to-be-forgotten morning of his abscission from the family arrives. To him and to them it forms a new era, as memorable as that of the Hegira to the followers of the prophet of Arabia. Father and mother, brothers and sisters, in sad silence, witness his departure, upon a journey he knows not where, to seek his fortune. The swelling of the big hearts left behind is known, not by words, but by the big tears that fell from them, as they gazed upon the young adventurer, till his image was lost to them in the far away perspective. Then, in mute eloquence all returned to their wonted cares and labors. Their first meals afterwards were eaten sparingly, imaginings of the departed one depriving them of appetite; and, during the first night afterwards, in their log hut, or cabin, or mansion, whichever it was, there was, as it were, an almost unearthly stillness, as if the messenger of the grave had there established his dominion.

In the early part of the present century, and in the last half of the previous one, scenes like this were of common occurrence among our hardy yeomanry. It is under such circumstances that the fond mother, however illiterate—however ignorant of the world and of polished society, presents a spectacle of simple dignity and natural affection that do honor to our nature. It is under such circumstances that young bosoms, with quick and strong pulsations, and with bright eyes dimmed by the crystal fountains within them, reveal to the spectator the hallowed power of communion between kindred souls. It is under such circumstances that natural affection, the undying symbol of that love which binds mortals to their Father in Heaven, presents to the world a chastened and purified passion, securing to itself a homage, in comparison with which the conventional formalities of modern high life sink into contempt and disgust. It is to the occasions which gave birth to these circumstances that our country is mainly indebted for its present prosperity and glory. Had they not arisen where are now Western farms, fruitful like ancient Eden, and where are now Western cities filled with living multitudes, and resounding with the rattling of machinery, and the buzz of every description of handy-work, the once dense and magnificent wilderness would still cast her dark and chilling shadows.

While the remaining inmates of his native residence were toiling with their accustomed tasks, the youthful traveller pressed onward, not even for miles looking backward. It was the broad world before him, and not the few acres left behind him ; it was the high pinnacle of fame, rising by successive gradations, like distant mountains, raising, one by one, their craggy cliffs or snow-clad summits, the most distant mingling with the blue ether, and not the valley or its tenants in his rear, which absorbed his kindling thoughts. He pressed onward till hunger and fatigue induced him, under the branches of a large tree, and by a cool, clear stream, to halt for rest and refreshment. He had seen no human habitation, nor heard human voice, nor human footsteps save his own, for hours. Upon a broad smooth stone here sat he down! At his feet the gushing fountain was ready to slake his rising thirst. Over his head the gentle breeze was floating by to cool his heated body. Here he sat and mused; for the first time in his life he felt himself to be alone—a mere unit in an immense expanse; for the first time his own destiny rested on himself, seemingly without any aid from mortal man! Till the present moment the rude outline of the self-made man had never been pictured to his imagination. Had he been sole heir of the world—had he been the first and the only one of his species, he could scarcely have been more deeply overwhelmed with the responsibility resting upon him, and with the august panorama opening to his view.

It might have been that he sighed! It might have been that tears rolled down his cheek! It might have been, that his bosom, like a troubled sea, was experiencing some inward commotion! Yet, his stern manhood, although in embryo, triumphed over desponding thoughts. He opened his sack and took a cold repast. He drank also from the gushing spring; and, then stretched himself upon the green turf, with his head on the stone, where he sat, for a pillow. Visions, however, of the future as well as of the past interrupted his repose; for while he slept, he dreamed; yes, dreamed, that he was in a spacious mansion, filled with beautiful furniture, and abounding in the richest luxuries; but, in the midst of such bursting joy, he also felt an impulsive pressure about his neck, almost like the crushing folds of a serpent, and a burning upon his face like that of scalding water; he dreamed, that the one was the embraces of his absent mother, and that the other was her feverish tears! On awaking from such a reverie, no one, who has had exciting dreams, need be told, the tumult of emotion that must have agitated him. Nevertheless, he was refreshed; he arose, and on he went.

It may be, that at the end of the second or third day, he reaches the city and seeks humble occupation. No matter what it is; possibly, he becomes an office boy and sweeps the floor and builds the fires; possibly he becomes an errand boy in a retail store, and thus renders himself familiar with the streets and the signs of business men, and with the general contour of fashionable society; possibly he becomes the youngest apprentice to some mechanical employment, and is a kind of servant to all his associates. Nevertheless the germs of his character take root, and send their branches upward. If he has a taste for mechanics, in a few years this lonely wandering boy becomes a full man, and is making contracts with capitalists for the building of blocks of houses or steam ships. If led to mercantile pursuits, in a few years he will be seen on Change, or in the bank, controlling discounts. We have known boys to start in some way similar to this, and to die worth hundreds of thousands of dollars, and a few even millions of dollars. Indeed, it will be found that most of our rich men began life in some small, humble way.

Or, it may be, that by accident, an old book, or the fragment of a book, had fallen in the way of our young adventurer, and that his mind thus became tinctured with scolastic aspirations. If so, some literary institution ; first, the grammar school, and then the university, was the object of his pursuit and the place of his destination, on leaving in the manner described, the home of his childhood. No sooner is this destination reached, than the culture of the mind is in full process. There is no slothfulness or indecision, there are no alternations between hope and despair. His career is onward, as it was at the spring, where he refreshed himself. It is immaterial whether his labors are in the subtleties of a dead language ; in the tropes of belles-lettres and the syllogisms of logic ; or, in the profound conceptions and the systematic demonstrations of deep science. Over them all he obtains a mastery. In a few years the uncouth farmer's son becomes a polished scholar, and may be seen and heard and admired, in our legislative halls ; in our courts of law ; and in the sanctuary of religion. Did not Henry Clay, and Silas Wright, and Daniel Webster, and Millard Fillmore, and Jared Sparks ; and, indeed a large proportion of our most eminent divines, lawyers, and statesmen—the living as well as those deceased—begin life much as here set forth ? Verily they did.

Was it not so with Franklin, Roger Sherman, and Andrew Jackson? Where are the ancestral annals of Martin Van Buren, Abbott Lawrence, De Witt Clinton, Robert Fulton, Winfield Scott, and Francis Wayland? Are they found in books of Heraldry? Our theory of human greatness is no dream life. The prospective delineation of it to the faithful adventurer may appear like a romantic shadow; but, when the vision becomes an element of history, the successful candidate for renown will give his testimony that his whole course was a stubborn reality. If there is any one thing indelibly impressed upon our own mental tablet, it is a three days' journey on foot nearly fifty years ago, made by us, with similar musings and toils and anticipations. The waves of oblivion may, like the waves of the sea, successively sweep away the records of human thoughts and human deeds till time shall be merged in the ocean of existence, but this one shall continue in freshness, unless memory give up her empire and the current of thought shall cease to flow in its accustomed channel.

It is a pleasure occasionally to review life, and to call to mind the incidents attending the progress made in the highway to eminence. It is not every one that reaches enviable distinction; and those who do, encounter perils and suffer hardships which require the most indomitable vigor. It has often seemed to us, that the successful laborer for wealth and for fame resembles in no small degree the traveller ascending the Alps or the high and precipituous mountains of our own country. When at the base the projecting cliffs, and here and there the almost perpendicular pathway seem to render the ascent impossible. On looking upward the mind is apt to become giddy, and the nerves to become unbalanced. Where there is one to make the intrepid effort, a thousand will prefer the ease and the quiet upon the rich bottoms below. The balmy fragrance and the rich verdure there within reach, almost without toil, save such as may be accounted a pastime, are more congenial to the taste and the mental and physical powers of the timid and the indolent. Neither felspar, or mica, or hornblende, or serpentine, or sienite, or porphyry, staple components in these uprisings of the globe, has much attraction for minds of that character. If gold were one of these components the inducement to examine them would be greater.

Hence, it is evident, that the scientific mountain sojourner must be a man of genius and enterprise, as well as of physical strength and endurance, in the prosecution of his labors. At first, he may experience less difficulty, the course being less precipitous, and winding around some small promontory ; by the aid of small trees and shrubs on which he can lay hold, he advances with encouraging success. But the higher he rises, the slower is his progress, and the more hazardous is his position. On the mountain side, in proportion to the elevation is the decrease of vegetable substance ; trees and shrubs are constantly become more sparse and individually more feeble, so that every now and then he is without their aid. Yet, he seizes, by effort, upon one after another ; sometimes so small as barely able to sustain his weight ; clinging also to precipice above precipice till the desired summit is more than half reached. Here, finding a table of a few rods square, he stops and refreshes himself, both with food and sleep. This is not unlike the half-way stage of human life ; a prominence in bold relief where the retrospect and the future may be comprehended by a single glance ; the one giving ability for the other.

Accordingly, our tourist here renews his strength, and brings into requisition his remaining energies. It is an occasion to paralise the feeble and to invigorate the powerful. He is already advanced so far that the oxen and sheep in the valleys below appear to him no larger than dogs and cats. The heights above him assume a more perilous and forbidding aspect ; yet, he resolves to move onward. Slowly he does so. New inconveniences now begin to arise. The atmosphere changes as if moving toward the polar regions ; and, he fails not to experience the frozen tempests of a northern winter. At one elevation the snow covers him ; at another a shower of hail is scattered around him and upon him ; and at each the thunder and the lightning display themselves in fearful grandeur. And his hands are so chilled and become so numb, that to secure and maintain himself holding to the stinted shrubs and projecting granite footsteps seems almost impossible. However, he does it ; his strength and ardor are proportioned to the emergency ; he reaches an elevation, in a few hours more, above the storm of elements ; then, with indescribable sensations, it is difficult to tell, whether of exulting complacency or of homage to the Supreme Being, he literally looks down upon the clouds and the gathering tempests ; and, at last, he stands erect upon one of the most sublime monuments ever raised to the Almighty's power and wisdom in the creation of the world. What a scene of magnificence is here spread out before him ! Whether east or west, north or south, no obstruction intercepts his vision, till reaching that broad circle, where the heavens and the earth appear to unite ! The men left in the nether distance appear to him like pigmies ; and, the stones lying broadcast around them like the dust of his feet.

Who does not recognise in his own career to eminence very much analogous to this sketch of the mountain sojourner ! It is by toil, and self-denial, and perseverance, that difficulties are overcome, and that the boon of desire is attained. The effeminancy from luxury, and the sluggish progress from indolence or inefficiency, in both cases would be ruinous to success. The path of the adventurer for fame is often made rugged by poverty. The struggling efforts to overcome it generate an energy rarely found in the lap of ease. It is by labor we become athletic ; and, it is by intellectual action, that the mind reaches its full capacity. The history of literary men prove this to demonstration. Rarely is there an exception to it. We remember an anecdote of the late Post Master General in the United States directly applicable to our purpose. It was related by Judge Collamer himself, at a commencement dinner ; and of course its authenticity is not be doubted. It is the more valuable, because the narrator, as in most similar cases, felt no disposition to disguise a truth, once an occasion probably of humiliation, but by his good sense converted into an occasion of pride and self-complacency. When in college he was so poor, according to this statement, he was induced habitually to go barefooted to the recitation room. The president of the institution probably not knowing it was the consequence of poverty, reproved him for it, as being out of character for a young gentleman. Young Collamer was both mortified and grieved at the censure, and by great effort obtained a pair of shoes ; but, to make them last as long as possible, he still went barefoot, except when appearing before his principal teacher. On these occasions, he would carry the shoes in his hand till he arrived at the recitation room door, when he would put them on ; and upon leaving it, as soon as outside of the door, he would remove them from his feet and carry them in his hand to his own room, as he had previously left it. With such an incident of character in a young man it required no prophetic acumen to predict the eminence to which he has risen. Such traits of character will usually lead to similar results.

It is proverbial, that a very large proportion of our distinguished men, like Webster and Van Buren, were born on a farm. The country is the great hot bed of genius. It would be easy to produce a long catalogue of those, who, like the boy above described, have risen to the highest rank in deeds of renown. Among the families of farmers in our country thus distinguished for producing great scholars is that of Noah Worcester, of Hollis, N. H. The facts annexed respecting him are collected from a memoir of one of his sons just published, and from our own personal knowledge. They are highly instructive ; and, it is cheering to the intelligent ruralist to reflect upon them.

Noah Worcester, the Hollis farmer, had four sons who became clergymen*; all four of reputable standing, and two of them, Noah and Samuel, rose to the highest grade in their time, both for professional talents, and as controversial writers. Samuel, the youngest of the four, was educated at college ; the other three, Thomas, Leonard, and Noah were literally self-made ; one went directly from the plough to the pulpit ; another went from the farm to the printing office, and from that to the pulpit ; and, the other from the farm to the shoemaker's bench, and from that to the pulpit. Samuel also worked on the farm till in the middle of his twenty-first year, when he purchased his time of his father, fitted for college, and gratuated with the highest honors, paying the expenses himself by keeping school. Both Noah and Samuel received the Honorary Degree of Doctor of Divinity ; the former at Cambridge and the latter at Princeton. Seventeen of the descendants of Noah Worcester of Hollis, have received a collegiate education. Of these Joseph E. Worcester, LL. D., the author of the Universal Dictionary and other valuable works, is one ; and besides these seventeen, six of his other descendants, without receiving a collegiate education, have entered the ministry. It is delightful to muse on such instances of mental development. They should in the family of the farmer be as familiar as household words. The writings and the memoirs of such men should become parts of all rural fireside literature. It is to such sources that we are mainly indebted for our national reputation. It is from them that we are mainly to receive impulses for a high destiny.

MORAL INFLUENCE OF SPRING.

THE vicissitudes of the seasons are among the most beneficent allotments of Divine Wisdom. All our instincts tend to harmonise with progressive development in material nature. We shrink almost involuntarily from whatever is monotonous, whether in the action of mental power, in the processes of human labor, or in the exhibitions of the material creation about us. Were we placed in a paradise, all in time would become dull and insipid. The bowers of ancient Eden, perfect as it was on coming from the hand of God, did not long satisfy our first parents. Nor, probably would they for a long period satisfy ourselves. No matter how rich might be the scenery ; no matter how limpid and pure might be its perennial springs ; no matter

how odorous might be its medicinal plants and trees; no matter how beautiful its foliage and its flowers; no matter how picturesque might be its landscapes; speedily there would be an oppressive satiety; speedily there would be a yearning for change —for some untried means to gratify the senses and to give buoyancy to the mind. It is not affirmed that there is any thing morally wrong in this feature of our nature. It probably leads to a better accomplishment of the purposes of our existence; to a higher and more refined circle of virtuous enjoyment. It counteracts a propensity to languor, and perhaps to indolence; with which the wide world, the earth and the heavens, are at variance. If there is any one characteristic, that, above all others, presents itself with ceaseless vigilance in the physical economy of our globe, it is that of a ceaseless activity and progress.

Were an attempt made for it, how difficult should we find it, to tell with which of the seasons of the year we should be most pleased, in case it were to be perpetual. Should we be dropped upon the earth, in mature existence, when autumn is bending under its annual weight of well filled storehouses; when hunger and thirst could be appeased without labor and anxiety; when the whole realm of earth's wide domain is teeming with desire to impart her treasures to the children of men; can it be imagined that we should evince no signs of discontent; that we should cherish no aspirations to witness the processes by which this abundance is produced! If we make observations upon the living world, we may feel assured that such would not be the fact. If we advert to the record of our thoughts, we may know, that a dull and unchanged perpetuity of every thing however excellent in itself, is not in accordance with our taste if with our reason.

So far as the seasons of the year are taken to illustrate this attribute of the human mind, it is doubtless true, that in each one there may be circumstances peculiarly adapted to fit man for his duties and his destiny. Summer presents to his imagination the wide world with all its vast designs and gigantic powers, struggling for great achievements, like man himself in the noon-

day of his existence ; autumn follows in its wake to teach him gratitude, and to inspire him with a sublime homage to the Author of his being ; nor is winter less important to him in teaching him by her own colorless shroud, that there is to be a crisis beyond which no one can advance. Thus, the revolving year with its varying seasons, is man's great schoolmaster. In her sublime teaching there is no heresy ; no false and delusive adulation ; it is all eloquence and impressive truth ; and those who follow it, with a willing heart, will reach a moral elevation belonging only to the wise and the good.

In becoming the disciples of this great teacher, especially in the spring season of the year, there is opened to our view a temple most gorgeously filled with all that can delight the eye and ear, and all that can impress the heart. Who can count the numner of ministering agents, that here inspire man with joy and wisdom ? It would seem that the whole spirit world is here in personification, speaking to the soul in every spire of grass, that from the ground rises upward, in every majestic tree, in every flower, and in every plant of this beautiful globe ! They steal upon him in every direction ! They start out to meet him from every fold in the rich drapery ! They embrace him from every niche and recess ! They greet him from behind each lofty column ! He hears their voice in every floating breeze ! In the note of every songstress that fills with their melody the dense forest as well as each verdant bough around the family mansion ! And he is continually inhaling the odors of a balmy incense from nature's high altar. Here is a magnificence compared with which the halls of princes are tasteless and mean ! Here is music more sweet and suasive than in the grandest artistic orchestra ! Here is eloquence more overpowering than human speech, whether in the Church, or the Senate !

Who feels no joy, no exultation, no renovated life, no new aspirations after heaven, from the return of spring ! When the green grass in some sunny nook first gives evidence of renewed existence ; when the swelling and the opening buds of the tulip, the apricot, the peach, the plum, and the cherry, present to our admiration their pure emblems of virtue, does the human soul experience no kindling ecstacy? Or, when a rich foliage, of a thousand shades and forms, comes forth and clothes the earth from every tree, and every shrub, and every plant, can the soul remain dead to the celestial graces which are to live in amaranthine beauty? Or, when the foliage is filled with the bright plumage of the many feathered tribes—when every leaf is quivering with an echo of their sweet anthems, shall no responsive breathings of joyfulness and adoration rise from the human bosom! When we witness all this loveliness, beauty, and animation in the return of spring, we cannot refrain from the conclusion that of all miracles, now and then, man is the greatest—a miracle of heartless philosophy; that of all paradoxes he is the most incomprehensible!

But with all our stoicism; with all our apathy to emotion, we cannot wholly resist these cheering and enlivening influences. Were we to do it, might it not be apprehended, that the purest susceptibilities of our nature have become extinct? Who can do it? Who does not observe, with an involuntary pleasure, at least, the first little rills of spring from the ice of a neighboring hill-side? Who does not mark, day by day, with lively interest, the gradual disappearance of the snow bank, on the cold aspect of his garden fence? Who does not listen to the first vernal concerts of the robin, the blue bird, the meadow sparrow, and the bobolink? Who gives no heed to the increasing playfulness of the lambs and calves, and the cheerful notes and pompous strutting of the poultry yard tenants, in a warm day of April? Few will do it. We do these things impulsively. There is a species of fellowship between man and other animals, and between them both and the vegetable kingdom. Man as well as they, needs the renovating influences arising in the spring time of this kingdom. The soul as well as the body becomes languid and more or less inanimate; and is restored to its appropriate energies, like the face of the earth, by such instrumentalities as we have here indicated. If animated by a spiritual essence, this essence is reached through corporeal organs.

The inspirations of spring are of the most gentle character. They are calculated to soothe as well as to enliven, rather than to terrify and overpower by august exhibitions. In one respect spring is like the Gospel dispensation, without terror, and teeming with beneficence. The dispensation preceding that of the Gospel was established by miracles of vindictive justice. Of these miracles were the general deluge, the dispersion at Babel, and the destruction of the army of Sennacherib. The latter dispensation was established by miracles of mercy ; among them were the feeding of the hungry, clothing the naked, giving health to the sick, sight to the blind, hearing to the deaf, and life to the dead. Analagous to these are the vernal ministrations of nature. We no longer witness the destruction of life from the convulsed elements of winter. Her ice bound mantle and her piercing blasts have disappeared. Her dominion has been subdued. Nor is there a rising terror in the bosom, from the lightning, and the thunder, and the hurricane of mid-summer. The season for their appearance is in the future. The rain of spring is in delightful showers. Her sunshine is genial and not scorching, and her winds are gentle breezes, like the Divine influences upon the soul. Or, if visited by a stiff north-easter, in or about the March equinox, it lasts but a few days, and is not unlike a box upon the ear of the heedless child from the petulant mother—always succeeded by a shower of smiles and a bountiful piece of bread and butter.

Even the hurry and commotion of spring in the country is not wholly unpleasant. To prepare the soil for the crops of the season does indeed leave the farmer little or no opportunity for recreation, or for protracted moralising on the agencies connected with his labors. Occasionally, till his seed is sown, he barely allows himself time to sleep. All other matters are to be deferred, aware that undue delays in this part of his rural duties would be attended with hazard. If he fail to sow, he cannot expect to reap. May not those who calculate on a spiritual harvest, learn wisdom from this fact in agriculture ; may they not see the necessity of attending in due time to that early culture, which alone can render them sure of the rich fruits of eternal life ? In both cases, most persons, occasionally at least, become sluggish. Hence, they need excitement. In the former case, the farmer is admonished to divest himself of rest and inactivity; to press forward ; to do with his might what he has to do ; and, as if to forget every thing else. In the latter case, all may and should be instructed by his example. This is one of the lessons inculcated when we reflect on the spring-time labors of the husbandman.

But there is another moral truth illustrated and enforced by the return of spring. In the winter all nature seems clothed in sadness. The beauties of vegetation have disappeared. The earth is hung in a drapery of bleak desolation. Had the change to be wrought by the return of spring never have been witnessed by us, who could realise or even believe such a change possible. And the seed we cast into the ground is apparently without vitality. Had it never been seen, who could imagine that in a few weeks, it would send forth a waving sea of verdure, containing all the elements of continued existence. St. Paul takes this fact to illustrate, not only the possibility but the reason to expect a similar resurrection of the human body from the grave. That doctrine is no more at variance with the soundest principles of philosophy than the resurrection of nature in the season of spring. If the hard kernel of corn is made to assume its original green vesture, why may not the human body that has once lain in the grave, be again clothed with all the characteristics of its first state ; bone be joined with bone ; muscle with muscle ; vein with vein ; and artery with artery ? Why may not the body again become erect, and again be endowed with locomotive power, as well as the new grown corn again be enabled to unfold its leaves and its flowers to the sun and the air ? Why not the human eye again be enabled to sparkle with intelligence and joy, as well as the new formed ear of corn again be permitted to adorn the stem which produces it ?

As no hope can be more precious than that of the general resurrection of our race from the grave, how interesting and delightful must be any instrumentality that can nourish and give stability to this hope! The spring of the year is such an instrumentality. The assurance of this truth is indeed unfolded to us in the Word of Life; but the conception of a truth so glorious, so congenial with our inherent sensibilities, and yet surrounded by so many obstacles to its recognition, requires all the adventitious aid, in any way to be had, to render it the anchor to the soul, so much needed by us. Let the doctrine of the general resurrection be discarded, and life itself is but an idle dream! Where else than in this doctrine is a destiny to be sought compatible with our mental endowments, and with our mental aspirations? Without it man sinks almost to a level with the brutes. With it he has fellowship with angels, and with all ethereal beings. Without it man is comparatively like the insect which glitters in the sunshine but a few hours or a few days. With it, there will be no end to his existence, and no limit to the expansion of his powers or the measure of his bliss.

THE PRINCIPAL VARIETIES OF SHEEP.

In the earliest records of the world, it is apparent that the value of the sheep was understood, and, that the care of it was among the first of human labors. In the life time of Cain and Abel, we are informed by the sacred penman, that this animal had an important relation to the primeval institutions of that period. The offerings of gratitude made by the first human family to the Author of their being, consisted of the first fruit of the ground, and the firstlings of the flock; and we are specifically told that the former of these brothers was a tiller of the ground and the latter a keeper of sheep. Nor is it an unnatural presumption, that this animal so soon became an object of such prominence; for the uses to be made of it were so manifold, its utility could not have escaped the attention of the most casual observer.

5*

It is well known that the most extraordinary results are produced in the form and covering of sheep by the climate, soil and food, which have influence upon it in different localities. Did we not know the fact we could scarcely realize that such results were possible. A distinguished writer* appropriately observes, that no animal varies more than the sheep, and none adapts itself so speedily to climate. It would almost appear that nature, convinced of its great utility, had bestowed upon it a constitution so pliant, as to enable it to accommodate itself to any point, in a wide scale of temperature. For though its natural situation as a wool bearing animal, like that of man, appears to be the wine countries, yet with him it has spread to every quarter of the globe, being impressed at every change with some peculiarity, alterable only by a change of situation, and varying, we might affirm, with the weather. Changes, occasioned by climates, are always limited to the fleece, horns, and disposal of the fat, and never extend to those parts on the permanence of which the animal depends for its station in the scale of being, as the teeth, the feet, and the digestive organs.

Under such circumstances, says Canfield, it cannot be expected that we can trace the origin of the different breeds of sheep. And as to the qualities and management of any of the ancient breeds, we know only what is furnished by the Bible, and by Greek and Roman writers; and so meagre are the accounts which they furnish of the different breeds, that any thing like a regular history of the sheep is entirely out of the question. But as the sheep has been widely disseminated throughout Europe, Asia, and Africa; as its young are easily tamed, and its milk, flesh, and pelts were extremely valuable to man in all ages, we may well suppose that it was one of the first quadrupeds which was domesticated. And as there is no animal which contributes more to the welfare and comfort of man than sheep, so, also, there is no one which requires more care and attention from him.

* Blacklock.

The sheep, in a state of complete domestication, appears equally stupid as it is harmless, and seems nearly to justify the observations of Buffon, who describes it as one of the most timid, imbecile, and contemptible of quadrupeds. When sheep, however, have an extensive range of pasture, and are left in a considerable degree to depend on themselves for food and protection, they exhibit more respectability of character. This is analagous to what we may observe in the human species. By domestication and tender care of the sheep it is rendered imbecile. So our children can be made effeminate and inefficient, by keeping them from occasions to develop their mental and physical powers; whereas, if allowed to derive the benefit resulting from such occasions, they become vigorous, able to take care of themselves, and even become distinguished among their fellow men. Is it not much so with the sheep? When trained to take care of themselves, a ram has been seen to attack and beat off a large and formidable dog, and even a bull has been felled to the ground by a stroke received between his eyes, as he was lowering his head to receive his adversary on his horns and toss him into the air. And when individual efforts are unequal to the danger, sheep will unite their exertions, placing the females and their young in the middle of an irregular square, the rams will station themselves so as to present an armed front on every side to the enemy, and will support their ranks in the crisis of an attack, harassing the foe by the most formidable and sometimes fatal blows. Such sheep too especially display considerable sagacity in the selection of their food; and in the approach of storms they perceive the indications with accurate precision, and retire for shelter always to the spot which is best able to afford it.

The elevated steppes of southern Siberia, and the mountain chains of central Asia, produce an animal, described by Pallas, under the name of Argali. According to a statement in the Farmer's Library, a few years since, two fine specimens of this animal, a male and a female, were placed in the British Museum, and may be regarded as among the most valuable and interest-

ASIATIC ARGALIC SHEEP.

ing of its zoological treasures. "Till we saw them," says Mr. Youatt, "we had no idea of so gigantic a sheep. Huge, massive, heavy, and powerful, is the Argali of Siberia! an ox, as it were in stature, but a wild sheep in form and characters. The male stands four feet in height at the back; and measured from the nose to the end of the short little tail over the head and neck, is seven feet nine inches; the circumference of the horns at the base is nineteen inches, and each horn, measured from its base along the curve to the tip, is three feet eleven inches. The horns are furrowed with deep transverse wrinkles, and are boldly spiral, diverging somewhat laterally. The limbs of the sheep are compact and well turned, the tail is short, and the ears are small. The female specimen is smaller in stature than the male, and the horns are considerably less, both as to length and weight. The horns of the male are triangular, and convex on the upper apex—those of the female are more compressed.

"The pelt is deep, close, and full. Its tint is a grisly brown. The limbs below the knee are whitish; the lips are grayish;

and the tail is also grayish. Probably the color varies at different seasons of the year, as is generally the case among wild sheep. The argali lives in troops, and is extremely active and vigorous, bold and resolute. In spring and autumn the rivalry of the males is excited, and they engage in desperate conflicts, striking each other on the head with such violence, that they often break off each other's horns, massive as these weapons are, and solidly as they are fixed on the skull. It is said that even foxes and other small animals may take shelter in the hollow of these horns, on their being separated from the dead animal, as they are scattered about."

In the valuable work from which the above is obtained, the same writer says, that upon the elevated plain of Pamir, eastward of Bokhara, which is 16,000 feet above the sea level, wild animals are met with in great numbers, particularly sheep of a large size, having horns three, four, and even six palms in length. The Roman palm was eight and a half inches. The shepherds, he remarked, form ladles and vessels of them for holding victuals. They are also used in the construction of fences to protect the sheep against wolves. These fences are peculiarly repulsive in their appearance. Col. H. Smith speaks of the Caucasian argali as having horns three feet in length, and so heavy and unmanageable, when lying on the ground, that he found it difficult to place them in such a position as to give a correct idea of their appearance when on the skull. Mr. Kotsbue says the Kamtchatkan sheep, which is supposed to be a diminished variety of the argali, is amazingly fleet and active, exhibiting itself on the loftiest pinnacles, and achieving, like the chamois, prodigious springs among the rocks and precipices, and consequently is not killed or taken without difficulty. In preparing for these leaps, its eye measures the distance with surprising accuracy. The animal then contracts its legs and darts forward, head-foremost, to the destined spot, where it alights upon its feet; nor is it ever seen to miss, though the point may be so small as to admit its four feet only by their being close together.

The Rocky Mountain sheep inhabit the lofty chain of mountains from whence they derive their name; from its northern termination, in latitude sixty-eight, to about latitude forty de-

ROCKY MOUNTAIN SHEEP.

grees, and perhaps farther south. They also frequent the elevated and cragged ridges with which the country between the great mountain range and the Pacific is intersected; but they do not appear to have advanced further eastward than the declivity of the Rocky Mountains, nor are they found in any of the hilly tracts near Hudson's Bay. They collect in flocks, consisting of from thirty to forty young rams and females, herding together during the winter and spring, while the old rams form separate flocks. Mr. Drummond informs us, that in the retired parts of the mountains, where the hunters had seldom penetrated, he found no difficulty in approaching the Rocky Mountain sheep, where they exhibited the simplicity of character so remarkable in the domestic species; but that where they had been often fired at, they were exceedingly wild, alarmed their companions on the approach of danger by a hissing noise, and scaled the rocks with a speed and agility that baffled pursuit. He lost several that he had wounded mortally, by their retiring to die among the secluded precipices.

Their favorite feeding places, are grassy knolls, skirted by craggy rocks, to which they can retreat when pursued by dogs or wolves. They are accustomed to pay daily visits to certain caves in the mountains, that are encrusted with a saline efflorescence, of which they are fond. Mr. Drummond says that the horns of the old rams attain a size so enormous, and curve so much forwards and downwards, that they effectually prevent the animal from feeding on level ground. Its flesh is said by those who have fed on it, to be quite delicious when it is in season, far superior to that of any of the deer species which frequent that quarter, and even exceeding in flavor the finest English mutton. Some naturalists have supposed that this variety of the sheep family is substantially the same as the Asiatic Argali, but of diminished stature. Others dissent from this opinion, not only on account of its size, but of a difference in the curvature of the horns. Those, who maintain it, imagine that some of the Argali originally passed Bhering's Straits on the ice to the American continent.

Captain Bonneville seems to imagine that there is a kind of congruity between the characteristics of the Rocky Mountain Sheep and the regions inhabited by them, they giving romantic effect to the natural scenery. Thus they bound like goats from crag to crag, often trooping along the lofty shelves of the mountains, under the guidance of some venerable patriarch, with horns twisted lower than his muzzle, and sometimes peering over the edge of the precipice, so high that they appear from the valleys to be no larger than crows. Indeed, it appears a pleasure to them to seek the most rugged and frightful situations, doubtless from a feeling of security rather than of contempt for their enemies prevented from approaching them.

In the early ages of the world flocks of sheep constituted a large proportion of the wealth of the people. In Palestine they were very numerous. It is stated that Job had twelve thousand sheep, besides oxen and camels. When the twelve thousand Israelites made an excursion into Midian, they brought away six hundred and seventy-five thousand sheep. When the tribes of Reuben and Gad made war with the Hagarites, their spoils amounted to two hundred and fifty thousand of that animal. The king of Moab rendered a yearly tribute of two hundred thousand sheep to the Jews; and at the dedication of the temple Solomon offered one hundred and twenty thousand. Travellers have seen immense flocks of sheep in the neighborhood of Aleppo; and, Dr. Shaw states that several of the Arabian tribes, who can bring no more than three or four hundred horses into the field, are possessed of more than as many thousand oxen and camels, and treble that number of sheep and goats.

Hair is an appendage of the skin of the mammalia. It consists of fine filaments growing from beneath the skin, to which it serves as a covering; it is nearly the same in its chemical composition as horn and feathers; it is kept flexible and moist by an oily secretion from the skin; it is furnished with blood vessels, like all the other organs of animals. Quadrupeds are more or less covered with it, and for the most part in the greatest degree where the cold is greatest. Man is slightly supplied with this universal defence; but he is enabled, by his reason, to adapt the hair of other animals to his use. When the hair of animals is very thick and strong, it forms bristles; when more fine, forms the commonly so called hair; when it is fine, and at the same time curled, it is termed wool. It is this curling property of the wool which renders it more suitable than any other species of hair for being woven into cloth. The fur of animals consists of a mixture of hair and wool, but the latter is often in a very small quantity.

The wool principally used for the purpose of forming clothes, is that of the domestic sheep; and we know that this substance has been employed from the earliest records of the human race. But the wool of various other animals is applied to the same purpose, as the camel, the lama, and the goat. Wool frequently loses its curling property, and passes into hair. In the warmer regions the fur of sheep is more hairy than in the colder, apparently because a less thick and matted covering is required for the protection of the animals. Hair is also found, and sometimes in large quantity, intermixed with the wool of sheep in cold and temperate countries. This intermixture of hair unfits the wool for many manufactures; and it is a process of art to separate it from the wool. By neglect in the treatment of the animal, the proportion of the hair increases; by care and more complete domestication, the quantity of hair diminishes.

The wool of sheep, like the hair of other animals, is periodically renewed, the older hair falling off, and a new growth taking its place. In the case of the sheep, this renewal of the wool usually occurs once in the year, and at the beginning of the warm season. It is at this period that we anticipate the natural process by shearing or cutting off the external part of the fleece. In some countries the fleece is not shorn, but is pulled off; and in certain conditions of climate and the animal, the wool remains for more than one year. Wool, like every kind of hair, grows quickly when cut. The wool of sheep is sometimes black or brown, and the wool of all the less cultivated animals tends more or less to a dark color. Some sheep, even of superior breeds, have black faces and legs, as in the English Southdown; and in all such breeds, there is a tendency to a mixture of black wool with the white. This is deemed an imperfection in wool; for the pure white wool receives a better color than that which is black or brown. Accordingly in attempts to improve the breeds of sheep one important object to be attained is the production of a white fleece. Some idea may be had of the magnitude and importance of the interest in the domestic sheep, that even in our own country the annual produce of wool now amounts to fifty millions of pounds; and that in Great Britain it is nearly three times that quantity. McCulloch says the value of woolen goods annually manufactured in that country is more than a hundred millions of dollars.

The cut connected with this paragraph is a correct portrait of a fat rumped Persian ram, which belonged to the Zoological

FAT RUMPED SHEEP.

Society in London. There are several varieties of this sheep, but they are substantially alike. The flocks of all the Tartar hordes, says Dr. Anderson, resemble one another, by having, particularly the males, a solid mass of fat formed on the rump, divided, as it were into two hemispheres, which take the form of the hips, with a little knob of a tail in the middle. Some of them have horns but others do not. Its covering is a mixture of hair and wool. Some of the breed weigh as much as two hundred pounds: and the mass of fat formed on the rump varies from a tenth to a fifth part of the entire weight. In the neighborhood of Caucasus and Taurida, the hind quarters are salted as hams and sent in large quantities to the northern parts of Turkey. It has been supposed by some writers, that this breed may be the same as that which was bred by the patriarchs in the days of Abraham and Moses. The sacred penman mentions that Moses took the fat, and the rump, and all the fat which was upon the inwards, and burnt them upon the altar for a burnt offering.

The sheep of Thibet, which are very numerous, are chiefly a small variety of the fat-rumped Persian and Abyssinian, with black heads and necks. Some are hairy, with short wool underneath, while others bear a long, soft, and fine wool. It is from the latter that many of the costly Indian shawls are made. Not a little of this peculiar wool finds its way to British India, and is there manufactured. This breed is found in its purest state, in the deserts of Great Tartary; no other variety being near to contaminate its blood. It reaches far into the interior, and northern parts of Russia, and is much disseminated in China, Persia, Hindostan, Asia Minor, and Eastern Africa, as well as in Thibet. In Palestine it is more numerous than any other breed; indeed the largest proportion of the sheep of Northern Asia being of this description.

The causes of the peculiar disposition of fat upon the tail and rump of different breeds of sheep, will probably ever be a mystery. Fat is a secreted tissue which intermingles with, and surrounds the muscular parts, and envelopes the viscera within the body. Ordinarily, it is dispersed throughout the body; but in many of the sheep of the above mentioned countries, it accumulates principally upon the rump and. tail. Professor Pallas conjectures that this character arises, in the fat rumped sheep, from their feeding upon the bitter and saline plants, found upon the borders of the Caspian and Black seas. And he asserts, that when they are removed from the places where these plants grow, the fatty excrescence becomes less. But says Canfield, as the fat-tailed and fat rumped sheep are varieties which are widely dispersed, it seems more probable that they may have been produced by accident, and may also have been perpetuated by accident, design, or fancy.

The fat-tailed sheep is very extensively diffused; it is found throughout Asia and a great part of Africa, as well as through the north-eastern parts of Europe. They differ like other sheep in the nature of their covering. In Madagascar, and in some

FAT-TAILED SHEEP.

other hot climates, they are hairy; at the cape of Good Hope, they are covered with coarse, hard wool; in the Levant, their wool is extremely fine. The proportion which the weight of the tail in some of these sheep bears to the whole carcase, is quite remarkable. The usual dressed weight of the sheep is from fifty to sixty pounds, of which the tail is said to make more than one-fourth part. Some of the largest of these sheep which have been fattened with great care are said to weigh one hundred and fifty pounds, the tail making one-third of the whole weight. The tail is described as being composed of a substance between marrow and fat, serving very often, in the countries to which the animal belongs, instead of butter, and is used as an ingredient in various dishes. While the animal is young it is deemed to be little inferior to marrow.

Russell describes two breeds of fat-tailed sheep about Aleppo; in one the deposit of caudal fat is moderate, in the other sort the tail is much larger, and it is this kind to which travellers allude in their descriptions of sheep with enormous caudal appendages. The sheep of this breed, which are fatted and attain the large size mentioned, are kept in yards, in order to prevent injury to their tails. He farther informs us that when these sheep are fed in the fields, the shepherds in several places of Syria fix by way of protection a thin board to the under part of the tail, and to this board are sometimes attached small wheels. Hence, with a little exaggeration, we have the story of the oriental sheep being under the necessity of having carts to carry their tails. True, there are writers of accredited veracity, who have so far endorsed such inflated statements, as to say they have seen the tails of these sheep weighing from seventy to eighty pounds; but if such were in rare cases the fact, it must have been the result of high feeding.

This tendency to fat in the prevailing sheep of oriental countries seems to adapt them in a peculiar manner to the use for which they are mainly designed. In Syria little meat excepting mutton is eaten, and excepting during a few weeks in the spring it is fat and well flavored. The lamb in the Spring is excellent. As mutton is almost the only animal food consumed in that country, a regular and abundant supply, especially for the large towns, is very requisite. It is calculated that sixty thousand sheep are annually consumed in Aleppo, the population of which amounts to about the same, thus making the consumption of one sheep a year for each person; that is, as it regards sheep alone, what the same population in an English town is calculated to consume. But as in England and our own country other meats are used to perhaps three times the quantity of mutton, the consumption of animal food in Aleppo is only about one-fourth part what it is in these countries.

The unctuous fat of the tails of these sheep is accounted a great delicacy, alike by the boors and the Hottentots of Southern Africa. In their primitive condition, while yet they claimed the country as their own, the Hottentots possessed immense flocks as well as herds, and pursued the pastoral arts with great success Nor are they now, in their changed and fallen state, less excellent as shepherds and herdsmen. Those entrusted with the flocks of their masters know each individual sheep, and have their attention so sharpened by practice, that if one out of several hundred sheep be missing its loss is immediately perceived. This faculty appeared to surprise Burchell, who paints a cavalcade of flocks returning home at evening, which must indeed have been a pleasing spectacle. It was, says he, an interesting sight to behold, a little before sunset, the numerous flocks streaming like an inundation over the ridges and low hills, or moving in a compact body, like an army invading a country, and driven forwards only by two or three Hottentots and a few dogs. At a great distance, the confused sound of their bleating began to be heard; but as they approached nearer and nearer the noise gradually increased, till the various cries of the multitude mingled with the whole air and deadened every other sound.

Russia is a great sheep country. In 1837 she exported more than six million pounds of wool to the British empire alone. This is mostly from the south part of Russia. In the Crimea there are immense flocks of coarse woolled sheep, with fat tails, the wool of which is white, black, or gray. A rich Tartar will frequently possess fifty thousand sheep, of which the wool is more or less valuable. It is well known that many of the Russian nobles derive a considerable portion of their revenue from their herds of horses; but their sheep are a still greater source of riches, a fact intimated by the circumstance that when the wealth of a noble is mentioned it is often estimated by the number of sheep which he possesses; some individuals are said to possess no fewer than a hundred thousand. These sheep are mostly of the ordinary Wallachian breed, noted for the huge size of the tail; with others of the Kalmuc variety, which carries a load of fat on each side of the croup, which even hides their short tail.

The Cretan sheep is said to prevail in Wallachia, Hungary, Austria, and the Western parts of Asia; but along the Danube it is, or rather was, the principal habitat. It is of the long tailed variety, though without any great tendency to a fatty enlargement of the tail. On the face, the hair is short and of a rusty black. On the body, the wool is white and long, perfectly straight —that is, has no spiral curve—thick set, and wiry, and is much mixed with hair. Its horns are very large, adding greatly to its striking and picturesque appearance. The horns of the male

CRETAN SHEEP.

rise almost perpendicularly from the skull, making a series of spiral curves in their ascent, while in the female they diverge, taking a lateral direction, and then ascending. But there is probably some little variation in the horns of this breed, as in those of most other breeds. This variety of sheep is said to be vicious and unruly, and of great strength. In certain characteristics it has considerable resemblance to one variety of the Persian sheep, and to the Black faced heath breed of Scotland. The pure Cretan breed of sheep at present is not very common even in the above countries, as it has been mixed, or partially superseded, by the introduction into them of the Merino.

The Iceland sheep, of which specimens have at different times, says Youatt, come under our personal observation, are of tolerable large size and strongly built. Their fleece consists of coarse hair externally with an under layer of close wool. Their horns are generally four in number, sometimes six or even eight, and this is the more remarkable as the Iceland cows and oxen are mostly polled. Many are the casualties to which the sheep of Iceland are exposed in their dreary country; they have to endure the storms of winter; they are exposed to the rushing descent of terrible avalanches, and to the overwhelming force of volcanic eruptions, to say nothing of the destruction of lambs by eagles, and occasionally by polar bears, drifted from Greenland on vast masses of ice. There are, however, no wolves, in Iceland.

To the Icelander the sheep is a most important animal; from its milk both butter and cheese are obtained; its flesh, when dried or salted, forms an article both for home consumption and for exportation; its wool enters into the material of almost every part of the Icelandic dress; of the skin fishing garments are made, and these, being smeared repeatedly with oil, so as to become saturated with it, are rendered quite water proof. The tenant of a farm pays his rent partly in wool and tallow. When butter is rare, sheep's tallow is the general substitute; and so fond of this substance are the children, that they may be seen eating lumps of it as if it was some sort of sweetmeat. One of the ways in which these primitive and hardy people reward their parish priests, is by keeping each of them a lamb for him during the severe winter season; they take it under their care in October, and return it to the minister in good condition about the middle of May. In Iceland, as in the eastern parts of the world, a sheep or lamb is the usual tribute of hospitality, and is the common present to the stranger and traveller, or killed to make him welcome.

The notices of the wild and the oriental breeds of sheep here

ICELAND SHEEP.

introduced are designed rather to gratify curiosity, natural to persons, on such a subject, in rural life, than to subserve the interests of agriculture. But to learn how the influence of climate, soil, and food, can produce so many varieties, from a single stock, as seen in the sheep family, and also in most of the domestic animals, is a subject worthy the attention of the philosopher as well as of the common observer. The principles here developed are at the foundation of all improvement in the breeds of farm animals. Knowing what changes have been wrought by external and incidental influences it is easy to calculate how to arrange and modify these agencies so as to produce whatever results may be described. This has been sufficiently exemplified by persons who have long been occupied in sheep husbandry, and who have reduced the principles, that have governed them in their labors, to a regular science. The stock breeder may reach the desired point of merit in animal economy, with nearly the same certainty, that the mechanic can produce a prescribed result in machinery if he exercise a corresponding degree of intelligence and vigilance.

Spain, from the earliest ages, has been celebrated for the production of wool of exquisite delicacy; and though, from the time when it was a Roman province to the present day, the country has undergone many changes, many revolutions, certain breeds of sheep have continued to maintain their ground, and at the same time their pristine celebrity. Strabo, in the reign of Tiberius, speaks of the beautiful woolen cloths made of Spanish wool, and worn by the Romans; and at a subsequent period, Columella exerted himself in the improvement of the Spanish sheep by the introduction of rams from Africa, and also of the more valuable stocks of ancient Italy. And, it is on record that some of the fleeces of the original Spanish flocks were black, while others were brown or of a reddish hue. The remains of these ancient varieties of color may still be discerned in the modern Merino sheep. The plain and indeed the only reason that can be assigned for the union of black and grey faces with white bodies, in the same breed, is the frequent intermixture of black and white sheep, until the white prevails in the fleece and the black is confined to the face and legs. It is still apt to break out occasionally in individuals, unless it be fixed and concentrated in the face and legs by repeated crossings and a careful selection. In the Merino south down of England the black may be reduced by a few crosses to small spots about the legs, while the Merino hue overspreads the countenance. The Merino hue, so variously described as a velvet, a buff, a fawn, or a satin colored countenance, but in which a red tinge not unfrequently predominates, still indicates the original colors of the indigenous breeds of Spain; and the black wool, for which Spain was formerly so much distinguished, is still apt to break out occasionally in the legs and ears of the Merino race. In some flocks half the ear is invariably brown, and a coarse black hair is often discernible in the finest fleece.

The Merino sheep are long in the limbs, but the bone is

small ; the breast and back are narrow, and the sides rather flat ; the fore-shoulders and bosoms are heavy, and the skin under the throat is loose and flabby, or indeed pendulous ; the forehead and cheeks are covered with a coarse long hair, but the lower part of the face is smooth and velvety ; the head is large, the forehead rather low. The male carries comparatively large horns, spirally contorted, the curvature being often very graceful. The females are mostly destitute of horns, and where these appendages are present, they are small. The wool of the Merino sheep is at once exquisitely fine, and admirable for its felting pro-

MERINO BUCK.

perties. As to length of staple, the breeds of different provinces somewhat vary. The average of fleece of the ram is from seven to eight pounds ; that of the ewe five pounds. Merino sheep, when fatted, usually weigh only from twelve to fifteen pounds per quarter.

Sheep husbandry in Spain is conducted on an extensive scale. Our farmers generally have no conception of the magnitude of this interest. The Merino flocks, after spending the summer in the mountains, descend to pass the winter in the milder provinces. Each flock, consisting generally of about ten thousand sheep, has a head shepherd, who is chosen as an experienced man, well acquainted with the nature of pastures, and the different diseases to which sheep are liable. This chief, or superintendant, has under him fifty shepherds, each one of whom is furnished with a good and powerful dog, as a defence against wolves, which are much to be dreaded. It necessarily follows, then, that the flocks are subdivided, each shepherd having his own peculiar charge, to which, under the direction of the principal, he expressly attends. Age after age these sheep have been conducted annually from the mountains to the plains, and from the plains to the mountains. This migratory system seems natural to this animal as well as some others. The Laplander pursues a migratory system with his herds of reindeer, and the Tartar with his flock.

Though the superiority of the wool of the Merino sheep of Spain was acknowledged for centuries over Europe, yet the idea of improving the native breeds by crosses with the Spanish seems not to have been entertained till a comparatively recent period. Wool was indeed largely exported from Spain into England, Germany, France, and other continental states, where the manufacture of fine cloth was carried on; but the attempt at rendering the inferior fleece of their own flocks available for this purpose, by a gradual intermixture of them with the Merinos, was long neglected. England had indeed her own sheep of great value, both short and long wooled, and therefore felt less than many other countries the necessity for any amelioration; perhaps even the great wool growers might doubt the possibility of improvement. It was in the bleak land of Sweden that the bold attempt was first made to naturalise the Merino sheep of Spain, and to improve the native race by judicious intermixtures. Individual effort was aided by the civil government, so that between eighty and ninety years ago there were in that country between sixty and seventy thousand pure Merinos, besides many valuable ones of a mixed breed.

Ere long Saxony followed this example of Sweden. The

wool of Saxony quickly became celebrated, and more than rivalled that of Spain in the market, and the woolen manufactures at the same time rose in reputation. It would appear, indeed, that the Merino succeeded better in Saxony than in Spain. In 1809 Saxony reared about 1,600,000 sheep of all kinds, and could boast of 900,000 partly pure Merino, and partly of the most improved and valuable crosses. These mixed breeds took the name of Saxon-Merinos. Other States of Germany were not slow to embark in this praiseworthy enterprise. Prussia, under the munificent auspices of Frederick the Great, who had there spent more than forty millions of dollars for agricultural purposes, became distinguished for her sheep husbandry. Her flocks of sheep have sometimes numbered in the neighborhood of 5,000,000, more than half of which are pure Merinos. Similar efforts were made in Austria. In 1775, the empress Maria Theresa imported three hundred Merinos from Spain and placed them on the imperial farm in Hungary. From time to time other importations were made from the same country, so that it is now calculated that Austria, including the Hungarian territories, has nearly or quite twenty millions of excellent sheep.

In the mean time France did not exhibit herself an uninterested spectator in the noble efforts for one of the most important branches of rural economy. The French Government, in 1786, purchased one thousand three hundred and seventy-six ewes and lambs in Spain. These were sent to the royal farm at Rambouillet, an establishment devoted to the improvement of domestic animals; and, like the Saxons, received all the attention which intelligence and wealth could bestow, and the consequence was soon manifested in their large size, and the increased weight and uniformity in the fineness of their fleece. Colman says that sheep, which he saw in France, and which were originally of the stock of Rambouillet, were, beyond comparison, the finest of the kind he had ever seen; and he expressed the opinion that they were of the best kind of sheep, for this country, that could be raised. They would weigh full twenty pounds to the quarter when dressed; their wool is of fine quality, and their fleeces extremely large and heavy. They are not so large or fat for mutton sheep as the Leicester or South Down of England, in which country mutton being a favorite food, is much more an object of demand than in the United States, but the superior fineness of their wool gives them to us a peculiar value.

Nevertheless, the progress of sheep husbandry in France has been rather impulsive. This in part is the result of the general

SAXON BUCK.

impulsive character of the people, and of the general tendency to revolution. Efforts to improve agriculture, and particularly the breeds of animals, should be systematic. There should be no vacillation of purpose, no transient or evanescent order to accomplish the desired object of pursuit. For reasons here intimated, or for other reasons less obvious, notwithstanding the success attending the efforts that were made in that country to improve the breed of sheep, much, very much, remains to be done. It was stated in 1811, twenty-five years after the establishment of the Rambouillet flock, that while there were in France thirty millions of the native breeds of sheep, there were only two hundred thousand of the pure Merinos.

It has already been implied in this brief sketch, that England was tardy in her attempts to naturalise the Merino, or to seek by its means the improvement of their own stock of sheep. The reasons for this are various. It is well known that without any such improvement their fleeces were of a high grade of excellence, if not as fine as that of some of her continental neighbors. Of course this reason originated in the superiority of her own breeds over that of many others. This was a ground of complacence, and thus smothered desire for any thing better. It was also contended that the Merinos would degenerate for the want of a Spanish climate, Spanish pasturage, and the long periodical journeys to which they were accustomed in their own country. True, it was known that in Germany, Denmark, and Sweden the experiment had been satisfactorily tried, so that this argument should have had little or no influence. But, in this one case as in most others, when men have once espoused any hypothesis, they are not likely to yield concession as soon as convinced. The pride of opinion is not easily subdued. For a long time they adhere to an original position. Moreover, the tardiness in question seemed to rest substantially on a more plausible assumption, to wit, that in England sheep are raised as much, and even more for the flesh than for the wool. When it is known how much larger the English sheep are than the Merinos, to say nothing of the supposed difference in the quality of the meat, it is not extraordinary that such a procrastination was occasioned.

Although the popular feeling in England was decidedly unfavorable to the cause of Spanish sheep, they had some zealous advocates; and, they were of a character to press forward in their aims to make a fair experiment. Sir Joseph Banks, eminent in his day as a man of science, and particularly as a naturalist, gave the enterprise the benefit of his name and of his ardent co-operation. Associated with him were several others, calculated from their talents and position in society to give it additional impulse. Among these were Lord Somerville and Mr. Bakewell; the latter being the gentleman who afterwards became so distinguished as a stock breeder. Nor was the scheme without royal patronage. George the Third, a most devoted agriculturalist, resolved that the Merinos should have a fair trial. Accordingly he purchased a small but choice flock and placed them on one of his farms. The success did not meet the expectations of those interested. But not discouraged, he made a second attempt, and applied to the Spanish monarch for permission to select some of the best sheep of the migratory breed and bring them direct from Spain. This request was promptly met, by a present of five rams and thirty-five ewes from one of the royal family. Still some difficulty arose in the care of them, so that the prevailing public sentiment continued adverse to the project. Such was the condition of the Merino sheep in 1791.

However, the few friends of improvement in sheep husbandry continued to make successive movements to secure a triumph; and, in 1801, their unwearied perseverance reached a crisis which scepticism could not resist. Hundreds who had previously ridiculed the idea that had animated the friends of the measure, now not only sent in their adhesion to it, but, as if to make atonement for past opposition, like most new converts to a doctrine, became the victims of undue zeal. Public sentiment is frequently like the vibrations of a pendulum, when first put in motion, not only passing from one extreme to the other, but actually compassing a space too wide to be maintained by the force of gravitation. So it was with the Merino sheep fever in England, during the first ten years of the present century. There was a wildness about it, that viewed in the retrospect, seems incompatible with common sense; limited, in its operation, not by the shores of the island, where it originated, but sending, as we shall by and by see, across the Atlantic, and to hundreds of our own citizens, a similar spasmodic impulse. In 1804, Merino rams sold there on an average for nearly twenty pounds sterling, and one of them for forty-two guineas. In 1805 a flock of rams and ewes were sold at an average of thirty pounds each. In 1808, one of the former was sold for over seventy-four pounds. And in 1810 a ram was sold for one hundred and seventy-three guineas, and several others varying in prices from that sum to one hundred guineas. The ewes also that year were sold at prices from sixty to an hundred guineas each.

When the world gets on stilts, though the strides be long, it requires no prophetic inspiration to teach us, that there will be a speedy cessation of progress. Our own experience teaches us that. So it has always been. So it was with the Merino sheep speculation in England. So it has been with a similar one in our own country. The effects of it produced a baneful influence for nearly the fourth part of a century. The reminiscences of that speculation are still remembered ; and to those who did not suffer from it, they furnish matter for amusement.

"In importance," says Colman, "sheep occupy a high place among the live stock of Great Britain. The wool finds a demand in the various manufactures of the country; and mutton and lamb make up an extraordinary portion of the food of the inhabitants. Size, thrift, or disposition to fatten, hardihood, early maturity, prolificness, quantity and quality of wool, are matters of great consideration in these animals. It cannot be said that all these properties have as yet been combined, in the highest degree, in any one kind of sheep ; perhaps such a combination is impossible ; but the efforts for improvement of the different races, and, in several instances, the success of those efforts, have been as remarkable as in the improvement of neat stock. There are, says he, no fine-wooled sheep in Great Britain. Size and fatness are the principal objects with the British farmer; and in the latter quality, it would be undesirable to attempt any further advance. Our limits do not allow any particular description of all their sheep, but a short notice will be given of a few of their principal breeds

Of the long-wooled sheep, the Leicester takes precedence of all others. This race of sheep owes much of its excellence to the sagacity and skill of the celebrated breeder both of cattle and sheep, Mr. Bakewell. It was his aim, by careful selection, to combine, if possible, fineness of bone, beauty and symmetry of form, tendency of disposition to fatten, with weight of carcass, and a good yield of wool. In all these respects, it is surprising what he seems to have been able to accomplish ; and for roundness and finish of form, flatness and width of back, shortness of neck, fulness of breast, width behind, and depth of fat upon the ribs, the best samples of them are most remarkable.

NEW LEICESTER SHEEP.

The success of Mr. Bakewell in breeding his sheep, and raising them to a high degree of perfection, is perhaps in no way more strongly evinced than in the fact that he let his first ram for the season, in 1760, for seventeen shillings and six-pence, and in 1789, he let one ram for one thousand guineas, and he cleared more than six thousand guineas, or more than thirty thousand dollars, the same year, by the letting of others.

The Lincolnshire, the Cotswold, the Dorsetshire, the Gloucestershire, the Oxfordshire sheep, are large, coarse-wooled, and coarse-boned sheep, which have their partisans in particular districts, and are much crossed and intermixed with others, but have not attained the enviable distinction of being improved, so as to form a distinct and extensively popular race. Their yield of wool is large, averaging six or seven pounds to a fleece, and in some instances more, and of variable price, dependent, of course, on the demand in the market for coarse fabrics. Some of these sheep, the Lincolnshire in particular, attain an enormous size. Instances are not extremely rare where they have weighed, when dressed, seventy pounds per quarter. And there is one well authenticated case, of one of this breed, slaughtered in 1836, the weight of the quarters being three hundred and four pounds, and which weighed when alive 434lb. The first time this sheep was sheared the fleece weighed sixteen pounds, and twelve pounds the second time. The Dorsetshire sheep have the peculiarity of producing lambs twice in the year. A sheep which will give two lambs in the year for the market, and her own fleece, is a profitable animal most surely.

The South Downs are an admirable race of sheep. Their average yield of wool is about four to five pounds, of a short staple, and of a tolerably fine, and extremely useful quality. Though they have a great disposition to fatten easily, and come to a good weight, such as twenty pounds per quarter, and often exceeding that, yet their fat and lean are well mixed, and the proportion of the one to the other in the same animal such as is desired. They have dark faces, short legs, and stand extremely well on the legs; are broad in the chest, round in the barrel, most compactly and strongly built; with flat backs, and broad and square behind; quiet and good tempered; much more hardy than the Leicesters, though in this respect inferior to the Cheviot and Highland sheep; capable of being driven, without injury, two, three, or more miles a day, and used often for treading the new sown wheat where the soil is thin; and doing the most ample credit to any care or kindness bestowed upon them. Their wool is much inferior in fineness to that of the Saxony or Merino; but for quality and amount of wool, for size and weight, for quality of flesh, and for general hardiness, it would be difficult to find a superior race of animals.

SOUTH DOWN SHEEP.

Jonas Webb, Esq., of Cambridgeshire, has been a breeder and a keeper of the South Down sheep for nearly a quarter of a century, and laid the foundation of his flock by a selection from some of the best flocks in the kingdom. Since he began his improvements he has never made a cross with any other breed; and no other individual has ever carried off more prizes at the various agricultural and cattle shows, where the premiums are always assigned by judges who are understood to be entirely disinterested, and without any knowledge of the parties to whom the animals belong. The average of wool upon his sheep of different ages, varying in number from one hundred and fifty to two hundred each season, is about eight pounds each.

Mr. Colman says the character of Mr. Webb's sheep is above all praise; and he expresses a desire that this breed be extensively diffused in the United States. As mutton, they are pre-eminent, combining with extraordinary fatness a fair proportion of lean meat, and in taste, deemed equal to the Highland sheep. We, as a people, have yet to acquire a taste for mutton. In this respect, we differ altogether from the English, with whom, in spite of all we hear about the roast beef of England, mutton seems every where to be a preferred dish. The immense quantities of poultry, likewise, which are brought to our markets, will stand in the way of other meats; yet our markets, especially in large cities, are likely to furnish a steady demand and an increasing one for mutton; and wherever they can be reached, it is believed that no breed of sheep are so likely to meet and constantly stimulate that demand, as the beautiful mutton of the South Down sheep.

The Cheviot and Black-Faced sheep of Scotland have peculiarities which entitle them to a brief notice. Both kinds are of moderate size, and good shape, weighing when dressed, from twelve to sixteen pounds to a quarter, and sometimes more. Their wool, especially that of the Black-Faced, is of very inferior quality. They are thrifty, and their mutton is of the best quality, commanding a high price from its resemblance in taste to venison, and is much sought after by epicureans. They are both well adapted to the cold and mountainous regions of the country they inhabit.

Lanarkshire is the great nursery, so far as Scotland is concerned, of the black-faced horned sheep, and they extend their range throughout the Grampian hills and their offsets, from their most southern to their most northern limits. In Lanarkshire the standing stock of these sheep has been estimated at one hundred and twenty thousand, of which the greater portion are breeding ewes, the wether lambs being sold to the farmers of the mountain districts. The males and sometimes the females of these sheep have large horns; the horns of the former are spirally twisted. They feed on the loftiest mountains, up to the verge. One shepherd has usually about five hundred of them under his care.

The Cheviot hills are a part of that extensive and elevated

range which extends from Galloway in Scotland, through Northumberland, into Cumberland and Westmoreland, occupying a space of one hundred and fifty or two hundred square miles. The majority of them are pointed, like cones ; their sides smooth and steep, and their bases nearly in contact with each other.

BLACK-FACED SCOTCH SHEEP.

Excepting at the very top they are fertile, but are bleak and exposed to the weather; and the snowy mantles which cover them in winter remain long in spring after the valleys have become verdant. On the upper part of the hill in Northumberland, which is properly termed the Cheviot, is the central locality of the sheep called by that name. They have been there from time immemorial. This breed, however, has greatly extended itself throughout the mountains of Scotland, and also into Wales, and the west part of England, and in many places has supplanted the black-faced breed. They are without horns, and with face and legs white.

Vast numbers of these sheep, says Spooner, have sometimes been overwhelmed by snow-storms, which in those lofty, exposed situations, descend with merciless severity. Many years ago, as tradition reports, in one winter alone, nine-tenths of the Cheviot sheep were entirely destroyed by the storms. The sheep seem possessed of an instinctive foresight of the approach of these storms, and will hurry to a place of protection, when the shepherd sees not a cloud. A graphic and interesting description is related by Mr. Hogg, the celebrated Ettrick Shepherd, of the snow storm in 1794, in which seventeen shepherds lost their lives, and sheep were destroyed by thousands; one thousand and eight hundred being found on the beds of the Esk alone, after the flood. The difficulties encountered and surmounted are described with the greatest interest; and though such severe storms occur but seldom, yet the losses are very heavy in ordinary bad seasons.

There can be no doubt that the business of sheep husbandry in the United States is destined to increase. The abundance of land well adapted to it, not less than a favorable climate, strengthens the idea, that the multiplication of sheep will equal at least the increase of our population. With these circumstances, well understood, and a rational presumption of good profits, the character of the American people renders such a conclusion morally certain. The views of Judge Beatty, copied from the American shepherd, on this subject, with a slight modification, are co-incident with our own. The returns of the census of 1840, says he, shows that the number of sheep at that time, in the United States, was about 20,000,000. Twice the number would probably not furnish more wool than would be needed by our population—that is—two sheep for each individual, if we were to manufacture all our own blankets, carpets, and every other description of woolen fabrics. The period is not very distant when this will be done, with the exception of some very fine goods. Upon this hypothesis, with the above population, 100,000,000 pounds of wool would be required for home consumption. And, should our population increase for forty years, as it has increased hitherto at a compound ratio of three per cent each year, we shall have in 1890 a population of about

60,000,000, and shall then need 300,000,000 pounds of wool each year. If the average yield of wool for each sheep be two and a half pounds, we shall at that date, about forty years hence, have in this country 120,000,000 sheep. Sheep husbandry, hence presents to the attention of the American farmer, a source of wealth not now perhaps deemed practicable.

Anterior to the present century, in many districts of our country, each farmer had a little flock, it may have been two or three sheep for each individual in the family. This number about furnished wool for domestic use, and the manufacture of it, by hand, occupation in the Autumn months, for one or two of the female members of the family.

Those living who recollect the perpetual whizzing of the wheel and the clatter of the shuttle throughout the day in most farm houses fifty years ago, cannot but be impressed with the changes that have been wrought in rural life, by the invention of machinery and the application of steam and water-power to the purposes of spinning and weaving. It is not easy to tell whether the change, on the whole, has added to or taken from, the aggregate of family content and happiness of the husbandman. The homestead is now ordinarily as silent as the mansion of the dead ; much the same excepting Monday, on the other days of the week as on Sunday. In the days of our grandmothers, it was all life and animation and commotion. Then there may have been no occasion to destroy the rats by poison ; for the noise of these domestic manufacturing implements must have driven them far from the premises. Then, too the incessant buzzing of the spindle must have rendered lullabies unnecessary to keep the babies in quiet ; and, even the flies if they happened to light on the periphery of this hand machine for stocking yarn, were in danger of being thrown off and having their necks broken. We remember them well ; and notwithstanding we rejoice in much of the progress which characterises the present age, it gives us pleasure now and then to take a retrospective glance upon scenes gone by.

The scattering or small flocks of sheep kept in this country, prior to the present century, by farmers for their own family use, were susceptible of great improvement. The object from the wool then was to keep the body warm, and not to make fine fabrics. Perhaps, had we never been taught the difference between fine and coarse wool, we should have been as well satisfied with the latter as we now are with the former. Refinement of taste in this matter, as well as in many others, may not have added to the amount of human happiness. No canons can settle such questions with any certainty. They are incidents belonging to human society, about which different persons vary in opinion ; yet, all must in a measure yield assent to them, or if not assent, submit to their dominion. The first persons that made substantial efforts to improve our sheep were Chancellor Livingston and General David Humphrey. It was several years before public opinion responded in their favor. However, the impulse given to woolen manufactures by the war of 1812 not only brought Merino sheep into notice, but as it had previously happened in England, led to an extravagance of opinion in regard to their value, which proved ruinous to individuals who engaged in the speculation, and brought the business into disgrace. Instances occurred in the few years subsequent to that war, when choice selections of these animals were sold, ranging from five hundred to fifteen hundred dollars each.

The following account of General Humphrey's agency in causing Merino sheep to be brought into the United States is from Mr. William Jarvis, also a very efficient and extensive operator in the same philanthropic enterprise. In 1801 Gen. Humphrey being then Minister Plenipotentiary at the Court of Spain, purchased two hundred of these sheep in that country, and shipped them for their place of destination. They arrived in the Spring of 1802. It seems to have been a custom at the Spanish Court, when a foreign minister was recalled, on taking leave a present was made to him of five or ten bars of gold, each bar weighing a pound or thereabouts. But as the law of his own country forbids any minister taking presents from a foreign government, he declined this overture and suggested to the Spanish minister, that royal license be granted to take out of the kingdom two hundred Merino sheep which would be a great gratification to him. This the Spanish Minister stated could not be done, but intimated to the General that if he wished to take them out no obstruction should be thrown in his way. The sheep were accordingly procured, and forwarded as already stated.

To give an idea of the progress of sheep husbandry in the United States, we will here mention a few of the flocks, the existence of which has casually come to our knowledge, the natural presumption, however, is that this knowledge has not extended to a tenth part of what may be found in different parts of the country, our attention never having been especially called to the subject. Doubtless, too, larger flocks exist than many here named; and that hundreds of individuals skilfully and successfully engaged in this branch of rural industry are unknown to us. The design of our present labor is simply to call public attention to one of the great branches of rural wealth. It may be supposed that the few facts here collected will effect that purpose.

Among the prominent sheep owners of our country is Judge Beatty, of Kentucky, to whom allusion has already been made. It is understood he has about 1000. The Hon. William Jarvis of Vermont has 160 pure Merinos, 100 pure Saxons, and 750 crossed between the two. William Brownlee of Washington County, Pennsylvania has about 3500; one half kept in his own county and the other half in Iowa. Charles B. Smith of Wolcottville, Conn. has over 300 pure blooded Saxons. John Johnston, of Geneva, N. Y. has nearly 1000 of the best breeds. Messrs Hull and Tilden of Lebanon N. Y. have 1100. Joseph Barnum of Shoreham, Vermont, shears over 600. Charles Colt of Geneseo, N. Y., has 250 full blooded Saxons. Samuel Whitman of West Hartford, Conn. has 275. Mark R. Cockrill of Nashville, Tenn. has about 1500; two-thirds fine wooled sheep, and the others long wooled, or mutton sheep. T. C Peters of Genesee County, N. Y. has 600; and D. B. Haight, of Dutchess County in that State has 250 Saxons and 30 South Downs

Talbot Hammond, of Burke County, Va., has over 1000, principally Merinos. A. B. Hodskins, of Walpole, N. H., has 400; about half pure Saxons, and the others Merinos, or crosses of the two. Messrs. Perkins and Brown, of Akron, Ohio, have 1300, Saxon, or mixed blooded of the Saxon and Merino. E. Kirby, of Jefferson County, N. Y., has 1500 of the best breeds. Jesse Eddington, of Virginia, has 3000, descended from Gen. Humphrey's stock and others equally good. Samuel Grant, of Walpole, N. H., has between 800 and 900. Stephen Sibley, of Hopkinton, N. H., has 300 of the Saxon breed. It would appear superflous to swell this list were we able to do it. In Illinois there were said to be 30,000 in 1849. In Vermont they are numerous. Messrs M. and A. L. Bingham have about 2000; Mr. S. W. Jewett there is well known for his interest in them.

It is said in the American Shepherd that there are Saxon flocks of sheep in Connecticut and New Hampshire which rival some of the best German; and that there are Merinos in most of the New England States, whose fleeces surpass in weight and fineness those of Spain at the present day, and equal the far-famed Rambouillets of France. The State of New-York has within her borders more than one quarter of all the sheep in the Union; and in the aggregate, the wool of her flocks is unsurpassed by that of any other State. The State of Pennsylvania, although she has fewer sheep by far than her soil is capable of supporting, yet on her western borders, especially in the county of Washington, she has flocks rarely equalled. Ohio, too, is advanced in the wool culture, and her flocks are of a superior quality; and many of the western prairies are being filled up with thousands and tens of thousands of these useful animals.

A few facts respecting sheep husbandry in South America cannot fail of being interesting. They were given to the Albany Cultivator by one of its intelligent correspondents. The fertile pampas, says he, in the interior of South America, have long been celebrated for the immense herds of cattle and horses reared upon them. So abundant are they, and so easily reared, that they are slaughtered in many places for their skins and tallow alone. Sheep, too, of the native breed, with coarse, hairy wool, have been so plentiful that their carcases were used for fuel in burning brick. The expense of transportation and the absence of timber and salt for barreling alone prevent us from the competition of their meat in our own parts. The attention of agriculturists there has been of late years turned to improving their stocks of sheep by large importations of Saxony from this country and from Europe. An English gentleman began the business with a stock of sixty Saxon rams and 300 ewes, and in the year 1835 he had increased the number to 45,000, and the grade was nearly increased to full blood. In 1837 he had 90,000, and intended to keep on until he numbered 200,000, which he has doubtless attained before this time. Others were copying his example. until the business bids fairly to outstrip that of cattle in a few years.

The facts here collected on sheep husbandry, and particularly on the most obvious varieties in the sheep family, are not designed to supersede complete and elaborate treatises on the subject, needed by every person devoted to this branch of rural economy. No one can reflect on the influences which produce these varieties both in the form and habits of the animal, and in the quality of the wool, without being assured that the improvement of it, almost to his own satisfaction, lies in the power of every person engaged in the care of sheep. The same principles apply to the improvement of sheep that apply to improvement in all animals; horned cattle, horses, and swine. The few facts, also, here given, it is to be hoped, will call the attention of many farmers to a department of productive industry congenial with their habits and within the range of their general operations, but as yet not so generally understood and appreciated by them as it should be. The facts, moreover, relating to the varieties found in the sheep tribe lead to a train of philosophical reflections and investigations well calculated to promote general mental development and literary taste, a matter of the greatest importance, especially with persons not enjoying the most liberal advantages for education.

Had this article not already exceeded the limits intended for it, a number of anecdotes might be introduced illustrating in an interesting manner the character of the sheep. As deficient as this animal is ordinarily supposed to be in mental endowment, it is well understood that in a few particulars such an opinion is quite erroneous. Cosset sheep may be trained to a variety of useful purposes. In large dairy establishments they are used for churning, far more advantageously than dogs, and at an hundredth part the expense of horse-power. The cost of keeping a few such sheep for churning would be but a mere trifle, and the yield of wool would be nearly the same as though they were not thus employed. Butchers, too, and other persons constantly receiving strange sheep, can place them under the care of an old cosset ram with as little liability for their being lost in straying away or suffering for want of feed, as though they were under the guidance of a shepherd. He, of course, is perfectly familiar with the good feeding ground and every avenue that leads to it. The strange sheep quickly find out his qualifications to be their leader, and will closely follow wherever he may go. Were he to go a mile, and in fifteen minutes to start back to the spot just left, they would all follow him. Were he thus to move from place to place ten or twenty times in a day, not one of them would be seen far in the rear. Were he to leap over a high fence, they would attempt the same; and if any were so unskilled in such feats of locomotion, as to fall back ten times on the ground, they would persevere till it was accomplished. Or if he were to scale a high wall, the whole flock would be in his wake instantly, not regarding the danger of broken legs. And at the close of the day all would collect around him for a night of repose.

The acuteness of the sheep's ear, says the Ettrick Shepherd, surpasses all things in nature that I know of. A ewe will distinguish her own lamb's bleat among a thousand, all braying at the same time. Besides, the distinguishment of voice is perfectly reciprocal between the ewe and the lamb, who, amid the deafening sound, run to meet one another. There are few things that have ever amused me more than a sheep-shearing, and then the sport continues the whole day. We put the flock into a fold, set out all the lambs to the hill, and then set out the ewes to them as they are shorn. The moment that a lamb hears the voice of its dam, it rushes from the crowd to meet her, but, instead of finding the rough, well-clad, comfortable mamma which it left an hour, or a few hours ago, it meets a poor naked, shivering—a most deplorable looking creature. It wheels about, and uttering a loud tremulous bleat of despair, flies from the frightful vision. The mother's voice arrests its flight—it returns—flies, and returns again, generally for ten or a dozen times before the reconcilement is fairly made up.

SALUBRIOUS AIR OF THE COUNTRY.

The process of breathing, and thereby sustaining animal life, is one of the most curious things in nature. The mechanism by which it is maintained, as well as the object accomplished by it, denote a wisdom of contrivance, as well as a perfection in the great theory of animal and vegetable existence, surprisingly wonderful. If there were no other evidence of the fact, this of itself would be sufficient to prove the universal agency of a Being, who made and presides over the material universe, of infinite wisdom, power, and beneficence. Not only is animal life kept in vigor by this process, but it is equally efficacious in giving vigor to the vegetable kingdom. This will more fully appear in the course of the remarks now to be made. Indeed, there is a well adapted harmony and mutual subserviency depending on this process, between the animal and vegetable kingdom. To appearance, neither could exist without it; and, it would surpass man's ingenuity to imagine any other mode of accomplishing these apparently simple, yet, in reality, vastly complicated designs in the wide realms of the material creation.

To understand this subject we must also understand the nature of the air which we breathe; the elements of which it is composed; and the respective functions of these elements. It is not intended to indite a philosophical lecture on pneumatics; but simply to glance at the few facts without which we can have no adequate conception of the processes in animal and vegetable economy to which attention is here invited. With these facts in mind, the beauty of these processes is apparent. A child may comprehend them; and in comprehending them he is led almost involuntary to a love of investigation, not only of them, but of all kindred topics.

It is to be observed, therefore, that the atmospheric air consists principally of two invisible fluids or gases, called oxygen and nitrogen. With them is combined a very small portion of hydrogen and carbon. Every animal has lungs or air vessels. These vessels in brutes are called lights, and in slaughtered animals are familiarly known to all. They resemble in structure common sponge; the interstices of the latter being readily filled with water, as the interstices of the former are designed to receive the air we breathe. They are located in contiguity with the heart, so as to bring the air received by them in contact with the blood, as it passes through the heart. As we open the mouth, the air rushes into it, and thence into the lungs, filling all these interstices, so that they instantly become swollen or expanded, like a bladder, or any air-tight bag, when we force the air into it. By a mechanical muscular action of the chest upon the lungs, as soon as the air has accomplished the object of its mission there, speedily to be explained, they are compressed so as to force from them the air before received, now become foul; and as soon as it is thus ejected, before the mouth closes, another current of fresh air rushes into it as before. Thus, at every opening of the mouth one current of polluted air is forced out of it from the lungs, and another current of atmospheric or pure air through the same channel rushes into them.

The air we breathe, or which we thus receive into the lungs, is worked over by a process similar to combustion. The lungs might not hence improperly be called a furnace to decompose the air, the same as a stove is a furnace to burn up or decompose the wood or coal placed in it for combustion and the generation of heat. Accordingly, the oxygen of the air, being separated from the nitrogen, when in the lungs, is employed to clarify the blood of its impurities, which are constantly accumulating, not very unlike the clarifying of coffee or any liquid by the application of a gelatine substance. The blood before being thus clarified is of a dark brown or blackish color, and thick or clotted. This dark color and coagulated consistence is occasioned by the carbon and other impure substances with which it had become impregnated in passing through the system. But when the blood is clarified or renovated by the action upon it of the oxygen in the lungs, it is of a bright red color, and then passes through the arteries to every part of the animal frame, yet, in its passage is constantly gathering up the impurities with which it was previously loaded. On reaching the extremities of the system, it passes into another set of vessels called veins, to answer the purpose of the backward track of a railroad, and thus it returns again to the head, dark and clotted as before, thus again to be purified by its contact with the fresh oxygen of the lungs. These processes of inhaling fresh or oxygenated air, or breathing; then of purifying the blood; and then of collecting the carbon and other poisonous substances of the animal system, are continued to the end of life; that is, if they were discontinued, the lamp of life would go out, as flame will be extingushed when the gas or oil which fed it is exhausted.

Thus to purify the blood, the oxygen is all extracted from the air conveyed to the lungs by breathing, and is literally burnt up; as much so as the fuel placed in a stove; and will no more answer for that purpose a second time, than the ashes from fuel already consumed in combustion would answer to make a new fire; or than the skins of grapes, after the juice had all been extracted, would answer to make wine; or that the excrements of animals would again answer for food, after all the nutritious element had been removed in its first use. Indeed we can no more use the air in breathing the second time than we can use our food the second time. The former in use becomes as foul as the latter; not only as foul, but as inefficacious for its legitimate agency. Hence no one can fail to perceive the necessity for a constant supply of pure air in breathing, and consequently in the preservation of life. To attempt living without it would be as absurd, as to attempt living without food. Moreover, we could live an hundred times as long without the latter as we can without the former; and, to mix arsenic with our food would be comparatively no more fatal to the vital principle, than to mix a poisonous gas with the air we breathe.

It is a well known fact, that we breathe eighteen or twenty times every minute; and, at each breath we inhale or take into the lungs about one pint of air, or over two gallons each minute. Thus in an hour an adult person consumes more than one hundred and twenty gallons, so that if he were enclosed in a hogshead containing one hundred and twenty gallons, before the end of an hour the whole of the air contained in it would be exhausted, and he would die for the want of the vital principle which pure air imparts in breathing. It is as well ascertained, that animal life depends on having a constant supply of atmospheric air, as it is that there must be a supply of food; and where this supply is deficient breathing will become difficult. It will be difficult also if the air is impure. The cases on record are numerous, where persons have suddenly fainted and died from entering deep wells, caverns, and vaults filled with noxious vapors. So they are of no rare occurrence where persons have died when sleeping in close rooms containing burning charcoal. The vapors thus inhaled are in reality the same of those ejected from the lungs in breathing. In the one case the carbonic acid gas is generated in a little iron or pipe-clay furnace; in another case, it is generated in the lung, already said to be analagous to any other furnace. This is the only difference.

Hence, if a prisoner were shut up in a cell perfectly air-tight, containing the cubic measure of twelve such hogsheads; or, if any one were to attempt sleeping in a room air-tight, of that capacity, in about ten or twelve hours the air would be so foul from use in passing through the lungs, that if life did not become extinct, breathing would be barely practicable. Or if four persons were to sleep in an air-tight room of the capacity of forty-five or fifty such hogsheads, in about ten or twelve hours they would all become incapable or nearly incapable of breathing. Or, if the cabin of a steam-boat, of the capacity of a thousand or twelve hundred of these hogsheads, and containing one hundred passengers, were without ventilation, and were to receive no fresh air, a similar effect would be produced on them all. On this account, it is evident, that all rooms for sleeping, and all public rooms, churches, lecture-rooms, and halls for amusement —containing a great number of persons, should be so situated and so constructed, that there may be a continuous escape of foul air as of ingress of that which is pure. In the location, therefore, of houses and sleeping rooms, and in the construction of all the apartments in a dwelling house, a judicious reference should be had to this subject. To have a sleeping room in a cellar, or elsewhere liable to be filled with polluted air, or excluded from a free ingress of pure air; or, in any situation, no matter how good, if the foul air generated in it cannot make its escape, or the pure air admitted into it is unnecessarily destroyed by heat from stoves or furnaces, denotes an ignorance of physiological science, and a reckless disregard of the preservation of health and of life, incomprehensible among those making the least pretension to good sense.

It may not be frequent, that the disastrous effect on life, happen from these causes, indicated as possible. It may be they do not frequently occur. If so, the reason is, not that our theory is false, but that enclosures are not made air tight. They could not easily be so made, if there were an attempt to do it; and moreover, the importance of a free circulation of air is so well understood, that there would be no attempt to do it. And in the country there is generally but little danger to be apprehended from such causes. There, houses are generally too open rather than too tight. It is in the city mostly, that we are to apprehend such evils. In the city little or no advantage can be taken of location, for obtaining an adequate supply of good air. The houses too, by all who can afford it, are so well constructed, there is little opportunity to obtain its free ingress, or to be relieved from that which is within, and is become unfit for respiration. Indeed, among the poorer classes, almost every yard of space in a house, from the cellar to the garret, is crammed with human beings, so that each one is nearly as badly situated, as though he were headed up in a hogshead. In many of the filthy locations of the city, in a single edifice, not larger than many farm houses in the country, may be found from fifty to an hundred persons, at least one to every ten cubic yards! Here they live; here they sleep; here are impurities of every description; and perhaps not the worst, the odors of their own evacuations, penetrating every niche and recess of the entire premises. Whence do these persons obtain the oxygen to purify their own blood? And, in addition to all other deleterious exhalations, from within and from without, how are they to be relieved from the two or three hundred pounds of carbonic acid gas, every twenty-four hours ejected from their own lungs? Positively, such a place would be a barbarous one for the execution of criminals!

It is estimated, that every full grown man emits from his lungs in breathing, each twenty-four hours, about three pounds of carbonic acid gas, so that in a city like New-York, of half a million of inhabitants, there is daily a creation of more than five hundred tons of this poisonous vapor, and in a month more than fifteen thousand tons of it, and in a year, more than an hundred and eighty thousand tons of it! What becomes of it all? There is no way to use it up and get rid of it, as in the country, as we shall, by and by, see. A portion of it is indeed carried off by the winds that successively pass over this congregated multitude of chimneys and spires; but not a little mixes with the air to be breathed, or settles down in the vaults and chimneys, and crevices between the walls, or rather becomes a permanent floating cloud, extending from one side to the other, like the pestilential vapors that hover above the grave yard. Could such a city be surrounded by an air-tight brick wall of seventy-five or an hundred feet in height, and then covered over by an air-tight roof, so that no fresh air could ever afterwards enter it, and none of its own foul exhalations escape it, the now living multitudes therein would soon become one mass of putrefaction, with as much certainty as did the inhabitants of Pompei and Herculaneum, when buried under a shower of volcanic eruption. There could in such an exigency, be no escape from such a catastrophe. That in such a city, there is thus generated such physical elements of self-destruction, is an undeniable truth; and the only reason we no oftener witness their destructive power is, because of the benignity of nature, in a concatenation of counteracting influences.

It is not denied, that in the country as well as in the city, men and women and children, and brute animals breathe, and in their breathing create a due quantum of the gas described; but it is not as in the city, a nuisance, an engine of destruction. Far from it. Instead of being an evil, here it is a prominent, efficient instrumentality, in promoting the purposes of the Supreme Being. Have any occasion to ask, what becomes of it, or what use is made of it? In giving an answer to the inquiry, we present one of the most beautiful harmonies of the material creation, that can be imagined. A mere glance at it cannot fail to fill the mind with admiration. If such exhibitions do not make men devout, no one can tell what would do it. On account of such exhibitions in rural life, constantly before the mind, it is maintained that the occupation of the farmer is peculiarly favorable to mental and moral culture. Where else can one, amid his every-day labors, continually find topics in his way, both for interesting scientific investigation, and for the development of pious thoughts?

The secret is this. Men and all animals, as has been described, breathe or inhale atmospheric air. Their life and their health depend upon it. On the other hand, trees and shrubs, and cereals, and flowers, and tuberous plants, every vegetable substance, also breathe, if it may be so called; that is, their leaves as constantly absorb this poisonous carbonic acid gas, prepared in the lungs of animals, as the animals are able to prepare and eject it. The principal distinctive attribute of vegetable life is derived from this source. Without this element of vegetable substance, the meadows and the fields would cease to be clad in their present waving beauty; the mountains and the forests would cast off their gorgeous panoply; and the valleys and the hill sides would become cheerless, if not desolate. While, therefore, the whole vegetable creation is continually giving itself up for the support of the animal kingdom, animals in return are as constantly furnishing that creation with the means for a rapid and vigorous rising into the perfection of its nature.

Thus too, other nuisances of the city, (for there they are so,) have no repulsive attributes in the country. They neither pollute the air nor become objects of disgust. Decaying vegetables, animal evacuations, and filth of every kind, are here viewed with complacence, nor do they but rarely become an offence to the senses. We spread them upon the fair bosom of the earth. Speedily they mingle with their native elements; and we know nothing more of them, till they return to us in a rich verdure, in blushing flowers, or a luscious fruit. The air, which they polluted and rendered noxious in the city, now in the country passes over them, and brings to us a most balmy and delightful fragrance; instead of conveying to the lungs as it then did, a fetid, poisonous gas, it now bears to the lungs a purifying essence, that imparts power to the muscular action of the system, and health beaming, with inimitable hues of loveliness and beauty on the countenance.

How sweet—how fragrant is the air of the country, when the trees are in full blossom, and after May-day shower! It carries with it a perfume, for richness, unknown to cosmetics. How delightful to snuff up the scent of the new-mown grass! Nor less does the rich ripe fruit impregnate the passing gales with their inviting odors. Under the influence of it, man seems to have a renovated existence, and the country to be invested with new charms. Here let childhood frolic and gambol in pristine simplicity. Here let youth attire itself in the dignity and the vigor of full manhood. Here let virgin innocence, with bright eyes and ruddy complexion, inspire that homage and love, which neither wane nor die, till crowned with the full grown honors of virtuous age. Here let those in the prime of their strength, like the first ancestors of our race, devote themselves to the dignified and healthful labors of tilling the ground. And here let the hoary-headed pilgrims for another world sustain themselves, in the freshness of their best days, till called to lie down in the slumbers of the grave.

BIRTH OF THE FIRST GRANDSON,

OR THE ADVANTAGE OF PLANTING ORNAMENTAL TREES.

Some time in February of 1822, we believe it was, we had occasion, from the capital of New Hampshire, to make a tour over the Green Mountains, into the western part of Vermont. Our object was to meet a professional engagement, that did not allow us much leisure for miscellaneous, social, or literary indulgences; and the peculiar season of the year was not favorable to the making of rural observations, so interesting in that State during all the seasons when her fields and pastures and her forests are green with grass or foliage. However, our habit through life has been to allow no opportunity to be neglected, where hints and facts in practical wisdom can be gleaned up. Indeed we never made an excursion in all our life among the plainest portion of our yeomanry, or spent a night at one of their houses, which we have done perhaps an hundred times, without learning something we did not know before, and without gleaning some hints upon the economy of human life, that could be used with others to the greatest advantage. Having finished our prescribed month of labor at Middlebury, which was the object of our tour, late in the afternoon of a mild day, about the middle of March, we commenced retracing our steps to our beloved granite hills. There was a full moon, and our ride, if a little lonely, no one being with us, was not unpleasant. At ten o'clock in the evening we reached the house where we were to spend the night; and, as it happened, where we spent the following day.

The house of our host was rather spacious, and with out-buildings and other appendages, exhibited evidence of substantial comfort and independence. It stood upon a concave declivity of the mountain, with a south-eastern aspect, so that the snow had mostly disappeared. Into this segment of a sphere the first rays of the morning sun shone with beauty, and falling upon our chamber window admonished us that the hour of breakfast might be nearer at hand than we had imagined. It was evidently the beginning of one of those delightful March days so much desired in maple districts to catch the sap for sugar. We were aware that in these districts the entire working population at this time were busily occupied; and we had supposed that the little commotion in the latter part of the night which disturbed our repose and made us late to rise, was a preparation to engage early in the sugar labors of the day. Nor were we undeceived respecting this fact till at the breakfast table. The name of our host we shall call Andrew Lawson, usually called in that region Squire Lawson. He had but two living children, a daughter, married, and living a mile distant; and a son named James, married the previous year, and, with his wife, occupying a portion of the home mansion with his father and mother. In the latter part of the night to them a son was born, which was the first born grand-son of Squire Lawson. The movements connected with this event we had suspected to be a preparation for the maple forests belonging to the homestead farm. It will be seen in the sequel what influence this event had in retarding our departure homewards.

At a rather late hour we were at the breakfast table; to wit: Squire Lawson and his wife, the grand-parents of the little stranger who had just made his appearance; James Lawson, the father of it; the family doctor, who still remained; and, lastly, ourself as the guest of the family. As usual on such occasions there was not much disposition in those most interested for conversation. There were only a few common place remarks, save by the doctor and ourself. On such occasions the heart is generally too full for free exultation. Tears rather than laughter or freedom of speech are usually the first manifestations of the joy then experienced. The anxiety and the anguish so recently felt had seemingly choked up all the avenues for the expression of the sudden rush of pleasure into the bosom. We have seen the father of the first-born, for a long time weep like a child. It seems a strange way of giving vent to a thankful heart. The philosophy of this is not of easy solution. Perhaps it is owing to the escape of the previously suppressed agony.

When this is gone, the channel is cleared for a stream of exhilerated and joyful emotion. To a cast of sadness on the countenance, succeeds quickly the bright shadows of renewed life and animation. To a manifestation of taciturnity, quickly succeeds even an impatience to eulogise the expanding hopes of the family. It is a little so, when the gate of the mill-pond is first raised. The first rush of water is turbid ; but anon the sediments pass off, and the flowing current becomes pure and transparent, like crystal.

When first seated at the breakfast table, from the appearance of James Lawson, a stranger might have been perplexed in knowing, whether his wife were shrouded in grave clothes, or had become a joyous mother, and himself the father of a hopeful offspring. As conversation increased, these obscurations of his swelling bosom began to disappear, as thin clouds in the morning disperse, when the sun begins to shine upon them. James was a plain young farmer, inheriting the sound common sense of his father, and had received what may be esteemed a good practical education. Such a man, if not possessed of all the flippancy and graceful social polish found in some departments of life, has a big soul ; and his perpetually gushing affections about the fire-side, are as constant as the pulsations of his own life. His love to his family is as unerring, as is the magnetic needle in pointing to the poles. When was the aged father or the aged mother ever abandoned by such a son? When did such a son, on becoming a husband or a father, ever fail to cherish with unwearied assiduity, a kind wife or an affectionate child; or to bow down in the overwhelming sadness of a broken heart, if called to follow one of them to the grave? We have never seen these noble traits of character better defined, than in rural life. Here are no pompous exhibitions of external affection, when the heart is cold and pulseless. Here we are to expect no conventional ostentation. Here all is simplicity, and, what is better, undoubted sincerity.

To have delivered a sermon at the breakfast table, at a season like the one indicated, would surely have been in bad taste, and productive of no good. It would have denoted on our part, a feeble knowledge of the means best adapted, to put into successful operation, the moral energies of the mind. Yet, at such a season, a few words may be so fitly spoken, as to be like apples of gold and pictures of silver. As the social elements of our young friend began to assume their proper equilibrium, an attempt was made, rather, however, in a colloquial, than in an austere form, thus to improve the opportunity. Each additional remark did evidently awaken new interest, and we are not without belief, that our effort was kindly received and duly improved. At any rate, there was every appearance that we are not mistaken. The grand parents were overjoyed at our remarks, as will fully appear before we conclude the narrative of this reminiscence; and, the new-fledged papa did not fail to appreciate something of the new relation in which he stood to the world.

As our remarks were in substance applicable to every similar exigency, in the domestic relations of life, we shall here give a brief abstract of them. To give them in the colloquial or conversational form in which they were uttered, might not be an easy task, after the lapse of thirty years; and it would occupy more space than we have to spare, and require more time for being perused, than consistent with the leisure of our readers.

It is a fact every man may understand, that on becoming a parent, particularly of a son, he is ushered into new relations, not only to his own family, but to the whole human family; not only to the age in which he lives, but to the never-ending future, as long as the race of man shall have existence on this globe. Instead of a broad deep river, in which the current of his affections may roll on with increased volume and accelerated motion to the end of time, without offspring, this current is confined by the little rill that becomes lost in traversing a bed of sand. Man, without offspring, is but the mere incident of his species. He is like the spark, that is blown from the burning mass, and in a moment is extinct and is forgotten. One moment he may exhibit a point of brilliancy; but the next moment it is gone forever, and that point is lost in the darkness of night. Let a man, without offspring, look through the telescope of time, upon the wide spread glory of the world in coming ages; let him, in imagination, read the history of a score of rising generations, and what will be there to make him rejoice that he ever had existence? Will he be able to appropriate any of that glory to himself? Will any of those countless multitudes trace back to him a proud lineage? Can he exult in the idea, that some of his own blood mingles with theirs? By no means. But a man with offspring may indulge the thought, that in a few centuries, the blood of a majority of the inhabitants, constituting a vast empire, may be part and parcel of his own; that the greatest generals and statesmen, and even kings, may trace back to him an unbroken lineage. May not a man then, on first becoming a parent, look with a dignified complacence on himself, never before imagined? So it appears to us. If not so, the most precious endearments of a social nature, are the meanest and most insiped of fictions.

In reflecting on a subject like the present, there is no necessity for these far-distant views. Our own powers of vision, without the aid of telescopic transportations, are sufficiently suited to the argument. The subject is even more impressive as in the case of young Lawson becoming a parent, when viewed simply in relation to his own family. What motive, in a young man like him, can one have for a life of toil, if there is no one after him, for whom he feels interest, to take possession of his wealth? And especially, as it sometimes will occur, that those allied to him in remote consanguinity, will almost rejoice in his decease, that they may receive what belonged to him. And even when alive and in the vigor of manhood, where will the man without offspring inhale fresh odor, that like pure incense, constantly rising about the fire-side, from the breath of his own buoyant children. Their affectionate prattle and their joyous glee will oftentimes make the fond father forget his toils, and inspire him with new energy while encountering the frowns of adversity. Is there no magic power in the caresses of a group of young children, hanging upon the knees and about the neck of the parent? Is there no hallowed fragrance in their oft-repeated embraces? We know there is; and, if there is ought in the retrospect of our own shadowy life, that we shall remember when on its last verge, it will be the infant nestlings and the paternal watchings therewith, of little ones, now grown to man's estate, and partly, in the dispensations of a wise and good Providence, called to exercise a similar agency towards those of another generation; a generation that is our own grand-offspring.

What we have here written will convey but a faint idea of all that was or might have been said, when surrounding the breakfast table of Squire Lawson. Before concluding, we said to the respected grand-parents: "Venerable friends, this to you must be a day of gladness. Such an event is always so, to the grand-parents; to you, it must be especially so. If this new-born child is to be blessed with a life like your own, you will, when you shall go down into the grave, have the pleasing anticipation, that for at least one generation more, the name of Lawson will be identified with this delightful farm; and, that within these very walls, may be born, live and die, other generations inheriting your own blood." The old lady wept; and from the eyes of her dignified spouse, we thought, a few tears stealthily fell. All was silence; each, by degrees, recovered self-possession ; then Squire Lawson, raising himself to a more commanding position, his elbows resting on the arms of his big chair, and with a firm, but rather choked voice, says : "James, I will tell you, what I want you to do, this very day : I want you to go up on the side of the mountain, and dig up two of the best sycamore or black-walnut trees you can find, and set them out somewhere near the house, as a memorial of the birth of that dear baby. You know that those large elms near the big barn were set out the year my father was born, on this very spot ; the sugar maples by the cow-house were set out the year I was born ; and the cedars by the big gate were set out the year you were born. When you have done this, James, we shall have trees marking the rise of four generations of one family, and what is better, in the same place. This is the family record I like best. I never look at the elms without thinking of my grand-father, who set them out ; nor of the maples, without thinking of my father, who set them out."

This was the first time we ever had seriously reflected on the value, as well as the beauty, of ornamental trees about the mansion, particularly as mementos and monuments of historic events. The remarks of Squire Lawson on their use, as a kind of family record, to us were new, and they have led to a new chapter in our philosophy. Had my own children been born in the country, instead of the dusty city, we would, succeeding to the birth of each, have caused to be planted a few trees, thus to be a standing memorial of the event. Let a family of children, where this is done, become grown up and scattered, as is common, except the eldest or the youngest son, who is to retain the homestead ; and when they make their periodical visits in after life, to the place of their birth, bringing with them those of another generation, how interesting will be to them such rural inscriptions to perpetuate the family chronology ! And, if one of a family be cut off by death, let there be planted a weeping willow, the token of grief ; or, an evergreen, the emblem of immortality ; so that when, in subsequent life, the survivors look upon them, they may be reminded, not only of the tears which humanity sheds upon the graves of our loved ones, but of the undying life to which they have gone. When we visit the place of our own nativity, one of the most cheering exhibitions, that seemingly bid us welcome, is the venerable broad-spreading and deep-colored foliage, under which, in a hot summer day, when a child, we were wont to build toy-houses. The human forms then most familiar to us are gone; and those that remain are so changed, as scarcely to be recognised. But the majestic trees, the work of nature's laboratory, still maintain their identity, changed only in having become more magnificent, from increased age, more massy trunks, and longer branches.

Few persons are aware how much a few well chosen trees will add to the beauty of a country residence. Here is an ornament available to the mechanic as well as the farmer. The cost of it is but a mere trifle. The few leisure hours at the command of all, devoted to this object, would add much to the elegance and pleasure of rural life. As our forests are yearly giving way to the axe-men, this is constantly becoming more apparent. Unless our wood lands are watched with more caution; and unless more attention is paid to supplying, by a new growth, the places wastefully deprived of trees, many portions of our country will present a most repulsive aspect of desolation! In no one particular have Americans evinced a more spurious taste, than in their ceaseless havoc of cutting down every thing like a pillar in the great temple of nature. The traveler, in passing through many of our towns, may observe where hundreds of acres, seemingly without any object, where the land was not required for culture, and where no benefit could have been realised from the timber, have been deprived of every green branch, even for the shelter of the birds, with which they were once covered. The mountain side, with bluff rising above bluff, like the steps of a stair-way, interspersed and covered with trees so far as there is soil for their nutriment, is not only a spectacle of grandeur, but is one of great beauty. Let the meditative sojourner, in passing over the Erie Railroad, when in the neighborhood of the Delaware Gap, survey the almost infinitely diversified mountain scenery that will there like enchantment rise up before him, and he will have increased conceptions of natural beauty.

Here in the season of autumn may be seen such a multitudinous assemblage of hues in the foliage as man had never before imagined. One might as well count the stars in a clear winter night as to count these hues. There is the pale green, the dingy yellow green, the reddish green, and the dark green of the fir which defies the power of cold, of frost, and of snow. There is the pale red and the dark red, the dingy scarlet and the crimson, the pale yellow and the dark yellow; and, all these so running into combinations with each other, that it is impossible to tell which has the predominance. It is usually said the newly opening foliage denotes the fresh beauty of children and youth, and that the seared leaf denotes the decay of manhood and the approach of old age. The illustration is most appropriate. And in the case we are describing may be seen, not simply the emblems of the two extremes in human existence—not simply of childhood and of man's final decrepitude; but, seemingly, every possible intermediate tint and blending of color, to denote, in like manner, every possible change in human aspect, from the cradle to the grave. Who cannot gaze with unwearied delight upon these mountain precipices, when covered with garments of such inimitable beauty! The landscape painter may spend his life in learning how to transfer upon canvas the images here constantly greeting the eye; but, never, never can be equal the reality! Yet, let an hundred pioneers, with axes and their brawny muscles, deprive these mountain ranges of their now uprising greenwood, and what a cheerless, paralysing shadow will they then cast upon the traveler! Even the wolf and the bear, as if spurning uch desolation, will depart to more distant regions, not thus defiled by such wanton desecrations upon nature's realm.

Our yeomanry, in making clearings, have generally made most unfortunate mistakes, in leaving so few forest trees about their dwellings, and even by the road side. Where few or none have been left, no time should be lost in supplying the place with a new growth. In this way another generation may be furnished with rural embellishments at present rarely seen.

This is the only atonement that can be made for the error committed. This is the only remuneration we can offer for the injury we have thus occasioned to those who are hereafter to occupy what is now in our own possession. To spend a few dollars in this way will be better for our children than double the amount invested in bank stock. A well chosen hickory, or chesnut, or maple, in twenty or thirty years after being planted, will be worth at least five dollars. It would be worth that at the present prices of timber; and not unlikely, that from the increasing scarcity of it, by that time, it will be worth double this sum.

If the inhabitants of any town within ten miles or so of a large city were to unite and plant good-sized and well-selected trees, two rods distant from each other, on both sides of every public road within its limits, the value of real estate, from this cause alone, in twenty years, might be doubled. It would present attractions that would draw to it, one after another, till every acre of it fit for a building lot and garden would be taken up. And within ten years from being planted, those trees would attract favorable notice. Throughout the whole length and breadth of any locality whose avenues were thus studded with handsome shade trees, there would be for riding or walking a kind of fascination admired and sought by all having knowledge of it. The passing traveller would here loiter and linger, that he might record in his note-book the pleasure he felt; and, the inmates of the city, in escaping the scorching heat reflected from a succession of brick walls and pavements, would seek shelter beneath the wide-spreading branches, as well to inhale the fragrance they impart, as to be fanned by the soft breezes which play among them.

It is now a little more than thirty years since we spent the night named at the house of Squire Lawson, and when was made the first recorded annals of a new generation to his family.

The circumstances of that visit are as fresh as ever in our long-taxed memory. The social impulses that then filled our bosom have not yet ceased their pulsations. We can now seem to behold the white locks of the venerable patriarch who presided in that plain but commodious mansion! We can now seem to behold the tears of joy that then ran down his furrowed cheeks! We can now seem to behold the unassuming dignity and the tender heart-gushing of his well-chosen companion! And, we can almost imagine ourself listening to the yet undying echoes of their mutual congratulations, that their name was destined yet to live on, and that even yet other generations under that name might successively rise up in the distant future, each in turn occupying the same homestead, and annually planting flowers upon the graves of a long line of beloved ancestors. If there is aught that can give dignity to human existence, it is a lineage thus perpetuated; a retrospect to each one composing it, full of honor, and in its progressive developments betokening no paralyzing dreams of waning hope. We would go far to see the mounds of earth covering such a succession of generations, and to pick the fresh flowers yearly rising thereon. In the affection that causes them thus to rise, there is a healthful fragrance not less delightful than that which the flowers themselves impart; and, both together give a charm to life and furnish an aliment for our social perceptions no where else to be found. In a measure, too, they furnish an antidote for that prevailing selfishness so deeply engraven upon the world.

Were we now to visit the spot associated in our mind with the incidents above related, what should we there see? Not indeed the venerated heads of the family of which we were a guest. For more than a score of years they have slept with the generations that preceded them. There are their graves! The grass is as green upon them as is the recollection of their virtues in the minds of their descendants! There stand, too, the beautiful sugar maples set out in commemoration of the birth of Squire Lawson; and not far distant two more weeping willows, grown to a goodly size, commemorative of the time when he finished his earthly stewardship. And there stand, also, the majestic elms which tell of the birth of his father, and the cedars that tell of the birth of his son James, and the black-walnuts that tell of the new generation that dawned in the family when we were there ; and, indeed, also, there is the beech, and the hickory, and the chesnut, and the pine, and the black birch, that tell of others since born and sharing in the family record—nor less to be observed, additional weeping willows and yew trees that have become additional records of the family mortality. What beautiful memorials of those who sleep in the dust ! Does the sculptured marble which tells the living world of deceased ones—the good and the great—so impress the heart of the beholder ? Do urns and cenotaphs possess such eloquence in calling into action the sensibilities of those who gaze upon them ?

The ownership of the same soil and the same mansion in the same family for successive generations, imparts a rational dignity far surpassing that of the most gorgeous heraldry ; and, none of the emblems of it compare in richness with the memorials we here recommend. These may be had by all. These may survive to reveal to unborn ages the deeds of those instrumental in their production, when all other records shall have perished. The trees we have planted are looked upon by us as the best legacy to be enjoyed by those who are to occupy the place now occupied by ourself ! The little fame we may have acquired from some successful efforts in literature and for the good of our fellow-men, cannot be of long duration. A few only leave a reputation that will survive the author of it a longer period than that spent in its acquisition. Thus it will be with our own. But the adornment of our lawns and groves and highways may not be in its full beauty till an unborn generation shall have risen up in our stead.

HISTORY AND PHYSIOLOGY OF THE HOG.

The hog is the only domestic animal seemingly of no use to man when alive, and was, therefore, as may be presumed, designed for food. The Jews, however, the Egyptians, and other inhabitants of warm countries, and all the Mahometons at present, reject the use of pork for food. The Greeks gave great commendation to the flesh of swine, and their Athletæ were fed with it. The Romans considered it as one of their delicacies. No proper experiments have been made in regard to its alkaline tendency ; but as it is of a gelatinous and succulent nature, it is probably less so than many others. Upon the whole it appears to be a very valuable nutriment. The reason is very obvious why it was forbidden to the Jews ; their whole ceremonial dispensation was typical. Filth was held as an emblem or type of sin. Hence, the many laws respecting frequent washings ; and no animal feeds so filthily as swine. Mahomet borrowed this prohibition, as well as circumcision and many other parts of his system, from the law of Moses. But it is absurd to suppose, as some do, that Moses borrowed any thing of the kind from the Egyptians.

The prejudices of the Egyptians against swine were most extraordinary. It would be difficult—indeed, impossible to account for them. It is said, that if an Egyptian even touched a hog at any time, except on the feast-day of the moon, when he was allowed to eat of it, he was obliged to purify his person and clothing, by plunging into the Nile. The swineherds formed an isolated race, considered as outcasts from society, being forbidden to enter a temple or marry into other families. It was an established custom among the Greeks and Romans, to offer a hog in sacrifice to Ceres, at the beginning of the harvest, and another to Bacchus, before the beginning of the vintage ; because it is said, that animal is equally hostile to the growing corn and the loaded vineyard.

All the varieties of this animal originate in the wild boar, which is found in most of the temperate regions of Europe and

Asia. It is smaller than the domesticated animal, and usually of a dark grey color, approaching to black. It is armed with formidable tusks, sometimes ten inches, or even more, in length; those in the under jaw curving upwards, and capable from their size, strength and sharpness, of inflicting the most dreadful

WILD BOAR.

wounds. Before these animals attain their third year, they are gregarious, and when danger is at hand particularly, they muster in numerous parties, and with great promptitude, at the signal of alarm. Uniting thus, they present so formidable an array, as speedily to disperse the enemy, few creatures, or none, daring to commence an attack against such a combination of strength and valor as they exhibit. When the wild boar is complete in growth, he depends upon his solitary exertions for his protection, is seldom seen in society, ranging the forests alone, rarely commencing an attack, as his food consists almost solely of roots and vegetables; but repelling one with all the fierceness of courage, and all the resentment of retaliation.

The wild boar is hunted by dogs, or else taken by surprise in the night, by the light of the moon. That the sport of boar hunting should have been followed, particularly in ancient times, when men delighted in perils, is not surprising. For it was replete with danger, to all engagod in it, men, dogs and horses; affording opportunities for the display of courage, strength and skill, and called up the energies of mind and body. It was practiced both on horseback and on foot, and the hunters were armed with strong sharp swords, and tough boar-spears. The animal was roused from his lair by dogs, and, during his retreat, was harrassed by the boldest of the pack, but not with impunity; from time to time would he turn, and by successive rapid blows, leave some of his assailants prostrate. At length his savage temper thoroughly roused, he would turn round and face the hunters now hard upon him. This was the critical time, the time for the hunter to display his skill and firmness. His object was then to receive the boar either on his sword or spear, aiming either between the eyes, at the breast, or immediately under the shoulder-blade. Sad was the fate of the unskilful or faint-hearted. If the hunter missed his thrust he was exposed to immediate danger, for the attack of the enraged beast was sudden and impetuous. Sometimes, indeed, the animal, by a sudden movement, would not only avoid the stroke, but contrive to seize the haft of the protruded spear, between his powerful jaws, one crunch of which, was sufficient to grind the wood to fragments; the next moment, unless another hunter came to his rescue, and struck with a better aim, the assailant was on the ground, ripped up in the abdomen by the sharp tusks of the infuriated animal. When the hunters were mounted, the boar struck at the flanks of the horse, raising himself on his hind legs, and not unfrequently man and horse came to the ground.

In England, and Scotland, a few centuries since, the wild boar chase was a sport in great repute with the nobles. Shakspeare describes it with graphic accuracy. And if in those countries men are now better employed, the sport has not become obsolete. It is still practised with great eagerness in India, and in those districts of Europe in which the animal still maintains his hold. As in the case of the wolf, it is valued only for its excitement, as no part of an old boar, except the head, is eatable. Young animals, and especially such as are under the care of the mother, are esteemed, but they afford no sport in the chase. The wild boar is now common in the extensive forests of France, Germany, Russia, and Hungary, as well as in Spain and some other countries. Even within forty miles of Paris, in the forests of Chantilly, the breed still lingers. The late Prince of Condé, who died in 1830, kept a pack of dogs for the express purpose of boar-hunting.

The following account of a boar-hunt, from Egan's Sporting Anecdotes, will be read with some degree of interest, from the fact that one of the greatest generals of the age was conspicuous in the incidents related. On the 30th of October, 1817, the hounds of the Duke of Wellington discovered a most enormous boar in the forest of Wallincourt, France, and a chase was immediately followed. The animal, on being disturbed, passed rapidly into the forest of Ardipart, which he completely traversed. Being then hardly scented by the dogs, he took to the plain, where he was vigorously pursued by hounds and sportsmen, and ere he could reach another wood was brought to bay. The animal then became ferocious, and destroyed all the dogs that approached him, when one of the aids to his grace plunged a spear into his side. This only rendered the beast more savage, and the Duke himself, seeing his dogs would be destroyed, rode up and with his spear gave him a mortal thrust. The animal made a desperate effort to wound his Grace's horse, but fell lifeless in the attempt.

It is not often, however, that the spear is used in modern boar-hunting in Europe. This weapon has given place to the rifle, and the sport is consequently deprived of danger, and, it may be added, of excitement. In Germany, the huntsman, armed with his rifle, goes into the forest accompanied by a dog trained to seek and rouse up the wild boar, the track of which he follows by the scent. When the boar quits his lair, the dog, giving tongue, pursues the game, but at a cautious distance, until he turns and stands at bay. The dog now, without venturing too closely, keeps up an incessant irritation, and dodges about, making feints of attack, until he leads the animal to expose his side to the hunter, who immediately fires, aiming at the shoulder. The boar generally falls, but, if only wounded, he commences a precipitate retreat, upon which a boar hound is let loose, who pursues him for miles, until the animal drops exhausted, or again stands on the defensive It is thus in the forests of Hanover and Westphalia, where wild swine are numerous, that numbers are destroyed, not only for amusement, but also for the sake of their flesh. It is from these wild hogs that the celebrated Westphalian hams are procured.

In the defence of their young, the females also are furious, and resolutely attack any animal, which excites apprehension in their behalf. This even the domestic sow will do when living at large in the woods. Mr. Lloyd, in his Field Sports of the North of Europe, gives the following account of the ferocity of a common sow which he met in Sweden. One day in the depth of winter, says he, accompanied by my Irish servant, I struck into the forest for the purpose of game. On this occasion we had no other guide than my pocket compass. Towards evening, and when seven or eight miles from home, we came to a small hamlet situated in the recesses of the forest; here an old sow and her progeny made a determined dash at a brace of very valuable pointers I had at that time along with me, who naturally took shelter behind us. My man had a light spear in his hand, similar to those used by lancers; this I took possession of, and directing him to throw the dogs over the fence, in the the angle of which we were cooped up, I placed myself between the dogs and their pursuers. The sow, nevertheless, still pressed forward, and it was only by giving her a severe blow across the snout with the butt end of the spear that I stopped her farther career. Nothing daunted, however, by this reception, she directed her next attack against myself; when, in self-defence, I was obliged to give her a home thrust with the point of the spear. These attacks she repeated three several times, and as often got the spear up to the hilt either in her neck or head; she then retreated slowly, bleeding at all pores. So savage and ferocious a beast I never saw in my whole life.

The wild boar, solitary and recluse, may be regarded as nocturnal in his habits; it is seldom, except when forced by the dogs of the hunter, that he stirs from his lair during the day. Night is his season of activity; on the approach of twilight he sets out in quest of food. In his search after roots, he ploughs up the ground in long furrows with his snout, nor does he refuse such worms or insects as he dislodges. Though not strictly carnivorous, the wild boar, like his domestic descendant, does not reject animal matters which chance may throw in his way. He is in fact omnivorous, and is endowed with an appetite and digestive powers suited alike for vegetable or animal aliment. Buffon mentions that he has found parts of the skin of a roebuck, and the feet of birds, in the stomach of the wild boar. Corn lands, bordering forests, haunted by wild swine, are subject to their nightly incursions, especially when the grain is ripe, or the ears full; while thus engaged in feasting amid the tall corn, which prevents their seeing the approach of any one, it is easy during moonlight nights, to come upon them by surprise, and shoot them before they are aware of the presence of the hunters. Vineyards are also subject to the ravages of wild swine, a fact which has been known from the earliest ages. It is probably on this account, as already observed, that the Greeks and Romans, at the season of harvest and of vintage, were accustomed to offer the wild boar in sacrifice to their gods.

The hog has generally been described as a creature of gross habits and unclean tastes, as having the senses of touch and taste obtuse, and even as being so insensible, that mice may sometimes burrow in his skin, without his seeming to feel pain from it. But these opinions are most unjust and incorrect. Far from being unclean, nature has furnished the hog with powerful organs of digestion, enabling him to derive sustenance from a variety of substances, and his voracity is only the result of the

OLD ENGLISH DOMESTIC HOG.

extent and perfection of his digestive and respiratory organs. Although one of the thick-skinned animals, the hog feels blows acutely, and manifests his suffering by loud cries. Indeed, that his sense of touch is dull, because of the thick layer of fat with which his body is enveloped, is most erroneous, for it is well known that the plexus of the nerves, which gives sensibility to the body, is exterior to this fat layer. So far from being insensible to pain, the hog even suffers under the irritation arising from the punctures of gnats, musquitoes, and other small insects, and endeavors to protect himself from their persecution, by rolling in moist places, and covering himself with mud.

Professor Low very justly says, the hog is subject to remarkable changes of form and character, according to the situations in which he is placed. When these characters assume a certain degree of permanence a breed or variety is formed; and there is none of the domestic animals which more easily receives the characters we desire to impress upon it. This arises from its rapid powers of increase, and the constancy with which the characters of the parents are reproduced in the progeny. There is no kind of live stock that can be so easily improved by the breeder, and so quickly rendered suitable to the purposes required; and the same characters of external form, indicate in the hog a disposition to arrive at early maturity of muscle and fat, as in the ox and the sheep. The body is large in proportion to the limbs, or in other words, the limbs are short in proportion to the body ; the extremities are free from coarseness, the chest is broad, and the trunk is round. Possessing these characters, the hog never fails to arrive at earlier maturity, and with smaller consumption of food, than when he possesses a different conformation.

In a wild state the hog has been known to live more than thirty years, but when domesticated he is usually slaughtered quite young. When the wild hog is tamed it undergoes the following, among other changes, in its conformation. The ears become less movable, not being required to collect distant sounds. The formidable tusks of the male diminish, not being necessary for self-defence. The muscles of the neck become less developed, from not being so much exercised as in the natural state. The head becomes more inclined, the back and loins are lengthened, the body rendered more capacious, the limbs shorter and less muscular; and anatomy proves that the stomach and intestinal canals have also become proportionately extended along with the form of the body. The habits and instincts of the animal change. It becomes diurnal, not choosing the night for its search of food ; and is more insatiate in its appetite, and the tendency to obesity increases. These changes of form, appetites, and habits being communicated to its progeny, a new race of animals is produced, better suited to their altered condition. The wild hog, after it has been domesticated, does not appear to revert to its former state and habits; at least, the swine of South America, carried thither by the Spaniards, which have escaped to the woods, retain their gregarious habits, and have not become wild boars.

The hog does not appear to have been indigenous to our

own country; but was taken hither by the early voyagers from the old world; each bringing them from their own country; and in the Eastern States especially a few of the breeds still retain traces of the old English character. From its nature and habits, the hog was the most useful and profitable of all the animals bred by the early settlers. It was their surest resource during the first years of toil and hardship. It arrived earlier at maturity, required less care, sought out, for the most part, its own food, was the least subject to accidents and diseases in a new situation, and, therefore, best repaid any portion of attention bestowed on the breeding and rearing it.

The prevailing idea is, that hogs are not capable of being taught; and, that they have a certain kind of obstinacy that is also an insuperable objection to any mental progressive development. All the offences they commit are attributed to an innately bad disposition. It seems to us that this is an erroneous conclusion. May it not proceed in no small measure from neglect or bad management? They are permitted, moreover, to live so short a time, they cannot learn much, were there no other reason. Indeed, they are rarely permitted to live long enough to attain anything like their full size. Would cattle or horses behave one iota better, were they treated as pigs usually are? They are the legitimate objects for the sport of idle boys, hunted with dogs, pelted with stones, often neglected and obliged to find a meal for themselves, or wander about half-starved. Can we wonder that, under such circumstances, they should be wild and unmanageable brutes? Here, as in many other things, man is but too willing to attribute the faults which are essentially of his own causing, to any other than their true source. We are disposed to become their advocate; to urge upon their owners more kind treatment; and, especially, that they be not consigned to the lowest grade in the animal economy. Can it be believed that a race of animals among the most beneficial to man, would have been created with animal physiological developments necessarily the most loathsome and disgusting that can be imagined. We cannot believe it. Seemingly it was a prominent design of the Deity, in forming the brute creation, to make such as were to be immediately useful to man, and of course to be constantly about him, to be also agreeable to him in their general characteristics.

THE CHINESE HOG.

We cheerfully recommend to our readers, Youatt's treatise on the pig. They will find much in it that is highly interesting, as well as useful, to the ruralist. He says, the Chinese and Japanese are great pig breeders, and make the art of crossing, breeding, and rearing swine, which furnish them with their principal animal food, an object of peculiar attention and study. Merchants who have resided for some time in China, and even travellers who have merely been able to bestow a superficial glance on the economy of that country, speak of the great care bestowed on this point; but no author appears to have given any details as to the course of practice adopted. Perhaps from the naturally jealous and uncommunicative disposition of the Chinese, they have been unable to acquire any; and perhaps few have thought it worth while to trouble themselves about so degraded an animal as a pig. However this may be, it is much to be regretted that information is so very scanty, for many valuable hints might probably have been thus obtained. If all our merchants who go abroad for residence, and especially our foreign ministers, were to be as observant and as vigilant, as one now and then is, in collecting facts and facilities for improving our agriculture, and particularly farm animals, the benefit to American farmers, would be immense.

In a work on China, by Mr. Lay, the naturalist, there are some amusing sketches of the pig. He says, between the natives of that country and their swine, there is a striking analogy. Thus, a Chinese admires a round face, and what might be called an aldermanic protuberance of the chest, with smooth curvatures; and, where opportunity admits, artificial means are used for increasing these auxiliaries to personal beauty in himself. The Chinese pig is fashioned on the same model. At an early period the back becomes convex, the belly correspondently pendent, and the visage shows a remarkable disposition to rotundity; nor is the resemblance merely personal; in the moral character there is an amusing similitude, contrariety and obstinancy being the prevailing characteristics of both men and brutes. And the same author pleasantly tells us that swine are rarely driven or made to walk in China, but conveyed from place to place in a species of cradle, suspended on a pole, carried by two men. But, says he, the difficulty of this operation is to get the animal into this vehicle, and this is accomplished by the cradle being placed in front of the pig, and the owner then vigorously pulling at porky's tail. From a spirit of obstinacy the animal darts into the place prepared for him. At the journey's end the bearers dislodge him by spitting in his face.

The cut pretty well represents the pure Chinese pig, and is also a fair likeness of many of the imported and their immediate descendants, seen in our country. They, however, necessarily vary somewhat in appearance, size, shape, and color, from the diversity in the style of breeding, and the various regions from which they are derived. The Chinese hog is generally deemed too small for common use, and requires to be mixed with larger breeds to produce the most profitable carcass for the market. For the purpose of refining the coarse breeds, no animal has ever been so successful as this. The hogs thus produced have fine bones, are short, and very compact, and have bellies almost touching the ground, with light head and ears, and fine muzzle. They also manifest great docility and quietness, are small feeders, and yield much meat for the quantity of food consumed. The Chinese hogs are of two kinds, black and white, and most of the above general remarks apply to each of them.

Mr. Youatt indulges himself in some humorous pleasantry in speaking of the hog, as found in Ireland. Here, says he, the hog is, in the fullest sense of the word, a domesticated animal. The Irish pig is born in the warmest nook of his master's cabin, reared among the children, and often better fed and more carefully tended than the ragged urchins who play around him; for the peasant will half starve himself and children, in order to have more food for his pig; and while the former have only potatoes, and few enough of them, the porker frequently gets not only a good meal of potatoes, but some porridge, or bran, or refuse of vegetables in addition. He is in fact the chief person in the household; on him the poor man reckons for the payment of his rent, or the purchase of the necessaries of life. Swine abound in all parts of Ireland; scarcely a peasant's cot but numbers a pig among the family; and the roads, lanes, and fields, in the neighborhood of every village, and the suburbs of every large-town, are infested or overrun with a grunting multitude.

Until lately, however, notwithstanding the value set on these animals, the real Irish pig was a huge, gaunt, long-legged, slab-sided, roach-backed, coarse-boned, grisly brute; with large flapping ears which almost shrouded the face; of a dirty white, or black and white color; with harsh coarse hair, and bristles that almost stood erect. It was also far from being a profitable animal, requiring a very considerable quantity of food, and when fat producing only coarse-grained meat. But since the facility of export has become greater, considerable improvement has been effected by the introduction of Berkshire and Chinese boars and sows, and crossing the old breed pretty extensively with these. Thus, the unwieldiness of size and coarseness of bone have been diminished, and greater aptitude for fattening communicated, which latter qualification is invaluable to the poor Irish peasant.

BERKSHIRE HOG.

The Berkshire breed of swine has been generally considered to be one of the best in England, on account of its smallness of bone, early maturity, aptitude to fatten on little food, hardihood, and the females being such good breeders. Those of the pure original breed have been known to attain an immense size, and weigh from eight to ten hundred pounds. One bred at Petworth measured seven feet seven inches from the tip of his snout to the root of his tail, and seven feet ten inches in girth round the centre; five feet round the neck, ten inches round the thinnest part of the hind leg, and two feet nine inches high; and, what was most remarkable in this monstrous hog, he did not consume more than two bushels and three pecks of ground oats, peas, and barley, per week. The first Berkshire hogs brought into the United States was in 1823. Other importations have been made from time to time, and no reason exists why the blood should not be as pure as in England. The Berkshires were favorably received, and at one time sold at high prices. Without doubt it is one of the best—if not the very best breeds in the country, although its merits are not generally appreciated.

The color of the Berkshires was originally a buff or sandy ground, with large black spots; and the feet, lower part of the legs, and a tuft on the tail, were buff. This color, in most of the modern race, has given place to white in the same parts. This variation, with the more important ones of early maturity and good feeding properties, has been ascribed to a Chinese cross, which has added the only characteristic in which they were before deficient. Although not so generally appreciated in this country as they should be, in the hands of skilful breeders, their merits are annually becoming more and more understood, and perhaps now take precedence with the majority of such farmers as have particularly given their attention to the subject. It is just to render to Mr. Brentnall, of Orange County, New-York, and Mr. Hawes, of Albany, in the same State, the credit of first introducing this breed of hogs into our country.

To show the success with which some of our pork amateurs have reared this breed of swine, a few specifications will be added. In 1845, Mr. Wakeman, of Herkimer, New-York, slaughtered a Berkshire hog, which weighed, when dressed, seven hundred and twenty-one pounds. His dimensions were—from the end of his nose to the end of his ham, six feet and seven inches; girth around the heart close to the fore leg, six feet and eight inches; around the middle, six feet and ten inches; height, two feet eleven inches. He was kept on grass through the summer, then a month on boiled pumpkins and potatoes, mixed with sour milk, and no meal; afterwards, for the most part, with dry corn. He was judged to gain three pounds per day, from the time he was put up till he was killed. This hog was the seventh cross on the native or common stock from Berkshire boars. In the same year, it is stated in the American Agriculturist, that Dr. Keever, of Ridgeville, Ohio, slaughtered a full-blooded Berkshire barrow, two years, two months, and seven days old, which weighed, on being dressed, net six hundred and sixteen pounds; and there was another in the same neighborhood, supposed to be about two hundred pounds heavier. On the same authority, the weight of a Berkshire hog slaughtered in Williamstown, Massachusetts, is given as seven hundred and eight pounds; the thickness of the heaviest of the clear pork being full ten inches. Mr. Spilman, of Carroll county, Kentucky, also in the same year, slaughtered nine Berkshire pigs, full blood, of one litter, twenty months and six days old, which, on being dressed, weighed three thousand four hundred and twenty-nine pounds. And more notable yet, the Sangamon Journal stated a Berkshire hog in that vicinity was slaughtered, when at the age of three years, which weighed, on being dressed, fourteen hundred pounds!

With such examples constantly presenting themselves, it is inexplicable that so many of our farmers pay so little attention to the breed of their hogs. The above were taken almost at random from agricultural journals, on which we could lay our hands, when at our writing table, opening only four or five of those nearest to us; and, the presumption is, were we to examine all the similar works we possess, we could find hundreds of similar cases. Multitudes of the old neglected breeds of the hog family are not worth being taken as a gift, if the receiver is to fatten them; for then the pork in all probability would cost him double per pound that it would to produce it from hogs of an improved breed, and moreover, also, being of an inferior quality. The secret of thriving or making money from agriculture lies chiefly in this; to keep cows that will pay for their keeping and for themselves in each year, while others keep such as will barely pay for their feed; to keep oxen of a character that one yoke will perform the labor usually done by two yoke; and to keep hogs of such breeds that pork can be made from them at about half the cost when made from inferior breeds. It costs but little more to feed prime animals than it does to feed those miserable caricatures too often seen upon a farm. It is analogous to the dissimilar modes of cultivating the soil. The cost of producing a premium crop on an acre of land is but a fraction in addition to the cost of getting from it only half a crop. Nearly the whole of this extra produce is clear profit.

This subject assumes a magnitude not before imagined, when it is considered that a pig or two is a staple means of subsistence with nearly every family in rural localities; and when it it considered that the aggregate of swine in the whole country forms one of the principal sources of our national wealth. The annual consumption of pork in the United States cannot be less than two hundred millions of dollars. This would be only about fifty dollars to each family in the year; or one dollar a week for each family. Every person with the most superficial knowledge on the subject must know this is insufficient. It is not half the allowance for penitentiary convicts. Only a dollar a week, or fourteen cents a day, for the pork to be used by a family of farmers! Thus stintedly our laboring yeomanry do not live! Thus they could not live! Now, is it a matter of trivial consideration, whether such a staple costs the consumer three or six cents per pound; or whether for the fifty dollars a family can have five hundred or a thousand pounds for its subsistence! To many it seems a trifle, whether a single pig, from the same feed, is made to weigh two hundred or four hundred pounds; but, when it is considered, that there may be slaughtered annually in the country eight millions of pigs, the subject becomes tangible and impressive. So it is in other matters. A single drop of water is viewed without interest; yet rivers and oceans consist of such drops! A single grain of sand is viewed without interest; yet mountains and the entire globe consists much of such grains!

Eight millions of hogs for annual slaughter in the United States is a low estimate, for since writing the above we have glanced at a statement in the Cincinnati Price Current, that the number of taxable hogs in the State of Ohio, some years has been about 2,500,000 ; and the presumption is that four-fifths of the number would each year be the current slaughter. The farmers in the Eastern States can form no idea of the extent of the pork trade of the West, says Mr. Edmundson, unless they personally inspect the slaughtering establishments that are to be found in Cincinnati, and other large cities. Twenty thousand hogs are very commonly slaughtered and packed for market in a single season, by one house ; and the whole number slaughtered in these establishments annually, west of the Alleghany mountains, average 2.000,000, weighing each two hundred pounds of nett pork. of which at least one-fourth are slaughtered in Ohio. It will here be noted, that this statement of Mr. E. does not include the hogs slaughtered on the farms where raised, whether for domestic use or for market ; and, beside these, many are transported alive to other places, so as not to be included in his statement.

The number of hogs packed at Cincinnati alone, equal 400,-000 head in a single season. During the month of December that city is crowded sufficient to produce suffocation, with droves of hogs, and draymen employed in delivering the barreled pork on board the steamers. Some fifteen hundred laborers are employed in the business from six to eight weeks, and in many cases it is kept in full operation both day and night, including Sundays, from the beginning to the completion of the season. The Sabbath is not at all regarded by those who are extensively engaged in the pork business. and a stranger spectacle could hardly be presented to a person brought up in a land noted for its steady habits, than to see many of the main business streets of the Queen City of the West, literally crammed with wagons, carts, and drays. employed in transporting hogs just from the hands of the butcher. from the slaughtering to the packing house, on the Sabbath. Indeed, this appears to be the great day for the bringing up and completing the week's work, among the principal pork packers of Cincinnati ; and as a very large proportion of the business is done in a densely populated portion of the city, it is not to be wondered at that disease and pestilence occasionally, at least, infest these portions of the city to an alarming extent. The filth and dirt, and impure atmosphere, in a large portion of the upper end of the city, can be better imagined than described, all of which are the products of hog slaughtering and packing establishments.

Within a few years past slaughtered hogs have been converted into a new branch of emolument. It is estimated that fifteen million pounds of lard and fat pork are used yearly in Cincinnati alone for making lard oil, nearly one-third being converted into stearine, a substance much resembling mutton tallow; the former is used in lamps instead of whale oil, and the latter for candles, instead of spermaceti. The entire carcass, excepting hams and shoulders, are subjected to a heavy pressure, and then by the aid of steam the fat is extracted. One establishment thus uses up six hundred hogs daily for this purpose. The whole quantity of oil produced from fat pork in this way is not known; but it is known that the business is annually becoming more extensive, and that it has already materially reduced the price of other oils used for light in families and elsewhere. In the Western markets it has nearly superseded the use of them; and in our Atlantic cities it is creating a powerful competition with them. This is the more important, as from the increase of our population and the growing sparseness of whales, their oil has had an upward tendency of price for a long period. Were this tendency not checked, ere long multitudes of the poorer classes would be unable to enjoy the pleasure of having their dwellings made cheerful by such a beneficent substitute for daylight. At the present time there are forty manufactories for lard oil in that city.

The Suffolk breed of pigs is with many a great favorite. Mr. Youatt says there is no better in Great Britain than the improved Suffolks. Among the crosses of the native Suffolk which he specifies are those of the Lincoln, the Berkshire, and the Chinese. A cross between the Suffolk and the Lincoln has led to a hardy progeny, which fatten well, and will weigh from

SUFFOLK HOG.

four to six hundred pounds. However, he gives preference to a cross between the Suffolk and the Berkshire, or Chinese. He says these are well-formed, compact, short-legged, hardy animals, equal in value to the best of the Essex, and superior in constitution, and consequently better adapted to the farmer. Those arising from the Berkshire and Suffolk are not so well shaped as those from the Chinese and Suffolk, being coarser, longer-legged, and more prominent about the hips. They are mostly white, with fine hair; some are spotted, and all easily kept in fine condition; and having a decided tendency to fatten, being of moderate size, with round bulky bodies, short legs, small heads, and fat cheeks.

The Rev. Mr. Rham, author of the Dictionary of Agriculture, says the Suffolk breed of pigs is, perhaps, on the whole, the most profitable of any in England. Several years ago the late William Stickney, of Boston, introduced into our country this breed of animals; and, although they have not become extensively diffused, the results thus far are highly satisfactory to all who have seen them. It was said by one gentleman, highly competent to give an opinion, on which reliance can be placed, that the pigs of this breed, at six weeks old, simply for being raised and fattened, are cheaper at the prices asked for them than the common country pigs at the low prices which they usually command. As the best ones are generally kept for breeding, the statistics of those slaughtered are not numerous. A few of these statistics will satisfy any reasonable person of their excellence. A Mr. McCormick, of New Hampshire, slaughtered one, seven months and ten days old, that weighed three hundred and twenty-five pounds; a Mr. Stearns one, six months and ten days old, that weighed three hundred and ninety-four pounds; and Mr. Knapp, of Northampton, Massachusetts, one, at fifteen months, that weighed five hundred and fifteen pounds; and a Mr. Fitcomb one, at nine months, that weighed three hundred and twenty-five pounds—also, two at seven months and thirteen days, which weighed two hundred and sixty-two, and two hundred and ninety-six pounds.

We feel disposed to say something that will in a measure relieve the swine family from the obloquy too often cast upon them. If hogs were kept to the age of many other animals, and the same pains were taken with them that we take with the ox and horse in training them to useful labor; or that we take with the dog in teaching him to watch and hunt, the success would, we have reason to believe, surpass any present imagination on the subject. When attempts are made for it, the hog has not by any means been found insensible to habits of docility or improvement. In proof of this, reference may be had to the learned pigs, as they are called, used in public shows, astonishing the spectators by their performances, such as arranging letters so as to form words, and other tricks indicative of the animal's tractability. Mr. Youatt incorporates into his work, on the Pig, the following narrative. Some thirty years ago, says he, it was mentioned in the public papers, that a gentleman had trained swine to run in his carriage, and actually drove four of these curious steeds through London. And, he adds, that at a much more recent date the market place of St. Albans, was completely crowded, on one occasion, in consequence of an eccentric old farmer, who resided some distance off, having entered it in a small chaise cart drawn by four hogs at a brisk trot, which pace they kept up a few times round the area of the market place. They were driven to the Woolpack yard, and after being unharnessed, were regaled with a trough of beans and wash. A gentleman present offered fifty pounds for the whole concern as it stood, but his offer was indignantly declined. In about two hours the animals were reharnessed, and the old farmer drove off with his unusual team.

Mr. Henderson gives an account, says Martin, in the Farmer's Library, of the manner in which himself and his brother, when boys, trained pigs for the saddle, and not only so, but for running matches; and, we do not think this to be a matter of much difficulty, for we have ourselves often seen pigs so docile as to allow themselves to be mounted by children to whom they were accustomed, and that without manifesting an insubordination or impatience. The most extraordinary instance, however, of the docility of the pig, and one which proves that it might become valuable for other qualities, were it duly educated, than those connected with its flesh, is given by Daniel in his Rural Sports. It is an account of a New Forest Sow which was broke in like a pointer to stand at game, and which, from the delicacy of her scent, she would find when the dogs had been unsuccessful. Thomas and Richard Toomer were concerned together, at the time in question, in breaking pointers and setters; some of their own breeding, and others sent to be broke by different gentlemen. In some instances these dogs were intractable or slack. At one time when impatient with want of success, a sow passed along, which on being fed by them evinced something in her manner that attracted their notice. The thought came into their mind that possibly she might be trained for this very purpose, so as to do as well as the animals to which they were devoting their attention. The idea was followed up. She was put into a course of training; and within a few days she would find rabbits and partridges with great precision. She daily improved, and in a few weeks, would retrieve birds that had run as well as the best pointer;

and, it was said that her nose was superior to that of the best pointer they ever possessed, and no two men in England had better.

This sow was a beautiful creature for one of her species. Her pace was mostly a trot, being seldom known to gallop, except when called to go out shooting. She would then leave home at full stretch for the forest. When called by a whistle or otherwise, she obeyed as regularly as a dog, and was as much animated as a dog on being shown the gun. She always manifested great pleasure when game, either dead or alive, was placed before her; she frequently stood a single partridge at forty yards distance, her nose in an exact line, and would continue in that position until game moved. If it took wing she would come up to the place and put her nose down two or three times, but if a bird ran off, she would get up and go to the place and draw slowly after it, and when the bird stopped she would stand it as before. The two Toomers lived about seven miles apart, and she would frequently go from the place of one to the place of the other, unattended, as if to induce them to go out for game. When her owner died she was five years old, and was then sold for ten guineas. She lived till she was ten years old, and became fat and slothful, but could point game as well as ever. She was not used, except for show to strangers. When killed, she weighed seven hundred pounds. Her death warrant was signed in consequence of her having been accused of being instrumental to the disappearance of sundry neighboring lambs.

Mr. Martin remarks, that in the case of this animal, however much the hog is usually despised, we have a most extraordinary example of docility and intelligence; and it furnishes evidence that, were the hog trained for to do so, it might, like the elephant, perform many kinds of useful labor. Nevertheless, for such purposes it is not generally needed; and consequently, except in a few isolated instances, its education is utterly neglected; all it has to do is to eat and sleep and become fat—its utility to man commencing at its death by the knife of the butcher. Yet, even under the disadvantages in which the pig is placed—debarred its liberty, prevented from exercising its natural instincts, and undisciplined in the slightest degree—it manifests both discernment and attachment; it recognises the voice, and even the footsteps, of its feeder, and is evidently pleased with his notice. Instances of the attachment of pigs to particular persons, and even to other animals, are on record. The senses of smell, taste in some cases, and hearing, are possessed by the hog in great perfection. It is a common saying that pigs can smell the coming storm; and certain it is that they are very sensitive of approaching changes in the weather. They become agitated, hurry under shelter, and during the continuance of the storm sometimes utter screams, run about with straw in their mouth, or carry it to their sty as if to add to their comfort and defence. Dr. Darwin noticed this peculiarity, and hence remarked, that it is a sure sign of a cold wind when pigs collect straw in their mouths, carrying it to their beds for warmth, and, as it were, stimulating their companions to do the same. And it is well known to all, that when cold they huddle together as closely as possible.

The natural term of a hog's life, says Gilbert White, is little known, and the reason is plain, because it is neither profitable nor convenient to keep that usually turbulent animal to the full extent of its time. However, he remarks, that a neighbor of his, a man of means, who had no occasion nor disposition to look very nicely into the cost of any matter to which his taste or curiosity might lead him, kept what was called a half-bred bantam sow till in her twentieth year, when she showed signs of age from the decay of her teeth, and the decline of her fertility. For about ten years this prolific mother produced two litters a year, of about ten at a time, and once above twenty at a litter; but as there was nearly double the number of pigs to that of teats, many died. From long experience in the world, this female was become very sagacious and artful. When she found occasion to associate with a boar, she used to open all the intervening gates and march by herself up to a distant farm where one was kept, and when her purpose was accomplished would return by the same means. At the age of about fifteen her litters began to be reduced to four or five; and such a litter she exhibited when in her fatting pen. At a moderate computation, she was allowed to have been the fruitful parent of three hundred pigs—a prodigious instance of fecundity in so large a quadruped. She was killed in 1775.

Vancouver very judiciously says, there is no animal in the whole economy of good husbandry that requires more attention as to breed, number, and supply of food, or will better requite the care and trouble of the farmer, than a well-managed and proper stock of hogs. These things, however, are too much overlooked, or rather disregarded, by farmers in general, though all are ready to agree that an overstock in other respects must prove fatal to his interests. Hogs are too frequently conceived to be a trifling and unimportant part of the stock upon the farm; whereas, if their first cost and the value of their food were duly considered with their improving value, it would certainly bear them out against some of the more costly animals, and challenge more care and attention than are usually bestowed upon them. In the first place, attention should be paid to the breed; but, let the breed be what it may, a well proportioned stock to every farm will most abundantly requite the care and repay the expense of the necessary food provided for them. As important as may be the breed, it is not less important that they should have an ample supply of feed, and that they should have local accommodations adapted to their nature and habits.

In addition to the profit of hogs, from the use of their flesh and lard, which are staple articles for food, in some form or other, in our own country and some others, too well understood to need description, their skins and bristles might be made, and are in some countries, sources of mercantile gain. The hide of this animal, when tanned, is of a peculiar texture and very tough. It is used in making pocket books, and for some ornamental purposes, but chiefly for covering saddles. The numerous little variations in it, and which constitute its beauty, are the orifices whence the bristles have been removed. As the common practice is to cook pork with the skin on, the quantity of leather made from it will always be of a limited amount. The bristles, too, of the best breed of hogs, are too fine and short to answer the end for which they are indispensable. The shoemaker would be obliged to resort to some new device for the accomplishment of his labors were he deprived of a coarse and stiff bristle with which to point his thread. The finer kinds will answer for brushes. The best bristles are taken from the back of the animal; and, it has been calculated that a pound of them may be obtained from a full grown hog. Mr. Youatt supposed that the bristles annually imported into England from Russia and Prussia must require the slaughter of about a million and three quarters of hogs. Those from the boar are best.

There is another circumstance connected with the keeping of hogs on a farm, never to be overlooked by the agriculturist. Their evacuations, both solid and liquid, may be rendered in some cases an equivalent for the feed consumed by them, especially if there is a constant supply in their pens of litter and other substances to be mixed with the dung and urine. This is too frequently neglected; but, to the farmer, it is of vast importance, if his lands require a large amount of manure. Those who keep extensive piggeries thus supplied with materials to be worked over into fertilisers will find an advantage from swine never to be disregarded; and, nearly equal to what is obtained from the horse-stable and cow-house. And even the mechanic, or small farmer, who keeps but one or two pigs may obtain, in this way, enough to enrich the soil of his garden. Decaying vegetable, muck, straw, and indeed almost every substance that can be named, are by hogs readily converted into the best of manures. One of the principal items in our own list of articles to be thus reduced into a useful material to be used in tillage is leaves from a neighboring wood lot. Two hogs will so reduce them to their original elements in a very short time, that half a ton every week would be none too much for them. When fresh they make the best of litter, always enabling the hogs to be clean; and, when unfit for that, may be removed and the place supplied by a new stock. After a rain in the fall or winter, when the ground is not covered with snow, leaves may be raked up and put into stacks of two or three hundred weight each, and there suffered to remain, to be used as required.

It is one of the rational and humane conclusions to which all sensible and Christian people should come, that an animal of so general and great utility as the hog, should be treated with kind regard. This should be done as a matter of interest to the owner; if not done, it must be apparent that his profits will be diminished in proportion to the neglect. Such a policy pertains to every description of productive investment and labor. It is a part of the economy of nature itself. It is a part of that regime under which man is placed. The earth requires his labor to be productive, as well as the enlivening influences of moisture, heat, and atmospheric action. Without this labor, and these influences, where would be our crops? We should seek for them in vain! And whence should we obtain the means for subsistence? Those means would be wanting. It is partially so in the animal kingdom. If many animals are enabled to subsist in a wild state, it is well known that in a domesticated condition they attain a higher degree of excellence, and high in proportion to the care bestowed upon them. Perhaps in the case of swine this is more apparent than with most other animals. Look at the wild boar and the old English hog! Look at the plump Chinese pig and some of our land sharks! Look at the well-fed Berkshires and Suffolks, and the various best crosses in this country and in England, and then at the half-starved alligators that herd about some poor farm-houses, and we shall see how much depends on skill and care in rearing these animals.

And in a moral view, not kindly to regard farm animals the most useful to us, is a reproach to be abhorred; a reproach incompatible with any claims for social virtue. Not to do it would be degrading to the instincts of a savage. The Arab who would plunder wandering pilgrims, when found traversing the barren sands of his birth-place, will never fail to caress the beautiful steed which does him service. The steed and his owner will even repose beneath the same shade and drink at the same brook. Were he to neglect or abuse that noble animal, he would fear the curse of the Arabian Prophet, and that the Crescent Banner would rise in judgment against him !

MONEY MADE BY FARMING.

One of the most hacknied abuses of agricultural labor is, that money cannot be made from it. This is the common slang of those who are too stupid or too indolent to secure to themselves from such labor the remuneration that would result from skill, and enterprise, and perseverance. These very persons get a living, somehow or other, from the soil ; it may be a living without luxury or elegance, though favorable to health and muscular vigor ; yet, were they depending for subsistence on many other occupations, and to manifest as little talent and application to judicious industry as they do on a farm, they would starve to death, or be compelled to take shelter in an almshouse. The mass of slanderers upon agriculture seem to imagine that getting a living upon a farm is not making money. They do not seem to consider that we do not eat silver and gold as we do beefsteaks, mutton-chops, and potatoes. They do not seem to realize that money is valuable only as the representative of property that can be used in social and domestic economy—as the means by which we may obtain what we need for food, or apparel, or the other things wanted in life.

What is the difference between the farmer and the mechanic in this respect ? It is simply this. The former raises his own bread-stuffs and tubers, and the latter makes shoes, or chairs, or ploughs, or wagons; then sells them for cash ; then with the money so received purchases those very articles produced by the agriculturist. The process is, simply, an exchange of labor ; the farmer works for the mechanic, and the latter works for the former, each depending on the other, in part at least, for what he wants. What is the difference between the farmer and the doctor? Very similar to that between the farmer and the mechanic. The one supplies the other with what he wants to eat, drink, and wear; and, in return, the last supplies the first with jalaps, and powders, and tinctures, and liniments—and when there is occasion for it, pulling his teeth, cutting off diseased limbs, and restoring to its proper place a dislocated bone. And what is the difference between the farmer and the merchant? It is pretty much of the same sort as in the two cases named. The farmer supplies him with beef, pork, lamb, veal, and poultry for his table; and, with wool, cotton, and flax, for his apparel. In return, the merchant hands over to the farmer money with which he pays his taxes, and procures for his family such articles as he cannot raise upon his farm. And besides supplying his own wants, the merchant takes all the surplus produce of the farmer, and sells it out as wanted to others, either to those at home or shipping it to foreign countries, paying him in cash for it. Thus, as the merchant produces nothing himself, he is the mere pedlar of the farmer and the mechanic, and gets his living by selling the several commodities which they produce, at a higher price than he pays for them. The farmer and the mechanic had better pay the merchant for being the factor and pedlar of what they produce, than to spend their own time in doing it. He can do this better than they can; and they can earn more in their respective vocations than in doing this.

This exchange of labor between the farmer and the three classes of persons named is one of the beautiful and beneficial features in civilized life and social economy. It is essentially the same between the farmer and every other class of persons. They are all living upon his labor, or are employed in completing the processes he has begun for sustaining the fabric of human society. Without him they could not subsist. If they were to fail of receiving his products, they all would cease to exist, or else they would be obliged to become farmers themselves. This is not mentioned to create odium against either class of persons in the community, but simply to show the relation between all classes, and the dependance severally of each upon any other one. It is strange, therefore, that it should ever have been imagined, that the occupation of the farmer is subordinate, or in any respect less reputable than that of any other persons. If a comparison were to be instituted, for which there is no occasion, the advantage would be the other way. It would be found that all others would be subordinate to him, and in some measure dependant upon him. So far as respectability is concerned, it may be proper to remark, that that depends not so much on the occupation as it does on the character and talents of the individual engaged in it. Any occupation, however respectable and elevated in itself, may be degraded and rendered comparatively disreputable, by a want of character and talents in the persons who have it in trust; and, on the other hand, an occupation that in itself is subordinate and seemingly without the elements of high consideration, may be elevated and rendered honorable when in the hands of those who have character and honorable position to associate with it.

But with all these facts, lucid as the light of day, and irrefragable as mathematical demonstrations, there are persons, who, to other slanders on agriculture, say no money can be made by it. The assertion is without truth. The inference proceeds from false premises. It originates in misguided conceptions. As far as any foundation for it exists, it is in the want of skill and industry applied to the soil, as already remarked. There is, indeed, connected with other occupations, particularly with merchandise, a show of money not seen about a farm. But it is to be remembered that a single merchant may have passing through his hands the produce of an hundred farmers. This produce becomes to him a kind of capital. He may be in debt for it, whereas, many may consider it his own. Thus it gives him credit. He raises money upon this credit, and hence may always be flush in funds. Being thus flush in funds, he may keep up an expensive family establishment, and spend a large amount annually for public objects, as well as for his wife and children. He passes off for a rich man; but not more than one in a hundred that do it, on having their estates closed by executors or administrators, or, as it frequently happens, by assignees, will be found worth a dollar. The others have eaten up all they earned, or else, as it too often appears, have lived on the money belonging to others, not having had enough of their own to purchase one of the little despised farms of the country. This is the secret of such apparent wealth; of such pompous displays of money and of family expenditure.

The process of doing this is quite simple. An individual having ample credit may purchase goods on long time to the amount of one hundred thousand dollars. Some of these goods he may sell without profit for cash; some on short time for small profits. This, it will be seen, may give him possession of available funds to half the amount of his purchase. It ends not here. Other purchases are made in the same way, from each one deriving an augmentation of available funds as in the first case, so that he may go on year after year, spending five or six thousand dollars annually upon his family, and the deception or the delusion, whichever it may be called, is not ordinarily discovered till a general revulsion, or some mishap of his own, drives him into bankruptcy, or till death closes the factitious, gilded drama. All this intervening time, the farmer is grumbling that the merchant makes all the money, and that he makes none; and, what is worse, the farmer's sons become half crazy to go off and play the game of getting rich; whereas, when the day of reckoning comes, it is found, that instead of wealth there is absolute poverty, and not only poverty, but destitution of disposition and capacity in all the members of the family to get a living as honest people should do it. Consequently they become genteel paupers, too proud and too ignorant to labor—the greatest nuisance ever permitted to have existence.

The business of buying and selling, or the occupation of a merchant, is not a little like a great lottery. If one person out of ten thousand who engage in it becomes rich, the fact becomes known far and near, producing a feverish excitement in all the young men who hear of it, wherever they are, to engage in some career that seems to promise a similar result, while the other nine hundred and ninety-nine, who lost their labor and all they had, are never once named or called to mind. So it is with lottery tickets; the individual who draws the highest prize becomes, as it were, a public character, known at least by reputation throughout the State where he lives, and his success stimulating multitudes of others to hazard money in the same game, while all who draw blanks or small prizes are never mentioned, and their want of success has no influence whatever in deterring these multitudes from similar disappointment. Now, let us suppose the institution of a lottery having thirty thousand tickets at ten dollars each, making a fund, when sold, of three hundred thousand dollars, one-third of it going to the object to be promoted by it, and to the expenses attending it, and the other two-thirds to be given in prizes, being probably the usual mode of apportionment. There is one big prize of one hundred thousand dollars; one of five thousand; ten of one thousand each; ten of five hundred each; twenty of fifty each; fifty of twenty each; and five thousand and five hundred of ten dollars each; thus having more than five blanks to one prize. Blanks and prizes are usually apportioned in this way.

The idea of getting this hundred thousand dollar prize puts thirty thousand persons into an instantaneous panic. As rich men do not buy lottery tickets, these persons are those of small means, perhaps devoted to manual labor. When the money to be distributed was in the possession of those who earned it, each one of the thirty thousand persons having ten dollars of it in his own pocket, it made no show—no one supposed such an amount existed. It was like the small sums in the hands of farmers scattered all over the country; not supposed to exist because there was no ostentatious exhibition of them. But when this three hundred thousand dollars is brought into one mass, and one-third of it is given to one individual to enable him to set up a rich man's establishment, it becomes a matter of general notoriety; all talk about it; all are in ecstacies to obtain such a prize, even spending the money for a ticket in another lottery, although it is wanted to procure bread for their families. In the same manner, if one person in trade obtains one hundred thousand dollars from trafficking upon the labors of farmers, all are agog about him, and thousands of farmers' sons get into a tilt to become merchants and get rich as he did, not thinking of the tens of thousands not worth a dollar if their debts were paid, nor are they deterred by this want of success from embarking in some perilous career.

It is a most curious thing in philosophy that the wealth, or the supposed wealth of merchants, should be so potent on the minds of those who witness it, when real wealth in the hands of our yeomanry, to a far greater amount, is wholly unnoticed. It is because one is chronicled in expensive ledgers and bank-books, and the other is carried about in the breeches pocket of those who own. In one case there is a display of fine broad-cloth, a patent lever gold watch, and a thousand dollar span of horses; in the other case there is nought of display but of coarse gingham, linsey-woolsey, and farm-wagons. Wealth in the hands of plain farmers is like an old picture in some garret, but when cleaned and oiled and put into an expensive gilt frame, is like the wealth of merchants, gazed at by every one in seeing distance. So it is with the wives and daughters of merchants; they are admired for their beauty, wit, and polished manners, simply, because they are dressed up in silks and laces, and are surrounded by rich sofas, ottomans, lounges, mirrors, and chandeliers; while the wives and daughters of farmers, possessed of infinitely more real personal beauty, are unobserved, or are pronounced coarse and vulgar, because they are attired in calico, and are not ashamed to be seen in the performance of domestic labor. It is a fact well known to all, who have the means of observation, that an exposure to the pure air of the country is peculiarly favorable to female beauty as well as health—to ruddy cheeks, bright eyes, elastic steps; full womanly developments, and sweet breath, like that of the delicious flowers whose odors are inhaled; while in the city girls grow up resembling plants secluded from the light and the air, pale, fragile, and infertile, and with breath foul and feverish. One might as well attempt to have a good flower and vegetable garden in his cellar, as to raise well-formed and healthy women shut up for life in the confined air of the city.

Agricultural wealth is overl oked, or not supposed to exist because it is in small sums; or, if known thus to exist, it is comparatively despised by those who wish to be thought rich, when in reality poor. Rarely does a farmer pretend to be worth more than he is—oftener does he estimate his property below its real value; whereas, the mass of persons in trade, especially those who are insolvent, use every means possible to be thought wealthy—sometimes living extravagantly and being profuse in charities, lest it should be suspected they are poor—in reality spending more than they would if they owned all they possess—from the fact that their business depends upon their credit, which depends on such a public delusion as originates in these systematic deceptions. Hence, an hundred young farmers who own each one thousand are looked upon as a species of paupers; and, although they are all industrious, they could, if they wanted it, obtain no more credit than one or two hundred each, unless they had endorsers, or would secure the payment by mortgage on their property; but let this hundred thousand dollars come into the hands of one man, and they all become tenants to him, and he can have credit of two hundred thousand dollars at least—double of all he is worth, although in the other case, they could obtain no more than about one-tenth of what they are worth. How absurd! Such humbuggery is the staple commodity in certain classes in society; the very ones that give tone to public sentiment!

The estimate is a moderate one, when we assume it as a fact, that in an agricultural community, there is on an average five hundred dollars to each individual, or three thousand dollars to each family of six persons. We entertain no doubt that this is at least fifty per cent. below the real amount; yet, at this valuation of rural wealth, with a population of fourteen millions to whom this wealth belongs, we have an aggregate of property amounting to six thousand millions of dollars! This is the estate of our American farmers, so despised by dandies and brainless women, as well as by insolvent mercantile gamblers. It would be difficult, if not impossible, with any data before us, to estimate the mercantile wealth of the country; and if we could do it, its value is depending on so many contingencies, and hence is so fluctuating, that what might be a truthful valuation in one year might the next year be fifty per cent. too high or too low. In a year of financial expansion there is a phrenzied excitement in all kinds of business, and prices of every kind range high. The minds of mercantile men become inflated, and they suppose themselves far richer than they were before. But in a year of financial contraction, or of panic from any other cause, trade becomes palsied, it may be there will be commercial revulsion and frightful bankruptcy, prices ranging low, and mercantile men estimating their property far less than they did before. With such vicissitudes farmers have only a remote connection.

Our remarks on this subject thus far are predicated on the common assumption, that money cannot be made by farmers only in very small sums, and by slow processes. This we consider an error. Our belief is, that agriculture, pursued in a proper manner, not only yields a competent living with small annual accumulations, but in proportion to the capital invested, and inversely to the hazards attending it, there may be large accumulations, equal, and even more than equal, to the profits of any ordinary merchandize. There have been so many cases of it, that there is no resisting the conclusion we have drawn from them. These cases, too, were not the result of accident, or of any novelty in the articles produced, or of any unusual conjunction of circumstances, not susceptible of application to the universal capabilities of the soil on the one hand, and the unvarying wants of the community on the other hand. Could all the cases of this description be brought together; could they be held up to the gaze and admiration of the community, like large mercantile accumulations; and if there could be any general concentration of the public mind to them, as in the other case, the fact would everywhere be conceded. Only a very small portion of them are ever presented to the public eye ; and even those brought to light through the agency of our agricultural journals and agricultural fairs, make, in many instances, but faint impressions ; either their truth is doubted, or else they are ascribed to some spasmodic effort, from which general inferences cannot be drawn.

It is now proposed to exhibit a few instances of the profit of agriculture illustrative of our theory. We take those of recent occurrence, because the impression they made on our own mind is the more fresh, and those so well authenticated, it is not possible that our readers can cherish a doubt concerning their reliability. The first one we shall select is that of D. D. T. More, of the town of Watervliet, N.Y., a few miles to the west of Albany. Here Mr. More has a farm of one hundred and eighty-five and a half acres, which, at the Fair of the New-York State Agricultural Society, in 1850, took the second premium. This farm has been in his possession and occupation for the last seven years, and during that period has presented one of the most striking examples of successful and profitable improvement ever known. Still, if the owner could accomplish so much others may do the same. If a farm in Watervliet can yield such profit, so can thousands of farms elsewhere, when similarly situated. Previous to the purchase of this farm by Mr. More, it had been for fifty years subjected to an exhausting course, under the leases of various tenants—the annual rent of the whole farm being but one hundred dollars, and that deemed too large a sum by the tenant—as the whole amount of produce was only worth from four to five hundred dollars a year. In fact, Mr. More bought the place in opposition to the advice of all his friends, who deemed it impossible that the land could afford him and his family a living. But, notwithstanding the soil was so reduced, that white beans was almost the only crop he could at first raise, yet his clear judgment and practical knowledge induced him to make the purchase, at sixty dollars per acre, and the result has more than realised his anticipations. He has not only brought his own farm to a fine state of cultivation, making, as will be seen, a little fortune already for himself, but his example has produced an entire change in agriculture in his own neighborhood, raising the price of land there to more than double what it was when he removed thither.

EXPENSES OF D. D. T. MORE'S FARM, 1850.

407 days of labor, at fifty cents a day, -	$202.50
Labor hired by the year and the month, - -	665.00
Wages of one girl a year and another one four mos.,	68.00
500 bushels of oats at forty-one cents per bushel,	205.00
Blacksmith's bill, - - - -	97.81
Bill for groceries, shoes, and dry goods, -	357.00
12 bushels of grass seed, at $2 25 per bushel, -	39.00
Bill for 10 bushels of clover seed, at $4.50, -	45.00
12 bushels of seed wheat, at $1.25 per bushel, -	15.00
Bill for seed corn and garden seeds, - -	10.00
26 bush. of seed buckwheat at 62½ cents, -	16.25
State and School taxes, - - -	31.22
Insurance in Mutual Ins. Co , average about -	4.50
Depreciation of farming tools, - -	100.00
Bill for tons of plaster, - - -	10.00
1,000 bushels of lime, at 4 cents, - -	40.00
Grains for cows, - - - -	119.17
Bill for hay purchased in April, - -	75.00
	$2,174.35

CASH RECEIVED BY D. D. T. MORE, ON ACCOUNT OF FARM, 1850.

363 bunches of asparagus, at eleven cents, -	$63.66
Received from produce of 5 acres of sweet corn,	257.33
610 bushels of rye, at sixty-nine cents, -	410.90
Melons, pumpkins, and citron melons sold, -	148.00
From 831 bushels of buckwheat, at 44 cents,	365.44
Received for raspberries sold, - -	31.25
Value of potatoe crop, - - -	100.00
1,240 bush. Indian Corn, at 65 cents, -	806.00

Milk sold from an average of 30 cows, 9 months,	1,629.81
From the sale of 64 bushels of wheat, -	64.62
Sold five tons of hay, - - - -	40.00
Sold five loads of straw, - - -	5.00
From 4,920 lbs. of broom corn brush, - -	250 00
From the sale of pigs, - - -	40.00
Value of straw not used on farm, - -	100.00
Corn stalks sold, not required on the farm, -	100.00
Twelve tons of clover-hay sold, - - -	72.00
Chickens and eggs sold, - - -	40.00
Pie plant sold, - - - -	11.25
Received for peaches sold, - - -	10.00
Twenty-five calves sold at one dollar each, - -	25.00
Received for pasturage, - - -	18.50
Work by teams and men during the State Fair,	117.30
Work done for S. Van Rensselaer, - -	114.00
140lbs. butter, at 18¾ cents pound, - -	26.25
Total receipts of farm for 1850, -	$4,852 51
Amount of expenditures for ditto, -	2,174.35
Nett profits of D. D. T. More's farm, 1850,	$2,678.16

The above is a statement given in under oath, to the New-York-State Agricultural Society, when applying for a premium. From the accompanying remarks made by Mr. More, it appears that the nett profit of an acre of strawberries was over two hundred dollars. From fifty square rods of land in an asparagus bed, besides supplying his family with this delicious esculent, he received for what was sold $69.66, being at the rate of about one hundred and sixty dollars to the acre, besides what was used by himself. The health of Mr. More, during the period of his owning this farm, has been so feeble that he has made no attempt at manual labor, which proves to be untrue what is frequently said, that if a farmer does make money, it is the result of his own personal toil. However, although Mr. More did not otherwise toil, he ever exercised the most vigilant supervision over the labor of others, showing the truth of the proverb, that the eye of the master will do more work on a farm than his hands. And, in addition to his own feeble health, he has had but little aid from his family, which consisted of a wife and five children, the oldest at the period being only fourteen years of age. And to show that he was not possessed of wealth or cash means to enable him to accomplish what he did on this worn out land, it is sufficient to add, that when he purchased it, he parted with all the money he had, or could raise, leaving him in debt, so that he had more annual interest to pay than the tenant had paid for rent. Besides, when he purchased, all the fences were rotted down, and the buildings were less than one hundred dollars in value. Of course the existing fences and buildings, which are very ample, are new, and, as well as the improvements of the farm, have been made from the profits of the farm. On the whole, this is a demonstration that money can be made by farmers, on a large scale as well as on a small one.

The next illustration selected is that of the Cypress Farm, so called from having in growth a native cypress tree, situated in Ontario County, N. Y., consisting of one hundred and fifty acres of land—sixty-six being for tillage, twelve in low meadow and pasture, sixty in timber, and the other twelve acres being waste land. This farm belongs to Elisha M. Bradley, and in 1850 was valued at $7,500 The expenses of managing the farm that year were, $1,242.50 ; and this sum included $525.00 for interest on the valuation, $300 for his own services, and $335 for hired help. The income of the farm that year was $2,702.37 ; in this sum was included $351.70 for wool sold—$319,30 for wheat sold—$275 for Indian Corn sold—$240 for hay sold—and $620 for cedar timber sold from swamp. Thus it will be seen, that this farm of Mr. Bradley, in the interior, without the advantage of a city market, yielded a nett profit of $1,459.87, equal to nine dollars and seventy-three cents per acre after paying interest and taxes, and the owner for his own time.

For a third illustration, the Blue Pond Farm is selected, situated in Wheatland, Monroe, County, N. Y., and owned by Rawson Harmon. The whole farm contains two hundred and forty-nine acres—seventy of which are swamp and timber land, and thirteen are covered by the pond from which its name is derived. The expenses of Mr. Harmon's farm in 1850, were $915,50; including $400.75 for hired labor—$251 for family clothing, furniture, and groceries—$97 for blacksmith's work—$52.75 for taxes—SIX DOLLARS for Agricultural Papers—and $21 for expenses in attending Agricultural Fairs. The receipts of the farm that year were $2,548.99; including $1,170.25 for wheat sold—$513.80 for wool sold—$348 for 24 bucks sold, being $12 each—and $84 for 28 ewes sold, being three dollars each. The nett profits of this farm that year will stand $1,633,-49—a pretty fair business we apprehend, after supporting the family, and encountering none of the anxieties and hazards of trade. One of the secrets of his success is, undoubtedly, in his free use of agricultural newspapers—the six dollars paid for them in his list of expenses, could have furnished him with six or eight, as a dollar is the common price, and the Plough and the Genesee Farmer, two of the best, cost only fifty cents each a year. Mr. Harmon's creed is to improve the mind as well as the soil This will always ensure success.

For a fourth case we take the farm of C. P. Holcomb, Esq. He is a lawyer, formerly of Philadelphia, we believe, but a good farmer. The farm of which we now speak consists of three hundred and twenty acres, and is situated in Newcastle, Del. Eight years ago it was purchased by him for the small sum of $2,400, it having been so run down and exhausted, as to be in the estimation of many, comparatively worthless. It was said, that when the purchase was made not enough of hay or straw could be found on it for a hen's nest. It has since been conducted by hired labor solely, the owner living on another farm, thirty miles distant. Yet, by a judicious expenditure of time, and a proper system of crops, the value of this farm has been raised to $6,000, besides paying all expenses, and leaving a

surplus of profits. It is now in a condition to continue to be profitable, although the former owner could not support his family upon the place. So much for the proper mode of farming. Mr. Holcomb's cows during the summer give an average of sixteen quarts of milk each per day, and that produces a pound of butter each per day. He has one cow that gives twenty-two quarts a day in the summer; and her milk is so rich, eleven quarts of it make a pound of butter.

In the American Agriculturist of September, 1850, is an account of Tyler Fountain's farm, well worthy the attention of our readers. It was published to stimulate farmers to a spirit of rural enterprise and improvement, as it shows them what can be done by proper efforts. Mr. Fountain's farm is in Peekskill, N.Y., and contains only eighty-three acres, of which, thirteen are wood and pasture. In 1849 he raised thirty-five tons of hay, estimated at $12 per ton; 425 bushels of Indian Corn, at 62 per bushel; 450 bushels of oats, at 40 cents per bushel; 60 bushels of rye, at 75 cents per bushel; 100 barrels of potatoes, at $1.50 per barrel; 2,200 pounds of pork, at six cents per pound; profits from his cows were $150; cattle sold, at $60; and apples sold, at $125;—the amount of all was $1,527. For hired labor he paid $150, so that there was remaining $1,377 for his profits.

Next in order we give the statistics of a farm in Newtown, Bucks County, Penn. It is owned by Mr. James C. Corning. The farm consists of 123 acres, and the products sold from it in 1848 were as follows; 516 bushels of wheat, at $620; 50 bushels of rye, at $40; 1000 bushels of oats, at 375; 1,037 bushels of Indian Corn, at $621; 4 bushels of Timothy seed, at $16; 100 bushels of potatoes, at $60; 500 bushels of apples, at $125; 70 tons of hay at $840; sheep and lambs, at $25; 14 calves, at $95; 20 swine at $240; poultry and eggs, at $125; and 3,708 pounds of butter, from February to October, at $948.58—in all $4,156.58. On the 12th of that year he had remaining on his farm the following stock—five horses, two colts, twenty milch cows, one bull, two heifers, ten sheep, and one breeding sow.

The expenses of the farm were not given by Mr. Corning, but if they were in the same proportion to the receipts, as in the previous ones, the net profits must have been large.

To diversify the character of this article, we here introduce, mostly in the language of the Indiana Farmer, a farming scene in the West. About eight years ago, says the writer, a Dutchman, whose only English was a good-natured "*yes*" to every possible question, got employment as a stable man. His wages, six dollars and board, that was thirty-six dollars in six months, for not one cent did he spend. He washed his own shirt and stockings, mended and patched his own breeches, paid for his tobacco by odd jobs, and laid by his wages. The next six months, being now able to talk decent English, he obtained eight dollars a month; and at the end of six more had forty-eight dollars; making in all, for the year, eighty-four dollars. The second year, by varying his employment—sawing wood in the winter, working for the corporation in the summer, and making gardens in the spring—he laid by one hundred dollars; and the next year one hundred and fifty-five dollars; making, in three years, three hundred and ninety-four dollars. With this he bought eighty acres of land. It was as wild as when the deer fled over it and the Indian pursued him. How should he get a living while clearing it? Thus he did it. He hires a man to clear and fence ten acres; he himself remains in town, to earn money to pay for clearing. In two year's time he has twenty acres well cleared and fenced, a log house and a stable, and money enough to buy stock and tools. He now rises another step in the world, for he gets married, and with his ample, broad-faced, good-natured wife, he gives up the town, and is a regular farmer. Thus, in five years, from a state of absolute destitution, he becomes the owner in fee simple of a good farm, with comfortable fixtures, a prospect of rural wealth, an independent life, and, by the blessing of Heaven and his wife, of an endless posterity.

The profits of a farm in Medford, Mass., owned by Mr. George E. Adams, are worthy of a place in this chapter. He is a young man who pursues his business for the profit; and the energy and judicious arrangements evinced by him are a sure guaranty that his reward will be most ample. His farm consists of 160 acres, and is chiefly devoted to the production of milk and apples. His stock consists of fifty-five cows, a yoke of oxen, and five horses. The annual sales of milk have amounted to $5,500, and in one year to $6,000. The apple orchard contains fifteen acres. The trees have been well managed—are large and generally bear abundantly. In some seasons he has picked 1,000 barrels of winter apples, and in one season sold them at two dollars per barrel. His Baldwin apples are usually sold for shipment to Calcutta. He has recently added a peach and pear orchard, not yet come to full bearing condition. Mr. Adams adopts a most enlightened policy in every thing tending to increase his products or to diminish the labor requisite on his establishment. His main barn is 160 feet long and 40 feet wide, with a cellar under the whole; and water is brought to the buildings by means of a small wind-mill, which works a pump, and keeps a large reservoir constantly filled.

It is not necessary to our purpose to make further additions of the same kind. There are in our country thousands of cases equally in point. They might be selected in reference to every branch of rural economy. The profits of agriculture are not confined to any one branch, or to any particular locality; they are within the reach of all, who use the means that may be used; premising, that some branches will be more productive in one locality, and others in another one; the preference to be adopted for either being easily determined on correct principles, by all persons of ordinary intelligence. For instance, it is apparent that articles liable to perish, before they can be sent to market, should never be cultivated remotely in the interior. Asparagus, strawberries, raspberries, and other things liable to decay or to lose their goodness speedily, should always be raised near the place where they are to be consumed, or where they can be taken to it by rail-road in a few hours. The same may be said of milk, unless it is to be converted into butter or cheese ; and, when produced for butter and cheese it should as evidently be produced in remote localities, where land and labor are cheap. If a farmer were to get up a dairy farm in the vicinity of a large city ; or if he were to have a sheep farm there ; or, if he were to raise cattle and horses there, where land is worth two or three hundred dollars an acre, it is apparent that those living where land is worth only ten or fifteen dollars an acre, could produce butter and cheese and beef and mutton and wool and horses at a far lower rate, and could of course undersell him in the market. Hence to him it would be a ruinous business, while to them it would be a profitable one. The same rule applies in multitudes of cases ; but, any one of common sense may make the application.

It is frequently said that merchants as a general thing cannot afford to do business on borrowed capital ; that the interest to be paid on this capital consumes so much of the profits of trade, and that the most they can ordinarily do, is to support their families as an offset for their labor, not expecting to grow rich or to lay up money from it. There may be exceptions, but they are few. In most cases where persons become rich in trade, it is from beginning in a very small way, working hard and living cheap, till something is saved from the profits, to be used subsequently as a capital in more enlarged operations ; or it is made, where the hazards are great, as in the purchase of lottery tickets, and where one here succeeds, thousands and tens of thousands, lose their all in the experiment. If a man with four or five thousand dollars were to go into a small or retail trade, build his own fires, sweep his own floors, and carry in the evening with his own hands, the goods sold to the doors of the purchasers ; and if his wife were to be her own maid and washwoman, and tailoress, and assistant shopkeeper to her husband, so as to save the expense of female servants, tailors, and clerks, no doubt he might acquire property, and now and then, as others have done, become rich. This is the way that Stephen Gerard, John Jacob Astor, Samuel Butler, William Gray, and William Bartlett did. But woe to that man who undertakes to become rich on trade with a five thousand dollar capital, if he does all the labor by hired clerks, depends on a professed amateur of the brush for shining boots, and sports with his horse and carriage, while his wife is decked off from morning till night in her silks, and has her house filled with a cook and chambermaid, with a family seamstress and a dress-maker, as well as with nursery woman and a man servant.

The question now naturally presents itself, how can a person best establish himself upon a capital of four or five thousand dollars. If he put this sum into stocks, or bond and mortgage, he cannot live upon the interest. With such an income, he cannot support a family. The idea of supporting a family on three hundred dollars a year as persons owning that amount of property generally wish to live, is absurd. Nor is this sum in trade worth much, if any, more than at lawful interest. In trade, interest and principal together would support a family only two or three years at most, living as most persons called merchants do live. But let an active, enterprising young man have five thousand dollars with which to purchase stock, and improve a farm, and he will be completely independent, and by the time he is forty, he will be comparatively rich. With such an outfit, industry, and good management on the farm and in the house, persons may lay up five hundred dollars every year with the utmost ease, especially when they commence and have no young children to support. Afterwards they will have the benefit of these accummulations, to aid them as their families become more expensive. Most of the individuals referred to had not half that sum with which to begin their farming operations; and yet they laid up on an average, more than double of five hundred dollars.

There is another beautiful feature in the pursuit of agriculture. In other pursuits there is danger of their being crowded. If crowded, disagreeable and unhealthy competitions result. Profits are reduced from a general reduction of prices, till at length there is difficulty in getting a living from any legitimate business. To obviate this the regular business is enlarged, or something having no kindred with it is attached to it or engrafted upon it. In both instances the result may be inauspicious. If the former is done, sales are not only made at reduced prices, but frequently on doubtful or long credit instead of cash or on short and good credit. All know that this is ruinous. If the latter is the alternative, the regular business is usually neglected. It is by no means difficult to anticipate the final consequence of either of these alternatives. But in a moral view there is something still worse to be dreaded. Each is apt to look upon his competitor as ruining his own business. Hence jealousies will arise, and not only jealousies, but ill will and discord. The pleasures of social life will be impaired ; and to a considerable extent, the sentiment of universal brotherhood, at least among those of the same occupation, or of rival and conflicting interests, will cease to exist. Instead of the warm hearts among the congregations of those who meet on change or in the marts of trade ; instead of the kind greetings with which all should there commingle, each is apt to cast his eye stealthily and suspiciously upon all of kindred pursuits and kindred interests ; instead of the spontaneous and affectionate uprising of the human bosom, which should be there as well as upon the altar of friendship, to impart its odorous breath to all around, there is diffused a poisonous vapor, impregnating the soul with the seeds of moral death.

In the pursuits of agriculture there is none of this deleterious tendency. In them there is no occasion for jealousy. There is ample space for all disposed to take part in them. The idea of more being produced from the earth than is necessary to support its inhabitants, never enters the mind. The apprehension is, that after all that can be produced there will be a deficiency ; that in some places there may be famine and starvation. And, whoever dreamed that the fundamental articles for sustaining life will ever be too cheap. When was ever beef, pork, and other meats—or breadstuffs—or potatoes and the kindred esculents too abundant to supply the wants of the masses, whose means for procuring them are scanty ? The prices of most of these staples have tripled, all have doubled, and some have quadrupled in our own life time ; and the tendency of these prices is still upward, notwithstanding all the improvements made in agriculture. To render them permanently less, no one dares to expect ; and, it is scarcely to be expected that upward advances will not be habitually witnessed. How can it be otherwise, especially in our own country, where population is rolling in upon us like the waves of the sea ? If the future is to be like the past in this respect, the time is not distant when millions in our country, as in the old ones, will scarcely be able to procure the means upon which to sustain life. To supply the increased demand for provisions to support our fast increasing population, there must be unwearied efforts to increase the fertility of the soil, and to augment the number of those who have the care of it.

Hence, among farmers there is no disposition to be alarmed when their numbers are multiplied, and when the crops are augmented. The more enlightened and successful of them, on the other hand, rejoice to observe all indications of improvement. When did a farmer ever get a patent for raising a great crop of wheat or corn ? If it is the result of any discovery of his own, he forthwith reveals the secret, that others may avail themselves of the benefit of it, causing it to be as free as the air of heaven. So in everything else. Each endeavors to stimulate his neighbors by his own success. On this account there is ordinarily among them a fraternal feeling, a general and harmonious co-operation in their labors ; all being co-laborers with the Supreme Being in supplying the wants and promoting the happiness of the human family. Is there no pleasure in this, to be even more than a balance against any extra toils or privations in the farmer's life ? We think there is ; and we challenge those in crowded occupations, where competition and rivalry are nearly ruinous to success, to produce in social life a feature so lovely, so elevating, and so benignant as that here set forth in rural economy !

THE VILLAGE SUNDAY SCHOOL.

The establishment of Sunday Schools, although comparatively of recent origin, has already done much in mitigating the evils of poverty. Indeed these schools were designed to elevate the poorer classes, by dispelling from them the ignorance to which they seemed to be allied; especially an ignorance of religion, without a knowledge of which they could be expected to rise only a little above the level of brutes. Where such a moral degradation was found, with it was also found a social degradation, but little less to be deplored, because it presented a broad gulph which few only would be enabled to pass, before they could enjoy the means for any general mental illumination. And, the fact has become proverbial, that intellectual without moral elevation is rather detrimental than beneficial to society; it is simply a qualification, not to do good, but to do mischief in the world; an engine, not to augment and render more stable the good already existing there, but to demolish it, and to erect on its ruins an empire of anarchy and blood.

Those not familiar with the state of society—or rather, with the poorer laboring classes of it—during the first ten years of the present century, and the last twenty-five years of the previous century, cannot realise the necessity that then existed for such an ameliorating agent for this benevolent work, as the Sunday School. The importance of general education was not then felt as it now is. At that period the public mind had scarcely begun to emerge from many of the shadows of the dark ages. During those ages learning had been shut up in cloisters or confined to privileged classes. Others did not generally deem it necessary; and the poor were unable to obtain it, however desirous they might have been. In manufacturing districts, where children were confined to labor, for twelve or fourteen hours each day, from one Sunday to another, and thus year after year, it became a natural consequence, that in such masses, the physical and mental powers lost their individuality, and were, to an alarming extent, rather moved like inert manner than like free and responsible agents. Had the social influences thus generated continued without abatement, even at the present day, there would be seen, especially in such localities, the most alarming exhibitions of human degradation and crime. How could it have been otherwise? The short intervals not devoted to confinement and toil in the six working days, seemed to render it a matter of necessity, that the remaining day in seven should be used solely for physical relaxation. Hence, the mind was wholly neglected, and there was no inducement in all these congregated multitudes to rise above the barriers that so closely and so firmly entrenched them.

The Sunday School was primarily designed to take the supervision of these children on their day of release from confinement, to give them the elements of a common education, and by causing them to undergo ablutions and to be decently clad, to inspire in them a feeling of self-respect, on which alone can there be any well grounded hope of thorough improvement. The design originated in the purest benevolence, and it embraced a scheme predicated on the soundest principles of common sense and philosophy. Rarely can we expect to render any moral service of value to an individual, and especially to a class of individuals, unless there exists with them a respect for themselves. If this feeling does not exist, the first effort to be made is to inspire it; for without it, our labors will neither be appreciated by those for whose benefit they are intended, or remunerative to those who perform them. At the period of which we are speaking, these children were generally ragged and filthy, and their parents were too poor or too indifferent to their comfort and respectability, to be instrumental in promoting for them any change. This change, as well as the labor of instructing them, was to be a work of charity. Accordingly, those who were to be the teachers were also to be the solicitors of the means, if not able themselves to furnish them, for rendering their pupils decent in appearance. Adults, if ragged and filthy, and especially if stupid and vicious, are naturally the objects of disgust and aversion, while children, having these repulsive attributes, are as commonly the objects of pity and assidious commiseration. This we all know from experience.

The spectator of the external changes wrought, especially by the first Sunday Schools, could not have failed of experiencing pleasure, though he might not have realized the internal change in progress. A contrast cannot be imagined more striking and impressive, than the presentation in the streets, or about the livery stables, or at the entrances to the garrets and cellars where they dwell, on a Sunday morning, of a few hundred children, as before described, and their appearance a few hours afterwards in the School Room, then clean and tidy and listening to the instruction of their teachers. We hesitate not to say, that in our life time we have frequently passed from street to street, beholding, in that season of their release from confinement, these wretched little personifications of poverty and ignorance, but not without painful forebodings of the destiny before them—of penitentiaries, and dungeons, and gibbets, the frightful retributions for a career of destitution and crime! And, it has been our delight, perhaps a thousand times—for we spent years, long since, in one of the receptacles of such children—in directing the instruction given them in the Sunday School department of a church; here they were comely and attentive, and not a few denoting, as time has since proved a reality, an inward aspiration to participate in the more elevated labors and enjoyments of a Christian community.

About the year 1815, we had the supervision of a Sunday School, composed wholly or nearly so of children in a manufacturing village, and habitually confined, year after year, in the cotton mills, for which it was then more distinguished than any other one in the country. These children were of all ages, from six or eight years to fifteen or sixteen, and had been collected mostly from the adjacent counties of the two States, on the contiguous borders of which the village was located. In some instances they were fatherless, but more generally the fathers were sick, or else had become the victims of intemperance, so that the responsibility of supporting and rearing them was cast entirely upon the mother. They were of course poor, and were placed in these establishments to earn money for the support of themselves and mothers, and also of brothers and sisters too young to labor, and, in not a few instances to our knowledge, of their drunken fathers. If there is aught in this world that can fill the bosom with righteous indignation, it is the sight of the heartless and brutal inebriate, who will thus take the hard earnings of his broken-hearted and sickly wife, and also of his own young children, that he may indulge himself in sloth and beastly dissipation. If there is aught that can fill the bosom with unwearied commiseration it is the toil and the robbery of such mothers and their tender offspring. We have no hostility to the establishments in which such children have found it necessary to labor. In numerous instances they have proved a great blessing to the community at large, as well as to individuals of the classes alluded to, for without them, many of these individuals would have become vagrants or public paupers. And while such establishments became the needful asylums of an untold multitude, thus reduced to poverty and the necessity for early toil, the Sunday School was planted by the side of them to save the inmates from that moral leprosy, almost a necessary consequence of such hardships, privations, and social influences as there seem an unavoidable element in their existence.

Our Sunday School was one of peculiar interest. Such schools were then in our country by no means frequently found. Hence the novelty of it, and the novelty of the labor to ourself, led to a devotion to it bordering on enthusiasm. It became a model or a pattern school; and as was natural, the notice taken of it by all around, additionally stimulated all engaged in it to secure for it a fidelity under other circumstances not always expected. The children were all decently equipped, not only for an appearance in the school room, but for the services of the church, which were immediately to follow. Indeed, for several of the first months, these children made the majority of the congregation for that service, being joined by a few only, save their parents and older brothers and sisters. With the exception of a single family, the service of the Episcopal Church was a novelty to that congregation. It is not our purpose in these specifications to indicate any disrespect to other modes of religious worship, or any inferiority in them for the legitimate ends to be accomplished in all divine worship ; but, simply, in the discussion of a more general subject, to make known the admirable working of it in the case now the subject of our remarks. In religious institutions, as in civil government, more depends for individual or public utility on the manner in which they are observed and maintained, than on any distinctive peculiarity possessed by any particular one. While well persuaded of the excellence of that to which we are attached, and in which we have long labored, we pretend not to think, much less to affirm, that persons attached to other forms are not equally sincere in their allegiance, and experience a satisfaction and moral culture not less to be desired.

Many of the children in this school partook largely in the ardor of their teachers. This was evinced by the punctuality of their attendance, their exemplary and praiseworthy deportment, and especially the amount of their recitations. Considering the little leisure time allowed them, the many hours they were held to service in the mills, every day, we believe from six in the morning till seven in the evening—it really seems impossible at this distance of time that some of the statistics of that school could have been founded on fact. Had we not been an ear and an eye witness, we might now doubt it. A stranger admitted to witness the exercises of it—to see the well-dressed and orderly children that composed it ; the intelligence that beamed from so many expressive eyes ; and especially the promptness and the accuracy that distinguished every branch of prescribed duty, could not easily have realised that these were a community of factory children, driven by poverty at this early age to earn their own living. Such, however, was the fact ; for it shows that the mental capabilities of all are essentially the same ; and, that so far as differences of character exhibit themselves in society, they are ordinarily the result of different degrees of culture, either mental or social. This fact, moreover, shows that not unfrequently factory children—or cotton bugs—as they are sometimes called—if favored by an attendance upon a good Sunday School, in connection with other helps for calling forth the better elements of their nature, will prove themselves, in the course of life, not inferior, to say the least, to the children of those whom they served. It is no uncommon thing that poor children become rich, and the children of rich in turn become poor. Society is in a constant state of fusion, new portions of it successively rising to the surface, while others are sinking from view to supply their place.

Such mutations in the social condition of man, especially in our country, are of the most frequent occurrence. Indeed, there is seemingly little or nothing in that condition which remains more than a score of years. We know but little of the rank of the ancestor from the position of the descendant. The instances are comparatively few where the son of the politician raised to the highest honors in the State is able to sustain in his family the ascendancy of the father; and, even the elevation from wealth with all the entrenchments cast about it, after the close of one generation, has but an impaired relic remaining behind; and in the second or third descending series, not a vestige of it is to be found. The country is the great nursery for supplying the places thus made waste. There, genius is constantly rising in freshness, and is sending forth to all the high ramifications of the body politic its long and vigorous tendrils One class of the youth leave the plough barefooted, and become errand boys, and then clerks, and then merchants and men of wealth in the city, while the sons of former merchants, from effiminacy, or extravagance, or indolence, after spending the property bequeathed them, become paupers. Another class of these youth in like manner leave the plough barefooted, seek some literary institution, and in due time are prepared to take, and do take possession of the most distinguished honors and responsibilities of the land. Even the poorest children and youth of the country—the offspring oftentimes of drunkards and forlorn widows, resort to the manufacturing village to obtain subsistence. Here the native energy, brought from their native hills and valleys, soon enables them to become adepts in the labors of their new vocation. A portion of the males are even able to excel in the arts of machinery and the varied routine of manufacturing, so that when revulsion, or any opportunity presents itself for it, they become themselves the principals in those very establishments, or raise up new ones surpassing those which preceded them.

Nor are these female youth doomed to spend life in such servile toil. They also rise ; as well as the males, they receive an impulse at the factory village Sunday School ; there the first mental germ begins to unfold itself ; there a taste is acquired for the decencies of life ; and there also they are stimulated to take part in the more desirable conjunctions within the sphere of female ambition. The culture of the Sunday School is succeeded by that of the Bible Class and the scientific lecture room, and then by self culture from the village library. Thus all the leisure hours, intermediate to the prescribed seasons at the spindle and the loom, are devoted to mental and social improvement. In a few years these once poor, ragged, and ignorant little girls, become well formed young women, and having laid by, as many of them do, hundreds of dollars, and being handsomely dressed, in the streets and at the church, they appear like ladies of the best class. Some of them become highly intellectual and polished in their manners. Afterwards, they marry gentlemen of enterprise and respectability — occasionally of wealth and high stations—thus from factory girls they become seated in parlors, and move and exert an influence in the first grades of life. In New England, especially, thousands and thousands, either from a state of poor orphanage, or worse than that, or from the fireside of the most straitened farmers, have worked their way till associated with all that is honorable and estimable in the most refined society. A volume might be filled with the most interesting memoirs of the factory girls of Lowell, Pawtucket, Taunton, Fall River, and Dover ; and an hundred years hence some of the best American heraldry may be traced back to these places.

Among the most interesting scholars in the Sunday School of which we were the head, was a little boy, perhaps ten years of age, when he first attracted our especial notice, who was scrupulously constant in his attendance, manlike in his demeanor, and indefatigable in his proficiency. He was one of the most faithful in his labors in the mill, but on each Sunday morning would recite half a dozen chapters from the New Testament, which he had so thoroughly committed to memory, with an accuracy and emphasis that might have ranked well with a good rehearsal of Shakspeare. In the course of one year he thus committed to memory, and recited at the Sunday School, the whole of the New Testament, the Prophecy of Isaiah, and most of the Psalms and Proverbs. Such an instance of assiduity naturally excited admiration. How a boy of that age, and constantly, for about twelve hours a day, devoted to manual labor, could do this, was a mystery. Inquiry was accordingly made. The result was as follows.

The family of this boy, in consequence of the drunken habits of the father, was, in a rural population thirty miles off, reduced to extreme destitution. He was beastly, sullen, and stupid. The mother was a woman of more than ordinary good appearance and intelligence, but, as usual under such circumstances, discouraged and heart-broken. There were several children, the eldest advanced to full womanhood, and the youngest in the cradle. The intermediate ones were of all ages, having succeeded each other, apparently, two years apart. Hither they were driven to obtain subsistence. The mother was housekeeper, the father did nothing but visit the dram-shops, and all the children of an age to work were kept in the cotton mill. The boy alluded to, with two or three others, joined the Sunday School, and he became so interested in it, that he procured a copy of the New Testament, from which he would cut leaf after leaf, as he wanted them, keeping one constantly spread out, in the inside of his hand, so that he could, without being observed by the overseer, glance his eye upon it when engaged in his work. This was his own device, as the children were not allowed to take a book into the mill; and, in his case might have been tolerated, inasmuch as it did not lead to a neglect of his employer's interest. In this manner our young pupil jointly pursued his labors and his Sunday School lessons, sometimes, also, getting up in the morning, and studying an hour by lamp-light, before the bell summoned him to his regular occupation; in this manner, too, he was enabled to accomplish all we have stated. Several facts connected with part of his early history might be narrated; but our limits will not admit of digression.

One advantage of the Sunday School, incidental to those of primary consideration, is the opportunity furnished for bringing to light and before the public, such instances of talent. Hundreds of such have in this way been gleaned from the greatest obscurity. Boys have been placed at school and furnished with a classical education, so that not a few of our professional men can trace back through this very channel, their progress to eminence, to the most humble conditions ever known to exist. Girls, too, of talents and moral worth have passed from the lowly cottage to a pupilage in the village Sunday School; then advanced to become teachers; and when performing duties as such, by frequent intercourse with the young pastor, or with theological students, frequently the male teachers in such a school, to become the wives of persons to whom their names would never have been made known, had it not been for their associations with such an institution. It is well known that the Sunday School room has often had a most bewitching influence in calling into action a pure and chastened affection between the sexes. As much as it has been celebrated for giving life and vigor to the young Christian, it has not been less memorable in moulding kindred minds into conjugal unity. Perhaps there is, or should be, a similiarity between the passion which binds two persons in wedlock, and that which binds all, male and female, to the throne of heaven!

At the age of about twelve years, the factory Sunday School boy, who committed to memory the whole of the New Testament, was placed under tuition for classical study. His proficiency was good, and he continued it till having passed through most of the books then required preparatory to the University. Owing to some change in the labors of his friend who had thus assisted him, his scholastic pursuits were relinquished for those of a mercantile character. In the latter he first labored with his accustomed fidelity, as a clerk ; and, then, as a merchant ; and, we apprehend, he still lives in an eastern city of our republic, respected by all who know him. It is many years since we have seen him, but we shall never cease to remember the honorable distinction he acquired at the village Sunday School, where he toiled for a subsistence ; and, how that distinction prepared the way for a subsequent career of business and respectability. A kind recognition on our part of distinguished merit in those who thus commenced the pathway of life, and one in which we were wont to share, is among the most pleasant reminiscences of former days May the time be far distant when those reminiscences shall fade away. They are better than gold, for gold is valueless when we suffer for the want of a pure conscience. They are better than fame, for fame is a mere empty breath, not imparting health to soul or body. They are better than life, for when life is ended they become an enduring memorial in the archives of heaven. In all cases it is more blessed to bestow a favor than to receive one ; and, in a case of this sort is a thousand-fold more so in the one than in the other.

Although Sunday Schools were originally established for the laboring poor, particularly in manufacturing districts, still, such has been their general utility, and so well adapted are they to the wants of a community, the rich as well as the poor, they are now found in connection with nearly every church in the land. As now generally conducted, the Sunday School is not so much designed to impart the elements of a literary education, as of a religious one ; to store the youthful mind with a knowledge of the great truths of the Gospel. Where there is any deficiency of this kind of instruction in the family—where parents neglect in a familiar and efficient manner indelibly to impress the minds of their children with the distinctive moral and didactic attributes of the Christian system, the Sunday School can alone supply that deficiency in family culture. Without this substitute vast numbers of children in Christian communities would seemingly grow up, aliens from the faith and the moral virtues of revealed religion. And where there is no manifest negligence in the family, by Christian parents towards their children, upon the subject here indicated, the Sunday School may not only become the handmaid of parental labor, but carry forward to a higher grade of religious attainment, both in doctrines and practice, what could, from the nature of the circumstances, only have received from that labor its first conformations.

It is a fact implying no dereliction of duty, that a vast majority of parents have not themselves had that mental training requisite to give their children a systematic course of instruction; and, of such as have had it, a few only have the leisure to do all, or even a moiety of what in this respect should be done. It would be unreasonable and absurd to expect, that persons can give, before having had preparation for the labor, judicious instruction in religion any more than in philology. It is seen, therefore, how valuable must be the agency of the Sunday School, as now conducted, in preparing the rising generation for a Christian life. Even if there has been no neglect of parents, it is about as necessary in preparing our children and youth for full membership in the Church, as the Grammar School is in preparing students for admission to the university. It is apprehended, nevertheless, that many parents are remiss in giving religious instruction to their children; that they seem to imagine all is to be done by the Sunday School, and nothing by themselves; that, since from a want of due qualification or of leisure time, they cannot do every thing in this important work, they will do nothing at all. It will be seen in the sequel of this memoir that more vigilance should be observed in fortifying the minds of our youth against the subtleties of infidelity than has been common. If not duly fortified, many a one will be lost to the Church—many a one will make shipwreck of their faith—many a one will be joined to the ranks of unbelief. This will appear from the case of Lydia Sheldon, another Sunday School pupil, with whose history we were acquainted.

Although there was a shade of wildness and romance in her character, not by any means attributable to the old-fashioned Scripture Christian name with which she had been endowed at the baptismal font, still she was a girl of rare talents as well as of great personal beauty. Partial as we are to that class of proper names, and hostile as we are to that irreverent and ridiculous fancy, which now-a-days shudders at the supposed vulgarity of them, and would sooner call cattle than children by such a cognomen as Beulah, or Kezia, or Machir, or Naomi, or Rachel, or Uzal, or Zilpa, we do not pretend to attach very great importance to them. Names are important, as furnishing a convenient mode of distinguishing one person from another; and, in giving them to our children, may be a delightful one in perpetuating the memory of esteemed friends, both dead and living. There may, too, be good taste in selecting such as are euphonious, or are easily recollected; but, beyond this, there is nothing of magnitude depending on them. They are mere matters of fancy; and more persons probably in the selection of them render themselves obnoxious to ridicule, than entitled to the exercise of good sense. We were always partial to the name of Lydia; and, our acquaintance with Lydia Sheldon, the sceptical, but subsequently pious Sunday School scholar, has embalmed it with increased affection.

The mother of Lydia, being left a widow, when this beloved child was but ten years of age, was induced to open a genteel boarding-house of limited extent, in Common-street, one of the most beautiful localities in Boston. This street being almost wholly occupied as private residences by the elite of that city, a boarding establishment there, especially when under the direction of a lady so accomplished and so well known as the Widow Sheldon—still young and essentially beautiful and cherished by the friends of her deceased husband—in his life time supposed to be rich, but on his death, found to be insolvent—was sure to receive the most ample patronage. By the boarders, all of the higher class, with abundant means, Mrs. Sheldon was treated like a sister, and had Lydia been an own child she could not have been more caressed and petted. The mother being much occupied with a supervision of the family, the daughter, when out of school, was almost wholly with the boarders, and saving at the table, or when her clothes were to be adjusted, they were seldom together; and what was especially unfortunate, Lydia's religious education was wholly neglected. The natural sweetness of her temper seemed almost to justify the apprehension that she needed not the culture required by most children. Nevertheless, her fine talents in connection with her peculiarly winning manners, made occasion for judicious culture even the more necessary. The more lovely the features the more brilliant the intellect, and the more fascinating the manners in an orphan girl of her age, the more likely was she to become tainted, when temptation presented itself, by some moral poison. Such in this case was the fact with Lydia Sheldon. The mother as well as the boarders generally, amid so much loveliness and so many bewitching charms, if they discovered some little defect of manners or thought, gave slight heed to it; they attributed it to some transient impulse, not calculated to leave behind any enduring blemish; and, especially, when viewed in contrast with its counterparts, was forgotten or undeserving serious consideration.

When Lydia was about thirteen years old, Dr. P., with two adult daughters, having had in everything but religion, the best education, obtainable at that time, became boarders in Mrs Sheldon's family. He was not then a practicing physician, but devoted his attention to general literature; and, although not a professed infidel, his sympathies were all that way, and he had in his library sundry sceptical books. His daughters, also, though privately, were a little, in this respect, disposed to listen to the inuendoes of their father. This was one of the unusual solecisms of female character. Not a year previous to this time they had lost a pious mother; and, that two surviving daughters should so soon have forgotten her counsels and her prayers—her pious resignation and her buoyant hopes of a blessed immortality—to us is incomprehensible. How a well educated surviving offspring—particularly a female offspring, can ever forget such counsels, or cease to avail themselves of the abiding consolation to be derived, at the sick bedside, from such prayers, such resignation, and such hopes, is more than we can explain. The Misses P. did seemingly forget these counsels, and did cease to avail themselves of that consolation. Knowing what we do of the warm susceptibilities of the female heart, an irreligious woman seems to us almost a paradox—sometimes we have thought her a monster! Knowing also the high mission assigned to her by the Author of her being, and the great trials she has to experience in the sphere of duties incident to that mission, we shudder at the very idea of her ever becoming recreant to the faith or the practice of the Christian. If there is no one to whom she can look for unerring guidance when the world all around her is dark as midnight; if there is none to whom she can repair with childlike confidence of succor when all earthly hopes and enjoyments are blasted and swept away, what a cheerless—what a paralysing destiny has been ordained for her! Depraved as the world is, bleak and desolate as it often presents itself to us, were it not for that conservatism found in female piety, sad indeed would be the picture of human life. Man may, and often does, long encounter courageously the storms that assail him; but, at last his spirit falters, and like the rent steamship on the tumultuous waves, he forever sinks amid the contending elements, and is lost and forgotten. Not so is it with woman! The world may, indeed, scowl and frown upon her, yet her well balanced piety enables her still to trust in a benignant Providence. She may, indeed, experience shipwreck; yet, she sinks not, she glides from wave to wave like the Indian skiff, or the modern life-boat, her faith and her hope sustaining her, till landed in the peaceful haven.

Lydia Sheldon soon became a favorite with Dr. P. and his

daughters, as she was with the other boarders. She was much in their apartment, and her inquisitive mind speedily led her to his library. One less intellectual might not have noticed the infidel books; but she noticed them and did not fail to read them. At first she was, as she afterwards confessed, shocked with the bold assumptions she there found. Nevertheless, like the fancied charm of the rattle-snake, they seemed to deprive her of all power to resist their influence; although for a period she abhorred their impious slanders, she knew not how to tear herself away from them. Her conscience, probably, reproved her for thus meddling with what her better impulses told her should not have been touched. Aware of her fault in this respect, she kept the whole matter a profound secret; daily reading them, privately, as she had opportunity, till she had at control all the popular arguments against the authenticity and truths of the Bible. True, she did not imagine herself an infidel, but her mind was too much poisoned to be an active Christian. She would sometimes for hours weep in her retirement that she had thus been entangled in these snares; still had not courage, either to her mother or her minister, to make known the dilemma in which she was held, or to seek for help in being extricated from it. Had she, in this matter, manifested the same frankness and simplicity, she had been accustomed to practice, she would have been saved from the peril to which she was now exposed—a peril ending in ruin had she not at last been relieved from it.

About this period Mrs. Sheldon relinquished her boarding-house establishment, and removed to a country village. Here were several young ladies nearly of the age of Lydia, belonging to the best families, but having finished their probation at the Sunday School. Simultaneously, or nearly so, Mr. Griffin, a young lawyer, established himself in it. A more beautiful spot could not have been found. Here was a select and well cultivated society sufficiently ample to make life agreeable. And there was enough of business to give it an air of invigorating activity, and prevent that feeling of ennui almost inseparable from a population wholly without business. However desirable that there be in every community a few individuals of refinement, wealth, and leisure, to look after objects of public interest as well as the poor, it is still desirable that the masses of all communities should be habitually devoted to some useful and active occupation. Without it, the mind becomes enfeebled, and the moral pulse beats low. The excitement of business is as needful to promote a good social temperature, as the wind is to drive away the sickly vapors rising from stagnant pools and undrained marshes. Such was the village where Mrs. Sheldon and her daughter took residence, to enjoy the little competence she acquired in the city. As yet she had learned nothing of the influence upon her daughter's mind from an acquaintance with Dr. P. and his family. Had she known it, her pleasure from the agreeable circumstances connected with her new residence would have been greatly lessened. As the sequel will show, it was best the secret of the daughter had not been imparted to her, though the principle of such concealments is ever to be condemned.

Young Mr. Griffin was beginning his business career with the finest prospects. His talents were of a high order; his moral character was unexceptionable; and his personal appearance and manners were also such as to make him friends and admirers. The Rector of the Church esteemed him a material acquisition to the interests of the parish; for in him, it was to be expected, he would ever find prompt and efficient co-operation. Notwithstanding the prejudice that often existed, especially in the country, against the legal profession; a prejudice not unnatural, and in some cases well founded; for it is well known that individuals have, formerly more than at present, made that profession a cloak for stirring up strife and oppressing the poor, without the manifestation of adequate redeeming qualities to be a fair offset; still it is an undeniable truth, that the members of that profession may, and should be, men of exemplary lives and of the purest social and moral principles. At the present time, such qualifications are the best guaranty they can have of success. Their position in society causes such traits of character to be in the highest degree efficacious upon all who witness them. Such traits, connected with their superior intelligence, and their prompt and systematic habits of business, renders it desirable that in every community, political, social, or religious, there should be more or less of persons belonging to the profession of the law. They are as necessary in the conventions and other business associations of the Church as they are in the senate, or in the business of their own particular sphere of labor. Nothing can be done wrong, where there is a shrewd and quick-sighted lawyer, without its being detected. It is a part of their duty to detect and expose misapprehension in small matters, as well as knavery and fraud in large ones. It is one of their very instincts to search out and unfold the lurking place of intrigue and dishonesty, and to present human conduct in the mirror of truth.

Our Church Rector had correct apprehensions of the advantages to be derived from the society and the services of the young gentleman, who had thus planted himself in the midst of his parish. It was accordingly proposed to him that he should take a class in the Sunday School. To this he assented, on the condition that a new class of more advanced scholars, should be formed, consisting of Lydia Sheldon and the other young ladies alluded to. The Rector was pleased with the suggestion and promised to use his influence to induce them again to become pupils. Under other circumstances they would have given a prompt refusal; but, feeling flattered with the compliment of such co-operation on their part being made the condition of securing the young attorney's charitable services, they at once consented to the proposition. The best of persons are sometimes, in a measure at least, induced to charitable labors by motives equally selfish. To do good to others individually, or to any general scheme of benevolent enterprise, without regard to the welfare, ease, and pleasure of those who are to be active in it, requires a higher degree of moral purpose than we ordinarily expect to find. Our motives at best are of a mixed character; even in our most charitable ministrations, it may be, that a tincture of ostentation or individual attainment may have had a lurking influence upon the better tendencies of our nature. Although it should, indeed, be the object in Christian morals to fix a high standard, yet, if we repudiate and denounce every deed not purely disinterested, it is feared much will be cast away that is now cherished with great assiduity.

Nor would it be unnatural, if these young ladies looked upon the proposed measure, as leading to a little social compact among themselves with Mr. Griffin, although in the Sunday School as their head and teacher, yet in reality or ultimately one of their number, as an equal. Such conjunctions of the sexes among those of their age are always viewed with favor. Why should they not be? Do they not lead to one of the most important designs of Providence? Through what channel is that design approached with a more chastened affection than the charitable services of which we are speaking? He had, indeed, passed through all the grades of classical and professional study, and was in the career for honorable distinction in the broad arena of life; nor did they probably consider themselves far in the back ground; they had as it were finished their education; their family alliances and their own personal accomplishments already reminded them that they might soon become the participa ors in the fame and the fortunes attained by those like himself. No doubt, when by themselves, commenting upon the new relations into which they were about to enter, and upon the new labors about to engross their attention, there might have been felt a little jealousy as to the success they would severally have on the social affinities of their proposed teacher, and consequently a little fun and raillery upon each other, in relation to that delicate matter. Be it so—it is one of the sweetmeats of young life—no one need be alarmed at its indulgence; even bashful young ladies would appear prematurely staid and matronly without this zest of youthful impulses.

In due time the new class in the Sunday School was duly organized; and the question came up, in rather a democratic way, what was to be done—what were to be the studies! The

10*

catechism had long since been made as familiar as household words. Besides, there was on the part of the pupils a feeling of dignity to be preserved, and had the idea been suggested of having them occupied with studies on a level with those of the children in the Sunday School, there might have been a speedy revolt. Mr. Griffin understood human nature too well, to hazard any such experiment, young as he was in his prófession. The best lawyers best understand human nature; and on this account they will glide along on the river of life amid curvatures, and shallows, and snags in perfect safety, while others without it are continually subject to disaster and ruin. He was prepared to meet the exigency. Having spent much of his leisure time in reading and reciting Shakspeare and some of the other best specimens of English literature, he proposed that the young ladies should, in a similar way, perfect themselves in readings and recitations from the Sacred Volume. At first they fancied nothing of the kind could be done satisfactorily; but, on hearing him read to them, with an earnestness and pathos becoming the subject, St. Paul's Defence before King Agrippa, it seemed to them they had never before known of that chapter. As he advanced in the Apostle's argument and plea, at every new point throwing his soul more and more into his enunciation, one after one became impassioned and wept, as if in reality before the court of Areopagus, where was speedily to be pronounced on the accused, the verdict of life or death. Accordingly this chapter was assigned for the first lesson in the new class. Each young lady was to practice by herself in reading, so that at the next Sunday all were to read it; and, thus, from reading, and from hearing each other read, the design was to enable them to read it with life and spirit as it always should be read; but as it rarely is read. We do not recollect, in the English language, a more masterly and overpowering appeal to a legal tribunal. No wonder that these young ladies hearing it read by Mr. Griffin, were so completely lost in the burning eloquence of the speaker.

The exercise thus introduced into this class of young ladies was peculiarly appropriate; for to read elegantly is one of the most desirable female accomplishments. With proper training and sufficient practice women may become the most elegant and effective readers. Their quick and deep feeling enables them to throw the sentiment expressed by the author so completely into their manner, their gestures and their looks, that the listener is completely engrossed with the subject, from the first to the last; and, with the bursting forth of every new and beautiful thought, there may be seen, as they advance, from the expressive curl of their lips and the penetrating flash of their eyes, the constantly kindling emotion that swells the bosom. It is to be lamented that so little attention is generally paid to this peculiar attribute in the female mind; for, it is susceptible of a culture, that gives a charm to female character of inestimable value. Let a beautiful and graceful woman with a fine clear voice, read the chapter alluded to; or St. Paul's description of Charity; or St. James' description of faith; or the miracle of the blind man restored to sight in the ninth chapter of St. John; or the description of the general judgment in the twenty-fifth chapter of Mathew; adapting her action to the language and the subject, and it will be seen with what certainty she can control, in all who hear, the elements of the human soul. Mr. Griffin trained this class of young ladies thus to read, and thus to recite select portions of the Inspired Volume, so that it was not long, before the Sunday School room was more crowded to witness their performances than was the Church to hear the sermon. A similar result would be produced in every Sunday School if a suitable teacher can be found to take charge of this branch of instruction. It was even surmised, that the Rector of the Church, well as he was accustomed to read the Daily Lessons of the Service, was incited by the example thus set before him, to improve his own manner of reading.

After a few months had been occupied so successfully and so satisfactorily to all interested in the exercises named, it was proposed that the young ladies should study the Evidences of Christianity. This branch of Sunday School literature has also been not enough appreciated. Even among the well educated of our religious congregations, it might be found that few only have paid that attention to the subject which would enable them to answer the interrogatories of the sceptic. This ought not to be. When infidelity, open or concealed, lurks in so many of the bye-paths of society, and especially when it, now and then, raises its banner in public places with dogmatic insolence, it is surely a part of decency for the friends of the cross to resist the insult with manly force. A failure to do this is construed into an inability to do it. Hence, Christianity is disgraced, and its friends are treated with scorn. Mr. Griffin himself, with all his familiarity with Blackstone, and Hale, and Shakspeare, and with all his ready knowledge of general literature, found himself under the necessity of brushing up his Theological polemics. Miss Sheldon very soon, at the recitations, commenced asking him such questions, as were suggested by her recollections of what she had read in the infidel books of Dr. P. The effect was as she had anticipated. He was perplexed in knowing how to answer; for he readily perceived, she had too much sagacity to be satisfied, or blinded, with any equivocal replies. Nor was she disposed to receive hypothetical explanations. She was quick to discover the difference between them and positive and well authenticated rejoinders. Nor was she ignorant of his conscious inability to remove her doubts; and, although she had, like all her associates, formed a high opinion of his superior mental endowments and moral excellence, like a shrewd tactician of her sex, the more she perceived his perplexity, the more she was disposed to multiply those little saucy and unmanageable interrogatories.

The fact was apparent, that the teacher was in an awkward dilemma. One of his pupils, at least, knew more about the evidences of Christianity than he did; or rather of the infidel exceptions to them. He made a confident of the Rector in the matter; and it was decided that he should be absent a few weeks from the school, ostensibly on business, but in reality for the purpose of making himself familiar with the writings of Leland and Paley and Watson on the subject, and that the Rector, in the mean time, should supply his place. Miss Sheldon was wise enough to suspect, as she subsequently admitted, the real cause of this change, and was willing that there should be a flag of truce till Mr. Griffin's return. When again at his post, better prepared for an attack, and even disposed to act on the offensive, she of course was more cautious in her onsets. However, no opportunity on her part was wanting to introduce inuendoes and exceptions to the received evidences of revealed religion. Generally he was able to quiet her suspicions; nevertheless, like a valiant soldier, she retreated cautiously, yielding the ground only where she was obliged to yield it. The result was, that in the course of a few months Miss Sheldon was fully confirmed in the truth of Christianity, and that the doubts she had been led to cherish rested entirely on superficial views. To one of her rare good sense and of her amiable social habitudes of mind, it was truly a consummation for which she expressed the most devout gratitude, as well as humiliation for her previous error in ever listening to such false and subtle delusions. It seemed to her providential, that her mother thus removed to this village. Had she remained in the city, under the influences which had caused her doubts, in time she might have become a confirmed infidel. It also seemed providential to her, that she was thus so unexpectedly placed in a situation to have her doubts removed, and to have her faith fixed on a basis so rational and so firm.

The above incidents show the importance of the Sunday School, and the necessity of having the evidences of Christianity, with competent teachers, made a part of the studies pursued. These labors of Mr. Griffin were continued under the most favorable auspices for nearly two years; the members of his class all becoming far better qualified for the duties of the Christian profession than they would otherwise have been. If all mothers were as well versed in the subjects on which these young ladies were instructed, few children, indeed, male or female, would become infidels. To impress the young heart with the sentiments of a true faith, and to train it to the habitual exercise of those sentiments is the appropriate duty of the mother. The father has not usually the opportunity to do it in the best manner. It should be done by one habitually with them, so that the most favorable occasions be taken for it. In the family, incidents are almost continually arising, by which young minds are rendered peculiarly susceptible to moral suasion. These are the occasions for this important work. A few minutes on such an occasion will lead to more enduring evidence of well directed counsel, than whole days formally and without reference to such contingencies, devoted to such culture. Besides, it is well known, that there is a kindness and a melting earnestness of manner in the mother, not always found in the other sex, however well cultivated their minds, or, however ardent their desire for the accomplishment of such an object. Here is the sphere for woman's appropriate action. Here is the theatre in which she effects great achievements. Here is the field where she acquires unfading laurels.

Mr. Griffin discovered in the convert he had made a quickness of perception and a mental vigor in harmony with his own estimate for high intellectual female endowment. Nor did he fail to discover social affinities and moral susceptibilities equally worthy of his appreciation. It was apparent also to himself, and to others, that he had been an agent in the hands of Providence, not only to rescue her from a most perilous condition, but to give her varied capabilities a development, which they might not otherwise have reached. The consequence upon his own feelings was a natural one. Philosophy in its expositions is universal. Had she been discovered by him, although a perfect stranger, in an upper apartment of a burning house, without means of escape, and he, with unstudied impetuosity, forcing himself through the flames to effect her rescue; or had she suffered shipwreck, and begun to sink in a deep abyss, and he, from the impulse of the moment, been led to plunge in after her and to save her, would he feel no untold emotion of friendship for her, --no undefined desire for a perpetuity of union with her? And, if thus snatched from ruin by such an one, would she fail to experience a reciprocity of affection? To deny, or to doubt this assumption would be to repudiate the teaching of universal experience, as well as the best philosophy with which we are acquainted. To stifle the uprisings of nature is no easy task. To say to the current of human passion, thus far shalt thou go and no farther, oftentimes is unavailing, unless aided by more than human power.

So it was in the present case. The facts of it were in union with all previous demonstrations. Almost before either imagined it, the Sunday School teacher and his pupil were unalterably in love. If either one of her companions would have been willing to have secured the conquest for herself, all had the magnanimity to acknowledge the fitness of it. To love succeeded matrimony; and in less than a year from bidding adieu to the school room, the once agreeable and fascinating Miss Sheldon, under a new name and with matronly dignity took possession of the nursery. The husband, prosperous in his profession, as well as blessed in his family, rose to eminence; and, if he acquired wealth, and shone as a star of the first magnitude in the national halls of legislation, it was to him a source of equal pleasure that his former pupil became distinguished as one of the most brilliant and accomplished ladies of the circle in which they moved; nor of less pride that she was one of the best wives and mothers. They are now grand-parents, with every earthly comfort; but they have never forgotten to cherish with fond remembrance the circumstances in which began their first love.

Such are some of the fruits of the village Sunday School in which we took part. May it survive, as it doubtless will, the longest life of its first friends and founders. May it continue, even to the end of our country's annals, as it has hitherto done, to spread abroad its genial influences—filling the Church with enlightened and devout worshippers, as well as the Christian household with all that can adorn and dignify social life. For one, we shall never cease to pray that peace and prosperity may be within her walls, and that the divine blessing may rest on all who do it favor.

To us the institution of Sunday Schools has ever been looked upon as among the best instrumentalities of the age, not only for elevating the laboring and poorer classes of society; but, for creating in all classes of society the finest embellishments of human character. It was originally intended to be merely a work of charity; an offering from the rich to the poor; but, in its development, it has illustrated the truth of the Divine aphorism, that the giver is more blessed than the receiver. Like all charities, it has made large the heart from which it proceeds. Giving to the poor calls into action those moral powers which make man better than he was before, and inspire him with a pleasure surpassing that of the one which participates in his bounty. The human bosom which does not glow with the inspirations of love and beneficence, is like a stagnant pool, whose waters are dark and turgid, and whose evaporations impart no new exhilaration or vitality to the realms of human existence. On the other hand, the human bosom which does continually glow with the inspirations of love and beneficence, is like the living spring at the mountain's base, whose waters are pure and transparent as crystal, and whose fertilizing powers impart renovating beauty, health, and vigor to every thing in contact with them.

It has been thus especially with Sunday Schools. They have purified the moral atmosphere of Protestant Christendom; they have enlarged the domains of Christ's visible kingdom; they have elevated the standard of the Christian character generally; and, to those in a particular manner who have been the agents in conducting them, there has been achieved a refinement and purity of Christian faith and morals which form a new era in the Gospel economy. Kind and unwearied as has been the labor of the Sunday School teacher; glorious as has been the regeneration in those for whom he has toiled; it is a truth, bold and effulgent in the annals of that labor, that the moral life which he caused in them to take root and to rise up in full perfection, has reflected back upon him the brilliancy of its own imperishable nature. The Sunday School teacher, by this very labor, has been advanced in the virtues of his vocation, from one degree of perfection to another, beyond what would ever have been vouchsafed to him, had he not performed it. To him it has been like physical labor, which always invigorates the constitution; every muscular effort creating strength for a new one; and not only that, but each additional one, being with augmented power.

The Sunday School teacher has sometimes appeared to us like the young man on a farm, who, near the door of his paternal residence, plants an acorn, or the kernel of a delicious peach or apricot. Anon the germ rises above the surface of the soil. He watches it day after day, furnishing it with all needful care and vegetable aliment. Its growth becomes rapid; its trunk is gradually rising upward, so that in a few years, he is able to sit under its long spreading branches, to be fanned by its green foliage, and at last to luxuriate on its rich fruit. The Sunday School has collected thousands and tens of thousands from the lower departments of society to become the brightest gems in the Church. After a due course of training, they take their places in the front rank of that army which is to struggle with the battlements of sin and unbelief. Some of their early teachers, when engaged in that pious service, were stimulated to leave a secular occupation and to devote themselves to the Christian ministry. Seemingly, before it could be realized, the young man is upon the watch-towers of Zion, and his former Sunday morning pupils become his life-guard, his standard bearers, and his main dependance, while marching onward to spiritual conquests. And not less important, some of the female teachers, and it may be successively in turn, some of the female pupils become mothers in the spiritual Israel. By what a moral force is such a phalanx bound together! With what unwavering fidelity do they advance to triumph and to victory!

With such an estimate on the agency of the Sunday School

in the cause of good morals and religion, we have from early life been an attentive observer of all its progressive stages. We have admired the young mechanic, the young farmer, and the clerk, who, after a week of toil and close application to business, and of course needing rest, rise early on a Sunday morning, spend an hour or two in looking over the lesson in which their pupils are to recite, and then repair to the Church to take charge of their respective classes. We have admired the young women, who, with equal promptitude, at the sound of the Church-going bell, relinquish that ease and quiet to be enjoyed in the serenity of the morning of that holy day, to devote themselves to the instruction of the ignorant poor. What motive can induce them thus to deny themselves self-indulgence, and to engage in a gratuitous labor, not without perplexities and the trying of patience, as well as unwearied assiduity? The motive must be a benevolent one; it must be a desire to do good to the young outcasts from the Christian fire-side; to train, for a life of religion and for heaven, those who have had no one thus to administer to them! To do this, requires a noble-minded, self-sacrificing consecration of one's moral and physical powers to the best of causes. Those who do it can rely on the Divine blessing alone for their reward, and the inward pleasure to be experienced in a labor of Christian love. This reward they will assuredly have; the latter when engaged in the labor, day by day, as the Israelites were furnished with manna in the wilderness; the former, when God in his mercy, shall deem it best for them, if not in this life, in that which is to come!

Nor less have we admired the children composing a Sunday School, with the delightful and balmy breath of the first morning in the week, rise from their beds and prepare themselves to receive the instruction to be given them; as freely given, as the air we breathe, or as the light in which we rejoice. What sight can be more pleasing than to see them emerging from their respective residences—some from a dreary basement—some from a dark pent-up garret—some through a dismal narrow alley from a cheap rear enclosure—and mostly from the abodes of penury. Speedily the side-ways of the street are filled with them. They are indeed, cheaply attired—some in muslin, and some in gingham, but all clean and comely—their eyes betokening an inward satisfaction at the purification and regeneration enjoyed preparatory to their participation in the delightful exercises of the Sunday School ; and their complexions becoming ruddy and fair from contact with the pure atmosphere. Without noise or tumult, they move onward, till reaching the village-green, when they begin to halt, and as it were, involuntarily, to cluster in groups, each cleaving to those of their own class, much the same as in the process of the chemical analysis of a compound substance, where each constituent assimulates with others according to its natural affinity, or as the soldiers of a regiment from a season of relief, at the beat of the drum, spontaneously fall into their own respective platoons and ranks.

Thus, upon this beautiful village-green, and under the several trees that shield it from the rays of the sun, they remain a few minutes for rest and friendly recognition, and it may be to recite to each other their lessons, and then repair with appropriate decorum, to their respective places in the Sunday School room. This is, indeed, but a miniature exhibition of human society ; but, who can deny there is in it moral beauty—who can deny there is rich and wholesome food in it for the meditative mind ? Often have we admired such exhibitions ! There we have seen little Lizzy, the cotton bug, with flaxen hair, in Miss Gildersleve's class ; yet, afterwards become an accomplished lady, being the wife of a rich manufacturer, and the mother of promising children ! There we saw little Ned, and little Tom, and little Dick, the sons of three drunkards, with earnest looks to become wise and good, catching with eagerness, instruction from the lips of their respective teachers ; yet, in less than forty years afterwards, the first became well known as the Hon. Edward Burnap, of our national senate ; the second, as Gen. Thomas Fillmore ; and the third, as Dr. Richard Ackerman, a learned professor in one of our universities ! And there we saw and admired little Jemmy Gilman, with black piercing eyes and full developments for mental vigor, although barefooted and evincing the poverty of a widowed and broken-hearted mother; yet, in the course of time, he was known as the Rev. Dr. Gilman, to whose eloquence crowded churches have been wont to listen with the same intense emotion, that he felt when receiving the words of truth at the village Sunday School.

HISTORICAL SKETCH OF THE HORSE.

The horse is known to most nations as one of the most beautiful, beneficial and submissive animals that live under the dominion of man. In gracefulness of form and dignity of carriage, he is superior to almost every other quadruped; he is lively and high spirited, yet gentle and tractable; keen and ardent in his exertions, yet firm and persevering. The horse is equally qualified for all the various purposes in which man has employed him; he works steadily and patiently in the loaded wagon, or at the plough; becomes as much excited as his master in the race; and appears to rejoice in the chase. Besides his invaluable services when alive, after death his skin is used for various purposes; the hair of the mane and tail is made into mattresses and cloth for chairs and sofas; and his flesh, although rejected by civilized nations, is much used by several rude tribes of the east. Nor is it to be passed over without notice, that the milk of the mare is converted into a spiritous drink of considerable strength by the Calmucs and other Tartars.

The horse, like the other tame animals, was no doubt originally wild, but his domestication happened at so early a period as to leave no record of the event, and it is now impossible to ascertain, with any certainty, in what country he originated. Wild horses, it is true, are found in various parts of the world, but in most cases it is impossible to say whether they are the remains of an ancient stock or are derived from the domesticated animal; though, as respects those found in the American continent, there is no doubt but that they were originally introduced by the Spaniards. Desmaret gives upwards of twenty varieties of the horse, and his catalogue is by no means complete. Arabia is generally claimed as the original native locality of this animal, and as the only source from which he is to be derived in the requisite perfection for the highest improvement of his race. However, it has been said, as an offset to this opinion, that at the commencement of the Christian era the horse did not flourish in Arabia; and, that the perfection he has since attained there is mainly owing to efforts made since the era of Mahomet.

The horse is vastly modified in his form and character by the physical condition of the country in which he is naturalised. If fed in a country of plains and rich herbage, he inclines to become large in his form; and such is the character of the horse upon the plains of Northern Europe, as of Holstein and England, and in portions of our own and other countries abound-in rich herbage. But, in an elevated country, where the herbage is scanty, the size and form of the horse vary with the circumstances in which he is placed. There he becomes small, hardy, and capable of subsisting on the scanty herbage with which the mountains supply him. No contrast between animals of the same species can be greater than that between the horse of the mountains and the horse of the plains. The pony of Norway, or the Highlands of Scotland, as contrasted with the huge horse of rich fens and prairies, presents such extremes of strength and size, that it is difficult to believe that creatures so different can be of the same species. Yet, all this great diversity is produced by difference in the supplies of food, as influenced by the effects of situation. Thus, the horse fed on the arid plains and scanty herbage of warmer countries assumes characters and a form entirely distinct from those of the large and massy animals fed on the rich pastures of temperate countries.

The history of an animal so beneficial and so much admired must necessarily be extremely interesting; yet, little is known of its early existence. Anterior to the flood, especially, it is not known for what purpose the horse was used. Every record of him was swept away by the general inundation, except that the ark of Noah preserved a remnant for the future use of man. That this animal did exist before the flood, the researches of geologists afford abundant proof. There is not a portion of Europe, nor scarcely any part of the globe, from the tropical plains of India to the frozen regions of Siberia—from the northern extremities of our own continent to the very southern point of it, in which the fossil remains of the horse have not been found mingled with the bones of the hippopotamus, the elephant, the rhinoceros, the bear, the tiger, the deer, and various other animals, some of which, like the mastadon, have passed away. In the majority of the cases, the bones are nearly of the same size with those of the common breed of horses at the present day; but in South America the bones of horses of gigantic size have been dug up and preserved.

There is something inexplicable in the early destitution of facts relating to the horse. In the list of presents made to Abraham by Pharaoh of Egypt, there are mentioned sheep, and oxen, and asses, male and female, camels, men servants and maid servants, but the horse is not in the list. The only way that this can be accounted for is on the supposition that this noble animal was not then found in Egypt, or at least, had not been domesticated there. The first allusion to the horse, subsequent to the flood, is a mere incidental one. This was fifteen hundred and ninety years before the birth of Christ, in the time of Isaac, when the son of Gibeon is said to have found mules in the wilderness—the progeny of the ass and the horse—as he fed the asses of that patriarch. The wilderness here referred to was that of Idumea. Whether these were wild horses that inhabited the deserts of that country, or had been subjugated by man, we have no means of deciding. Nearly a century after this, when Jacob departed from Laban, a singular account is given of the number of sheep and goats, and camels and oxen, and asses, he possessed; but no mention is made of the horse. Hence, also, it may be inferred that the horse was either not known or was not used in Canaan at that early period.

The first direct and positive declaration in reference to the horse, in sacred or profane history, is found in the records of the famine in Egypt. It is familiarly known to all, that Joseph having risen to the highest office under the monarch of that country, availed himself of the cheapness of corn during the years of plenty, and had thus accumulated great quantities in the royal granaries, which he afterwards sold to the starving people for money, as long as it lasted, and then for their cattle and horses. But this affords no clue to the purpose for which the horse was used. The presumption is, it being named in connection with the cattle, it might have been used for the same purposes they were used for. From specimens of Egyptian architecture of that date, it appears that chariots, even on state occasions, were drawn there by oxen. A few years after the famine another incident is recorded which casts additional light on the subject. When Jacob lay on his death-bed he called his sons around him, giving them his parting benediction, and prophesied what would be their fate and character. In speaking of Dan, he said, "Dan shall be a serpent by the way, an adder in the path, that biteth the horse's heels, so that his rider shall fall backward." From this it is seen, that the horse was then used for the conveyance of burdens upon his back. Job, too, spoke of the horse and his rider. He also spoke of him as used for the purposes of war; saying, "he hurries on to meet the armed men; he mocketh at fear; he turneth not his back from the sword; that he smelleth the battle afar off, and heareth the thunder of the captains and the shouting; that the quiver rattleth against him, the glittering spear and shield, and yet that he regardeth them not."

At the time the Israelites forsook Egypt, it is said that Pharaoh pursued them with six hundred chosen chariots and all the horses and chariots of Egypt, and all the horsemen, to the Red Sea. The breeding of the horse, and his employment for pleasure and in war, were forbidden to the Israelites. They were commanded to hamstring all that were taken in war. This animal is occasionally mentioned in the early history of the Israelitish commonwealth; yet, no definite duty is assigued to him, and it is said of the monarch of that time, that he shall not multiply horses to himself. The reason for this prohibition is not easily understood. The country was not indeed favorable to their production, yet it would seem there must have been some other reason not made known to us. In the time of Solomon, however, this restriction was so modified, five hundred years from their departure from Egypt, the horse was domesticated among them; and then so rapidly did the animal increase, that he is said to have had a thousand and four hundred chariots, and twelve thousand cavalry, and stabling for forty thousand horses. The greater part of these horses are supposed to have been imported from Egypt.

The Persian horses became celebrated; but until the reign of Cyrus they were few in number and of inferior quality. That monarch, whose life was devoted to the amelioration and happiness of his people, saw how admirably Persia was adapted for the breeding of horses, and how necessary was their introduction to the maintainance of the independence of his country. He therefore devoted himself to the encouragement and improvement of the breed of horses. He granted peculiar privileges to those who possessed a certain number of these animals; so that at length it was deemed ignominious in a Persian to be seen in public, except on horseback. At first the Persians vied with each other in the beauty of their horses, and the splendor of their clothing; and incurred the censure of the historian, that they were more desirous of sitting at their ease, than of approving themselves dexterous and bold horsemen; but under such a monarch as Cyrus, they were soon inspired with a higher ambition, and became the best cavalry of the East. The native Persian horse was so highly prized, that Alexander considered one of them the noblest gift he could bestow; and when the kings of Parthia would propitiate their divinities by the most costly sacrifice, a Persian horse was offered on the altar. It is reported that Bucephalus, the favorite war-horse of Alexander, would permit no one to mount him but his master, and he always knelt down to receive him on his back. Alexander rode him at the battle of Hydaspes, in which the noble steed received his death wound. For once he was disobedient to the commands of his master—he hastened from the heat of the fight; he brought Alexander to a place where he was secure from danger; he knelt for him to alight, and then dropped down dead.

ALEXANDER TAMING BUCEPHALUS.

The following is the story of Alexander taming his horse Bucephalus: "A very spirited horse had been sent to Philip, Alexander's father, when the latter was a boy. This horse was taken out into one of the parks connected with the palace, and the king and many of his courtiers went to see him. The horse pranced about so furiously, that everybody was afraid of him. He seemed perfectly unmanageable. No one was willing to risk his life by mounting such an unruly animal. Philip, instead of being thankful for the present, was inclined to be in ill humor about it. In the meantime, the boy Alexander stood quietly by, watching all the motions of the horse, and seeming to be studying his character. Philip had decided that the horse was useless, and had given orders to have him sent back to Thessaly, where he came from. Alexander did not much like the idea of losing so fine an animal, and begged his father to allow him to mount the horse. Philip at first refused, thinking the risk was too great. But he finally consented after his son had urged him a great while. So Alexander went up to the horse, and took hold of his bridle. He patted him upon the neck, and soothed him with his voice, showing him, at the same time, by his easy and unconcerned manner, that he was not in the least afraid of him. Bucephalus was calmed and subdued by the presence of Alexander. He allowed himself to be caressed. Alexander turned his head in such a direction as to prevent his seeing his own shadow, which had before appeared to frighten him. Then he threw off his cloak, and sprang upon the back of the horse, and let him go as fast as he pleased. The animal flew across the plain, at the top of his speed, while the king and his courtiers looked on, at first with extreme fear, but afterward with the greatest admiration and pleasure. When Bucephalus had got tired of running, he was easily reined in, and Alexander returned to the king, who praised him very highly, and told him that he deserved a larger kingdom than Macedon.

Bucephalus became the favorite horse of Alexander, and was very tractable and docile, though full of life and spirit. He would kneel upon his fore legs, at the command of his master, in order that he might mount more easily. A great many facts are related of the feats of Bucephalus, as a war-horse. He was not willing to have any one ride him but Alexander. When the horse died, his master mourned for him a great deal."

It is in some of the Grecian sculptures, that we first see the bit in the horse's mouth, but it is not always that we do see it; on the contrary, there is frequently neither bridle, saddle, nor stirrup. It was, however, frequently necessary to make use of cords or thongs, in order to confine the horse to the place at which it suited the rider for a while to leave him. These cords were fastened round the animal's neck, and may be seen in several of the ancient figures. According to some writers, the occasional struggles of the animal to escape from these trammels, and the strength which he exerted in order to accomplish his purpose, first suggested the idea of harnessing him to certain machines, for the purposes of drawing them ; and it is evident, that soon after this, it must have occurred to the horsemen, that if this rope were put over the head, and over the muzzle, or perhaps into the mouth of the animal, he would be more easily fastened, or led from place to place, and more securely guided and managed, whether the man was off or on his back. Hence arose the bridle. It was probably at first nothing more than the halter or cord by which the horse was usually confined. An improvement on this was a detached cord or rope, with prolongations coming up on both sides of the mouth, and giving the rider much greater power over the animal ; and after that, for the sake of cleanliness, and to prevent the wear and tear of the rope, and also giving yet more command over the animal, an iron bit was fitted to the mouth, and rested on the tongue, and the bridle was attached to each end of it. This was the common snaffle bridle of the present day, the iron being jointed and flexible.

In the records of the horse at that period, no mention is made of saddles, such as are used in modern times; but by way of ornament, and partly for convenience too, the horses were covered with beautiful cloths, or with the skins of wild beasts, secured by a girth or surcingle. Thus the horse of Parthenopius was covered with the skin of a lynx, and that of Æneas with a lion's skin. In their religious or triumphal processions the housings of the horses were particularly magnificent, being frequently adorned with gold and silver and diamonds. Rich collars were also hung round their necks, and bells adorned their crests. The trappings of the young knight in the days of chivalry did not exceed those of the Grecian warrior on days of ceremony. The stirrup was long unknown. The adoption of that convenient assistance in mounting the horse was of singularly late date. We have no evidence that it was in use for more than a thousand years after the commencement of the

Christian era. The heroes of ancient times trusted chiefly to their own agility in leaping on their horses' backs, and that whether standing on the right or the left. They who fought on horseback with the spear or lance had a projection on the spear, or sometimes a loop of cord, about two feet from the bottom of it, which served at once for a firmer grasp of the weapon, and a step on which the right or the left foot might be placed, according to the side on which the warrior intended to mount, and from which he could easily vault on his courser's back. And the horse was sometimes taught to assist the rider by bending his neck or kneeling down.

There was a peculiarity in the Greek mode of riding, at least with regard to the cavalry horses, and, sometimes, those used for pleasure. Two or three of them were tied together by their bridles, and the horseman, at full speed, leaped from one to another at his pleasure. This might occasionally be useful; when one horse was tired or wounded, the warrior might leap upon another; but he would be so hampered by the management of all of them, and the attention he was compelled to pay to them all, that it never became the general way of riding or fighting; nor was it practised in any other country. The Greeks must have carried their management of the horse to a very high state of perfection; and the Grecian horse must have been exceedingly docile, when exhibitions of this kind could take place. It was, however, to the draught of the chariot that this animal was principally devoted in some other countries, and among the Greeks in the earlier periods of their history. No mention is made of a single horseman on either side, during the ten years' siege of Troy; but the warriors all fought on foot or in chariots.

The chariots were simple in their structure, open at the back, and partly on the sides; and containing the driver in the front, and the warrior standing on a platform, usually somewhat elevated. These vehicles seem rarely to have been brought into collision with each other; but they were driven rapidly over the field, the warrior hurling his lances on either side, or alight-ing when he met with a foe worthy of his attack. These chariots were not only contrived for service, but were often most splendidly and expensively ornamented. They were the prize of the conqueror. Sometimes they were drawn by three horses; but the third was a spare one, in case either of the others should be tired or wounded. Some had four horses yoked abreast—such was the chariot of Hector. The charioteer, although at the time inferior to, or under the command of the warrior, was seldom or never a menial. He was often the intimate friend of the warrior; thus Nestor, and even Hector, are found among charioteers. When not the personal friend of the warrior, he was usually a charioteer by profession; and drove where he was directed.

Occasional reference is made in the histories of that time to chariots with armed instruments in the form of scythes, projecting from the axles of the wheels, by means of which whole ranks might be mown down at once. They were confined, however, to the more barbarous nations, and were used neither by the Greeks nor the Romans. They were advantageous only on tolerably open and level ground; and it not unfrequently happened that, affrighted by the clamor of the battle, or by wounds, the horses became ungovernable, and, turning on the ranks of their friends, threw them into complete disorder. They were on this account laid aside, even by the barbarians themselves; and, in process of time, war-chariots of every kind fell into disuse, and the higher classes of warriors were content to fight on horseback, where their personal strength and courage might be as well displayed, and discipline could be better preserved.

Almost to the period of the Christian era, and long after that in many countries, the use of the horse was confined to war, to the chase, and to public pageants. The first employment of the Egyptian colonists, when they landed in Thessaly, was to rid the forests of the wild cattle, and other dangerous animals, with which they were then peopled. In the central and southern parts of Greece, the country was more open, and the wilder animals were scarcely known; but in Assyria and Persia, and in every country in which the legitimate prey of the hunter was found, the horse was employed in the pursuit. And in process of time, in order to decide the comparative value of different horses, or to gratify the vanity of their owners, and to give more effect to certain religious rites and public spectacles, horse races were introduced. The most celebrated of these exhibitions was that of Olympia, in Peloponnesus, held every fourth year, in honor of Jupiter. The young men flocked thither, from every district of Greece, to contend in every manly exercise—hurling the javelin, leaping, running, wrestling and boxing. The candidates were persons of unblemished reputation—the contest fairly and honorably conducted, and the conqueror crowned with laurel, or with gold, was received in his native town with acclamations of joy. Nearly a century, however, passed before the attraction of the exhibition was increased by the labors of the horse. Connected with these labors were numerous formularies, some of which to us appear frivolous, and others barbarous.

The Romans, from the very building of their city, paid much attention to the breeding and management of the horse; but this was more than seven hundred years after this animal had been imported into Greece, and his value and importance had begun to be almost universally acknowledged. Horse and chariot races were early introduced into Rome. The chariot races fell gradually into disrepute, but the horse races were continued to the time of the Cæsars, and the young men of the equestrian order were enthusiastically devoted to this exercise. There were not, however, any of the difficulties or dangers that attended the Grecian races. They were chiefly trials of speed, or of dexterity, in the performance of certain circles, now properly confined to theatrical exhibitions. The rider would stand upright on his steed, lie along his back, pick up things from the ground at full speed, and leap from horse to horse in the swiftest gallop.

The Arabian horse is justly celebrated for his swiftness and agility, and in the power of endurance he has no superior; but his most distinguishing characteristics are sagacity and docility. He is the companion of his master, and fond associate of his children, whom he allows to play around, and play upon him, like a good-natured house dog; and he moves round in the tent among them with great caution. The Arab deserves commendation for the tender care and affectionate regard for his horse, as in his kind treatment he sets a noble example, worthy of the imitation of the civilized world. Instances have been known of this horse travelling one hundred and twenty miles with hardly a respite; running four hundred miles in the space of four consecutive days; and five hundred and twenty miles in six days, and then after resting three days, returning the same distance in five days. The swiftness of these animals renders the idea of catching them, when running wild, by chasing them with dogs or otherwise, wholly impracticable. Consequently, when a wild Arab was to be taken, he was to be entangled by snares, carefully concealed in the sand The poor creature being so terrified, frequently falls to the ground and is easily secured. The wild Arabs are now nearly extinct, the high prices paid for them having induced the natives to have nearly exhausted their existence in the desert.

Arab horses are of moderate size; usually between thirteen and fourteen hands in height. Their common color is dappled grey, and sometimes dark brown, with short and black mane and tail. To the wandering Arab this animal is of indescribable importance. It has been supposed that the great powers of endurance in these horses is owing to their having been bred in the arid deserts of that country, and consequently, to the hardships there experienced when colts. Mr. Youatt says this is an error; that the Arabs select for their breeding places some of those delightful spots, known only in countries like these, where, though all around may be barren and dry there is pasture unrivalled for its succulence, and its nutritious or aromatic properties. The powers of the young animal are afterwards developed, as they alone could be, by the mingled influence of plentiful and healthy food, and sufficient, but not severe, or cruel labor, except for making trial of their powers. The Arab habitually washes the legs, tail, and nostrils of his steed every morning and evening, and again after a long ride. The mane and tail are left in their natural state, not often combed, lest they should be thinned, which would diminish their supposed beauty.

ARAB HORSE.

Arabs give preference to mares rather than horses; finding them better able to endure fatigue and privations consequent upon long journeys over the desert. Besides, a large number of them can be kept together without danger of their quarrelling. On this account it is difficult to induce one ever to sell a mare, on any terms; and there is a law to prohibit the exportation of one from the country. Instances have been known that these mares have been sold for two thousand pounds each, and horses for half that sum. It would be difficult to imagine a more interesting spectacle than that of the tent of a wandering Arab at night in the desert. After a day of wearisome toil and privation, all are here collected together for rest and refreshment.

The proud courser is stretched out upon the ground, and some half dozen little dark-skinned, naked urchins,may be seen scrambling across her body, or reclining in sleep, some upon her neck, some on her carcass, and others pillowed upon her heels; nor do these children ever experience injury from their gentle playmate; she recognises them as the family of her friend and patron; and towards them all is manifested the most exemplary kindness of disposition. A nursing mother of our own species could not be more careful of her own offspring than is the Arabian mare of the children of her owner.

The following anecdote illustrates the sagacity of the Arabian horse, as well as his ardent affection for his master. An Arab chief, with his tribe, had attacked, in the night, a caravan of Damas, and plundered it. When loaded with their spoil, however, the robbers were overtaken on their return by some horsemen of the pacha of Acre, who killed several, and bound the remainder with cords. In this state of bondage they brought one of the prisoners, named Abou el Marck, to Acre, and laid him, bound hand and foot, wounded as he was, at the entrance to their tent, as they slept during the night. Kept awake by the pain of his wounds, the Arab heard his horse neigh at a little distance, and being desirous to see, for the last time, the companion of his life, he dragged himself, bound as he was, to the horse, which was picketed at a little distance. "Poor friend," says he, "what will you do among the Turks? You will be shut up under the roof of a khan, with the horses of a pacha or aga; no longer will the women of the tent bring you barley, camel's milk, or dourra, in the hollow of their hand; no longer will you gallop free as the wind of Egypt in the desert; no longer will you cleave with your bosom the waters of Jordan, which cool your sides, as pure as the foam of your lips. If I am to be a slave, at least may you go free. Go—return to your tent, which you know so well; tell my wife that Abou el Marck will return no more; but put your head into the folds of the tent, and lick the hands of my beloved children." With these words, as his hands were tied, he undid, with his teeth, the fetters which held the courser bound, and set him at liberty; but the noble animal on recovering his freedom, instead of bounding to the desert, bent his head over his master, and, seeing him in fetters and on the ground, took his clothes gently in his teeth, lifted him up, and set off at full speed for home. Without ever resting, he made straight for the distant, but well-known, tent in the monntains of Arabia. He arrived there in safety, and laid his master safe down at the feet of his wife and children, and immediately fell down dead with fatigue. The whole tribe mourned him; the poets celebrated his fidelity; and his name is still constantly in the mouths of the Arabs of Jericho.

Arabian colts are usually weaned when a month old, and are then fed on camel's milk for an hundred days. At the expiration of that period, a little wheat is allowed; and by degrees that quantity is increased, the milk continuing to be their principal food. This mode of feeding continues another hundred days, when the foal is permitted to graze in the neighborhood of the tent. Barley is also given; and to this some camel's milk is added in the evening, if the Arab can afford it. By these means the Arab horse becomes as decidedly characterized for his docility and good temper, as for his speed. The kindness with which he is treated from the time of his being foaled, gives him an affection for his master, a desire to please, a kind of pride in exerting every energy in obedience to his commands, and, consequently, an apparent sagacity which is seldom found in other breeds. In that delightful book, Bishop Heber's Narrative of a journey through the Upper Provinces of India, the following interesting character is given of him. "My morning rides," says the bishop, "are very pleasant. My horse is a nice, quiet, good tempered little Arab, who is so fearless, that he goes without starting close to an elephant, and so gentle and docile that he eats bread out of my hand, and has almost as much attachment and coaxing ways as a dog. This seems the general character oi the Arab horses. It is not the fiery dashing animal I had supposed, but with more rationality about him, and more apparent confidence in his rider, than the majority of English horses."

When the Arab falls from his mare, and is unable to rise, she will immediately stand still, and neigh, until assistance arrives. If he lies down to sleep, as fatigue sometimes compels him, in the midst of the desert, she stands watchful over him, and neighs and rouses him, if either man or beast approaches. The Arab horses are taught to rest occasionally in a standing position; and a great many of them never lie down. The Arab loves his horse as truly, and as much as the horse loves him; and no little portion of his time is often spent in talking to him and caressing him. An old Arab had a valuable mare that had carried him for fifteen years in many a rapid weary march, and many a hard-fought battle; at length, eighty years old, and unable longer to ride her, he gave her, and a scimitar that had been his father's, to his eldest son, and told him to appreciate their value, and never lie down to rest, until he had rubbed them both as bright as a mirror. In the first skirmish in which the young man was engaged, he was killed, and the mare fell into the hands of the enemy. When the news reached the old man, he exclaimed, that life was no longer worth preserving, for he had lost both his son and his mare, and he grieved as much for one as for the other. He immediately sickened, and soon aftewards died.

The wild horse is found in various parts of Tartary; but no where can it be considered as the remant of an original race, that has never been domesticated. The horses of the Ukraine, and those of South America, are equally the descendants of those that had escaped from the slavery of man. The origin of the horses of Tartary has been clearly traced to those that were employed in the siege of Azof, in 1657. Being suffered from want of forage, to penetrate into the desert in order to find subsistence, they strayed to too great a distance to be pursued or recalled, and became wild, and created a new breed. They are generally of a red color, with black stripe along the back. They are divided into numerous herds, at the head of which is an old stallion, who has fought his way to the crown, and whose preeminence is acknowledged by the rest. On the approach of apparent danger, the mares and their foals are driven into a close body, in front of which the males are ranged. There are frequent contests between the different herds. The domesticated horse, if he falls in their way unprotected by his master, is instantly attacked, and speedily destroyed; but at the sight of a human being, and especially mounted, they all take flight, and gallop into the recesses of the desert. The young stallions, as they grow up, are driven from the herd, and are seen straggling about at a distance, until they are strong enough to form herds of wild mares for themselves.

The Cossacks are accustomed to hunt the wild horses, partly to keep up their own stock, and partly for food. A species of vulture is sometimes made use of in this affair. The bird pounces upon the poor animal, and fastens itself on his head, or neck, fluttering his wings, and perplexing and half blinding him, so that he becomes an easy prey to the Tartar. The young horses are generally tamed without much difficulty; they are, after a little while, coupled with a tame horse, and grow gentle and obedient. The wild horses thus reclaimed are usually found to be stronger and more serviceable, than any which can be bred at home. The flesh of the wild horse is a frequent article of food, as already remarked, among the Tartars; and, although, they do not, like the Indians of the Pampas, eat it raw, their mode of cookery would not be very inviting to persons of an epicurean taste. They cut the muscular parts into slices, and place them under their saddles, and after they have galloped thirty or forty miles, the meat becomes tender and sodden, and fit for their table. At all their feasts, the first, and last, and most favorite dish, is a horse's head, unless they have a roasted foal, which is the greatest delicacy that can be procured by them. Some of the Tartar and Kalmuck women ride fully as well as the men. When a courtship is taking place between two of the different sexes, the answer of the gentler sex is thus obtained. She is mounted on one of the best horses, and off she gallops at full speed. Her lover pursues, and if he overtake her, she becomes his wife; but it seldom or never happens that she is caught, unless she have a partiality for the pursuer.

In South America, although constant war is carried on against them, there are innumerable herds of wild horses. All persons who have crossed the plains extending from the shores of La Plata to Patagonia, have spoken of the numerous droves of them there met. Some affirm that they have seen ten thousand in one troop. They appear to be under the command of a leader, the strongest and boldest of the herd, and whom they implicitly obey. A secret instinct teaches them that their safety consists in their union, and in a principle of subordination. The lion, the tiger, and the leopard are their principal enemies. At some signal, intelligible to them all, they either close into a dense mass and trample their enemy to death, or, placing the mares and foals in the centre, much like the Tartar wild horses, they form themselves into a circle and welcome him with their heels. In the attack, their leader is first to face the danger, and when prûdence demands a retreat, they follow his rapid flight. In the thinly inhabited parts of South America it is dangerous to fall in with any of these troops. The wild horses approach as near as they dare; they call to the loaded horse with the greatest eagerness, and if the rider is not on the alert, and has not a considerable strength of arm and sharpness of spur, his beast will divest himself of his burden, take to his heels, and be gone forever.

Lord Byron, in his Mazeppa, beautifully describes one of these troops of the wild horse. It is too graphic not to interest and delight the reader. We give the following extract.

A trampling troop—I see them come,
In vast squadron they advance!
I strove to cry--my lips were dumb.
The steeds rush on in plunging pride,
But where are they the reins who guide?
A thousand and none to ride!
With flowing tail and flying mane,
Wide nostrils—never stretched by pain—
Mouths bloodless to the bit or rein,

And feet that iron never shod,
And flanks unscarr'd by spur or rod—
A thousand horse, the wild, the free,
Like waves that follow o'er the sea.
On came the troop . . . .
They stop—they start—they snuff the air,
Gallop a moment here and there
Approach, retire, wheel round and round,
Then plunging back with sudden bound;
They snort, they foam, neigh, swerve aside,
And backward to the forest fly.

WILD HORSES.

The following is a description of catching wild horses on the prairies of Texas, by the Spaniards, who call these animals mustangs. The pursuer provides himself with a strong noosed cord, made of twisted strips of green hide, which, thus prepared, is called a lazo, the Spanish word for a band or bond. He mounts a fleet horse, and fastens one end of his lazo to the animal, coils it in his left hand, leaving the extending noose to flourish in the air over his head. Selecting his game, he gives it chase; and as soon as he approaches the animal he intends to seize, he takes the first opportunity to whirl the lazo over his head, and immediately checks his own charger. The noose instantly contracts around the neck of the fugitive mustang, and the creature is thrown violently down, sometimes unable to move, and generally for the moment deprived of breath. This violent method of arrest frequently injures the poor animal, and sometimes even kills him. If he escapes, however, with his life, he becomes of great service to his master, always remembering with great respect the rude instrument of his capture, and ever after yielding immediately whenever he feels the lazo upon his neck.

Being thus secured, the lazoed horse is blindfolded; terrible lever, jaw-breaking bits are put into his mouth, and he is mounted by a rider armed with most barbarous spurs. If the animal runs, he is spurred on to the top of his speed until he tumbles down with exhaustion. Then he is turned about and spurred back again; and if he is found able to run back to the point whence he started, he is credited with having bottom enough to make a good horse; otherwise he is turned off as of little or no value. This process of breaking the mustangs to the bridle is a brutal one, and the poor animals often carry the evidence of it as long as they live. After service during the day they are hoppled by fastening their fore-legs together with a cord, and turned out to feed. To fasten them to one spot in the midst of a prairie, where neither tree, nor shrub, nor rock, is to be found, is quite a problem. But that is accomplished by putting on a halter, tying a knot at the end, digging a hole about a foot deep in the ground, thrusting in the knot, and pressing the earth down around it. As the horse pulls generally in a horizontal direction, he is unable to draw it out. When a number are caught, they are generally driven to market, where they are purchased for three or four dollars, branded, hoppled, then turned out to take care of themselves till needed. At some future time they may become valuable articles of export.

It has long been an impression that the ordinary duration of a horse's life is much shorter than it ought to be, and that the excesss of mortality is the result of carelessness or ignorant management. The great error consists in regard to the temperament and general constitution of a horse, as altogether different from those of a human being; whereas, they are much the same in all important respects. Disease, arising from excessive fatigue, overheating, and exposure to the air, want of exercise, improper diet, both as respects quality and quantity, and from many other causes, affects the horse and his master alike, and neglect, in either case, must terminate fatally. Indeed, when a man or a horse has acquired, by a course of training, a high degree of health and vigor, the skin of each is an infallible index of the fact. It has often been remarked in England, that the skin of the pugilist, who has undergone a severe course of training, when he prepares himself for the fight, exhibits a degree of beauty and exceeding fairness that excites the admiration as well as the wonder of the spectator. So with the horse—his skin is the clearest evidence of the general state of his health. Even the common disease of foundering is not peculiar to the horse, but is merely a muscular affection, to which many men, who have overstrained themselves at any period, are subject. In fact, the medical treatment of the horse and his rider ought to be the same; and we confidently believe that if this principle were acted upon with a moderate share of attention and resolution, the average age of this animal would be much longer, and the profit derived from his labors proportionably greater.

Many anecdotes have been told of the memory and the sagacity of the horse. A most interesting volume might be made of anecdotes of farm animals, that would be amusing and would furnish an excellent moral. We here give the following one; the facts furnished by a venerable and intelligent gentleman, who lived in one of our cities on the Eastern seaboard, illustrative of the character of these two traits in the character of the animal named.

At the close of the Revolutionary War, when every thing

was unsettled and in disorder, an acquaintance residing on the Boston road, some thirty or forty miles from New-York, lost a valuable young horse, stolen from his stable in the night. Great search and inquiry were made for him, but no tidings of him could be heard, and no trace of him could ever be discovered. Almost six full years had now elapsed, and the recollection even, of the lost animal, had nearly faded from the mind. At this period a gentleman from the East, in the course of business, was travelling on horseback on this road, on his way to Philadelphia. When within four or five miles of a village on the road, the traveller was overtaken by a respectable-looking gentleman on horseback, a resident of the village, returning home from a short business ride. Riding along side by side, they soon engaged in pleasant desultory conversation. The gentleman was immediately struck with the appearance of the traveller's horse. And every glance of the eye cast towards him, seemed to excite an interest and curiosity to look at him again, and to revive a recollection of something he had seen before: and soon established in his mind the impression, that for all the world he looked like the horse he had lost some six years before. This soon became so irresistably fixed in his mind, that he remarked to the traveller,

"You have a fine horse, sir."

"Yes," he replied, "an exceedingly valuable and excellent animal."

"What is his age, sir?"

"Well, I suppose him to be about ten or eleven years old."

"You did not raise him then?"

"No, I purchased him of a stranger, a traveller, nearly six years since."

"Do you reside in this part of the country?"

"No, I reside in the Bay State, and am on my way to Philadelphia, on business. How far is it now to New-York?"

"Well, sir, I really regret to interrupt you, or put you to inconvenience—but I am constrained to say, I believe you have in possession a horse that I must claim."

The traveller looked with surprise and amazement, and replied,

"What do you mean, sir?"

"I believe the horse you are on, in truth, belongs to me Five years ago, the past Autumn, a valuable young horse was stolen from my stable. Great search was made for, but no tidings of him ever came to hand. In color, appearance, and movements, it seems to me he was the exact counterpart of the horse you are on. It would be hardly possible, I think, for two to be so near alike. But my horse was an uncommonly intelligent, sagacious animal. And I will make a proposition to you, that will place the matter in such a position, that the result will be conclusive and satisfactory, I think, to both of us. We are now within a mile of my residence, which is on the road, in the centre of the village before us. When we arrive at my house, your horse shall be tied to the East post in front of my door. The horse I am on, to the West post. After standing a short time, the bridle of your horse shall be taken off—and if he does not go to a pair of bars on the West side of the house, and pass over and go round to the East side of the barn, and pull out a pin, and open the middle stable door and enter, I will not claim him. If he does, I will furnish you conclusive evidence that he was bred by me, but never sold—that he was stolen from me just at the conclusion of the war, about the very time you say you purchased him."

The traveller assented to the trial. The horse was hitched to the post as proposed—stood a few minutes—the bridle was then taken off—he raised his head, pricked up his ears—looked up the street, then down the street, several times—then deliberately and slowly walked past the house and over the bars, and to the stable door as described, and with his teeth and lip drew out the pin, and opened the door, and entered into his old stall. We hardly need to add, he was recognised by the neighbors, who fully attested to the facts stated by the claimant, and that the traveller lost his title to the horse.

There is, says Mr. Youatt, much uncertainty with regard to the origin of what is called the thorough bred English race horse. By some it is traced through both sire and dam to Eastern parentage ; others believe it to be a native horse, improved and perfected by judicious cross with the Barb, the Turk, or the Arabian. But if the pedigree of the present English racer be required, it is traced back to a certain extent, and ends with a well known animal, The Flying Childers. What might have been the particular excellence of the English horse at the time of the Norman invasion, is not known, but it is certain that the

FLYING CHILDERS.

Saxon cavalry, under Harold, were speedily overpowered by William, at the battle of Hastings, which at once secured the throne to the conqueror. History first informs us of the improvement in British horses by importations from abroad during this reign, which consisted of a number of Spanish stallions. These were supposed to be strongly imbibed with the Arabian blood, which had been brought over to that country by the Moors, who had founded the empire of the Saracens in the Peninsula, three centuries before. Subsequently, importations were made from Flanders ; and the improvement of their various breeds was still pursued. with more or less judgment and zeal, by other British monarchs, till they reached their highest grade of merit during the middle of the last century.

Flying Childers, Eclypse, Highflyer, and other race horses familiarly known to persons who have made these animals a study, have probably exceeded in speed any thing before achieved; while the draught-horse, the roadster, the hackney, the cavalry, and the hunter, attained a merit at that time which some judicious authorities claim has not been since increased. It is even asserted, that some of the more serviceable breeds have been seriously injured by too great an infusion of foreign blood; while the almost universal absense of long heats on the turf has tended to the improvement of speed rather than bottom in the race horse. Mr. Youatt remarks that the racer is generally distinguished by his beautiful Arabian head; tapering and finely-set-on neck; oblique, lengthened shoulders; well-bent hinder legs; ample muscular quarters; flat legs, rather short from the knee downward, although not always so deep as it should be; and his long and elastic pastern.

The first Flying Childers was the descendant of an Arabian horse, and was owned by the Duke of Devonshire; and there was another called Bartlett's Flying Childers, from the same stock; and from these two descended a host of fine animals among the most noted of which was Eclipse. The two Childers took their name from Mr. Childers their breeder, and the one owned by the Duke, was the fleetest horse of his day. He was at first trained as a hunter; but the superior speed and courage which he discovered caused him to be transferred to the turf. Common report affirms that he could run a mile in a minute, although there is no authentic record of the fact. However, it is certain that he ran on one occasion three miles, six furlongs, and ninety-three yards in six minutes and forty seconds, and on another occasion, four miles, one furlong, and one hundred and thirty eight-yards in seven minutes and thirty seconds.

In the year 1775, a horse called Bay Malton, the property of the Marquis of Rockingham, ran four miles in seven minutes and forty-three seconds, which was seven seconds less than the same distance had been run before. Twenty years afterwards there was a beautiful horse, the son of Eclipse, and inheriting a great portion of his speed without his stoutness. He won almost every mile race for which he ran, but never could accomplish a four mile one. It is recorded in the Farmer's Library, that one of the most severely contested races ever known, took place at Carslile, England, in 1761. There were no less than six heats, and two of them dead heats. Each of the six was honestly contested by the winning horse, so that he ran in good earnest twenty-four miles without injury. Eclipse was the grandson of Bartlett's Childers, and when a colt, upon the death of the Duke of Cumberland, his owner, was sold to Mr. Wildman for seventy-five guineas. He sold an interest in him to Colonel O'Kelly, but the price is not known. Soon aftewards O'Kelly becoming assured of the great power of the animal, purchased the remaining interest in him for eleven hundred guineas. This horse soon acquired such a reputation for speed that no one was permitted to run against him. He was never beaten, nor ever paid a forfeit, and he won for his owner more than twenty-five thousand pounds. Afterwards, as a stallion, being taken from the race course, he earned double that amount for his owner; and, what is still more extraordinary, he produced three hundred and thirty-four winning colts, and these netted to their owners more than one hundred and sixty thousand pounds. Among the extraordinary feats in horse speed and horsemanship the following deserve notice. In 1741, a Mr. Wilde, in Ireland, engaged to ride one hundred and twenty-seven miles in nine hours. He performed it in six hours and twenty-one minutes. He employed ten horses, and, allowing for mounting and dismounting, and a moment for refreshment, he rode during six hours at the rate of twenty miles an hour. In 1745, Mr. Thornhill exceeded this; for he rode from Stilton to London and back, and again to London, being two hundred and thirteen miles, in eleven hours and thirty-four minutes. This amounts, after allowing the least possible time for changing horses, to twenty miles an hour for eleven hours, and on the turnpike road and uneven ground. And, in 1762, Mr. Shaftoe, with ten horses, and five of them ridden twice, accomplished fifty miles and a quarter in one hour and forty-nine minutes. In 1763, he won a more extraordinary match. He engaged to procure a person to ride one hundred miles a day for twenty-nine days, having any number of horses not exceeding twenty-nine from which to make his selection. He accomplished it on fourteen horses; but on one day he was compelled to ride a hundred and sixty miles, on account of the tiring of his first horse. However, Mr. Hull's Quibler afforded the most extraordinary instance on record of the stoutness as well as speed of the race horse. In December 1786, he ran twenty-three miles round the flat at Newmarket, in fifty-seven minutes and ten seconds.

While the writer, so frequently quoted or referred to in the present sketch, is favorable to a becoming development of the power of this noble animal, he severely repudiates the cruelty occasionally exercised upon him. Several instances of it have become matters of public record. Of the number he enumerated, the following are extracted:—an American horse, was, in 1829, matched to trot ten miles with a Welsh mare, giving her a minute's start. He completed the distance in thirty minutes, being at the rate of rather more than nineteen miles an hour, and beating the mare by sixty yards. This is a fair matter of speed, at an endurable distance; but when this same horse, some time afterward, was matched to trot thirty-four miles against another, and is distressed, and dies the following night; when two hackneys are matched against each other, from London to York, one hundred and ninety-six miles, and one of them runs one hundred and eighty-two of these miles and dies, and the other accomplishes the dreadful feat in forty hours and thirty-five minutes, being kept for more than half the distance under the influence of wine; when two brutes in human shape match their horses, the one a tall and bony animal, and the other a mere pony, against each other for a distance of sixty-two miles, and both are run to a complete stand still, the one at thirty and the other at eighty yards from the winning point, and, both being still urged on, they drop down and die;—when we peruse records like these, we envy not, says he, the feelings of the owners, if indeed they are not debased below all feeling. Mr. Youatt also refers to an individual, who, in 1827, rode on a small gelding, ninety-five miles in company with the Limerick coach; and to another, on a Galloway, less than fourteen hands high, who started with the Exeter mail, and reached that city a quarter of an hour before the mail, being one hundred and seventy-two miles, at the rate of more than seven miles an hour; as instances of great cruelty. The narrator says he saw this pony, a few months afterwards, strained, ring-boned, and foundered—a lamentable picture of the ingratitude of some human brutes towards a willing and faithful servant.

One of the most remarkable cases of bottom, and even speed, in ordinary labor, known in the horse, is that of a mare raised in Hopkinton, N. H., and brought in the fall of 1832 to Rockport, in Massachusetts, thirty-two miles from Boston, being then between five and six years of age. She is of a deep bay color, about fourteen hands high, and weighs, in her best condition, something less than eight hundred pounds. For nearly nineteen years she belonged to different livery stables in Rockport, and was used as a let horse. The following statements are made by her present owner, B. Haskell, but are confirmed by the former owners, persons of undoubted credibility, so there can be no doubt of their truth. Rare as are such instances, we once knew a small black horse belonging to a livery stable in Pawtucket, R. I., whose performances, according to the best of our recollection, were about equal to those of this mare. And, we have owned a horse, within a few years, reported by the former owner to have travelled one hundred miles in twelve consecutive hours; and, from our own experience with him, did not discredit the representation, but looked upon it as probable. And about thirty years since, we owned another one, after which we travelled in a carriage, seventy-five miles a day, of twelve hours, more than once, without the slightest apparent inconvenience to the animal.

Whenever a horse was wanted for a journey, at Rockport, that required speed and endurance, this mare was invariably selected for the purpose. To carry two men in a chaise, from that place to Boston and back, a tour of seventy miles ; to drag a carry-all, with four persons, to Salem and back, forty-two miles, were common daily performances with her. She was known to have seen Boston and Rockport as many as five times in six weeks, besides her ordinary labor during the intermediate portions of the time. In her prime, she was often hired by a stone contractor, who lived two miles out of the village of Rockport. To his house it was usual to send her in the morning, from whence he would drive her to Charlestown bridge and back—not always alone, and return her to the stable in the evening, making the distance of her travel upwards of seventy-five miles. This contractor assured Mr. Haskell, that he once drove her to Charlestown in three hours and a half, stopping at Salem about half an hour, and returned the same day. This was at the rate of twelve miles an hour. About the same time, she travelled from Ipswich to Rockport, eighteen miles, in an hour and a quarter.

When about eleven years old, she had seen so much hard work, and had so many hard drives, that it was thought she must have been worn out ; and she was accordingly offered for sale at seventy-five dollars, without a purchaser. Subsequently, however, she was sold for one hundred dollars ; and, since that sale, it has cost nearly five hundred dollars to keep a horse in the stable by the side of her. Three horses, valued at upwards of one hundred dollars each, subject to the same service with her, and often favored at her expense, were completely used up. One of them returning from Salem, in company with her at noon in the hottest day of the summer of 1840, the mare having the hardest of it, died within an hour after his arrival, while the mare was in the harness the next day, as if nothing had happened. In January, 1843, she went with a sleigh to Salem in an hour and three-quarters, stopped there one hour and a quarter, and returned in an hour and fifty-five minutes—forty-two miles in three hours and forty minutes. Not long before this period, she carried four persons in a heavy carry-all to Salem, in the forenoon, returned with them late in the evening, and at midnight was harnessed to the same vehicle, and sent back in urgent haste, to return early the next morning with two persons. Thus, in scarcely more than twenty-four hours, dragging that carry-all, with its load, she went eighty-four miles.

DRAUGHT HORSE OF NORMANDY.

Among the best horses of New England, are those called Morgans, in not a few particulars resembling the Arabian horse. They are well known and esteemed for activity, hardiness and docility; are adapted for all work; good in every spot, except for races on the turf. They are lively and spirited, lofty and elegant in their action, carrying themselves gracefully in the harness. These horses have size in proportion to their height; bone clean, sinewy legs, compactness, short strong backs, powerful lungs, strength and endurance. A mixture of the Morgan blood, though small, may be easily known from any other stock in the country. There is a remarkable similarity prevailing in all of this race. They are known by their short lean heads, wide across the face at the eyes; eyes lively and prominent; open and wide in the under jaws, large wind pipe, deep brisket, heavy and round in the body, broad in the back, short limbs in proportion to size, broad quarters, a lively quick action, indomitable spirit, move true and easy in a good round trot, and are fast on a walk. Their color is a dark bay, chesnut, brown, or black, with dark, flowing, wavy mane and tail; their height is about fifteen hands; the head is well up, and they move without a whip. Like the Arabian, they are high mettled, but uniting the playfulness and good humor of the pet lamb with the power and courage of the war-horse, whose neck, in the language of Job, is clothed with thunder, mocking at fear, and rejoicing in his strength. These horses have sometimes been sold as high as two thousand dollars.

An able writer has given the following desirable characteristics in the horse. A very general account, he remarks, only can be given of the proper conformation of the animal, for it varies essentially with the breed, and the destination to which it is designed. There are some points, however, which are valuable in horses of every description. The head should not be disproportionably large, and should be well set on—that is, the lower-jaw bones should be sufficiently far apart to enable the head to form that angle with the neek which gives free motion, and a graceful carriage to it, and prevents its bearing too heavily on the hand. The eye should be large and a little prominent, and the eyelid fine and thin. The ear should be small and erect, and quick in motion. The lop ear indicates dullness or stubbornness; and when it is habitually laid too far back upon the neck, there is frequently a disposition to mischief. The nostril, in every breed should be somewhat expanded; it can hardly be too much so in the racer, the hunter, the roadster, and the coach horse, for this animal breathes only through the nostril, and would be dangerously distressed when much speed is required of him, for the nostril could not dilate to admit and return the air.

The neck should be rather long than short. It then enables the animal to graze with more ease and to throw his weight more forward, whether he is in harness or galloping at the top of his speed. It should be muscular at its base, and gradually become fine as it approaches the head. The withers should be somewhat high in every horse, except, perhaps, that of heavy draught, and it does not harm him, for there is larger surface for the attachment of the muscles of the back, and they act at greater mechanical advantage. A slanting direction of the shoulder gives, also, much mechanical advantage, as well as an easy and pleasant action, and a greater degree of safety. It must not, however, exist in any considerable degree in the horse of draught, and particularly of heavy draught. The chest must be capacious, for it contains the heart and the lungs, the organs on which the speed and endurance of the horse depend. Capacity of chest is indispensible in every horse, but the form of the chest admits of variation. In the wagon horse, the circular chest may be admitted, because he seldom goes at any great speed, and there is, comparatively, little variation in the quantity of air required; but in other horses is often fearful.

The quantity of air expended in a gallop is many times that required in hard work. Here we must have depth of chest, not only as giving more room for the insertion of the muscles, on the action of which the expansion of the chest depends, but a confirmation of the chest which admits of that expansion. That which is somewhat straight may be easily bent into a circle when greater capacity is required; that which is already circular admits of no expansion. The whole seems to be implied in the general rule, that the loins should be broad, the quarters long, the thighs muscular, and the hocks well bent and well under the horse; or, in other words, where there are these developments the other points will exist almost of course. We talk much of the beauty of the human figure, and of its appropriate conformation, to answer particular physical functions; but the principles of beauty in the horse, and of his appropriate conformation to particular physical functions, are equally well defined and well understood by those who study his physiology.

Without the horse, it may be asserted that man could not have reached his present pitch of civilization, nor have been able to overcome the numerous obstacles to comfort and happiness. The want of these animals was one of the principal causes which rendered the aboriginal inhabitants of this country so inferior to their invaders; and the decided superiority of the white over the Indian was owing almost as much to the horse as to the knowledge of firearms. In fact, next to the want of iron, the want of horses is, perhaps, one of the greatest physical obstacles to the advancement of the arts of civilized life. Even with the progress that has been made in the arts—with the facilities now existing through the agency of steam, for traveling, for the conveyance of burdens, and as a substitute generally for animal power, were we henceforth to be deprived of the horse, there would be a blank in the means for human comfort but little realized. Between this animal and such comfort there has become a relation which cannot be destroyed without leaving a chasm in society that cannot again be filled. No wonder, then, that the horse has been and is still so highly prized. We are taught to do it almost by instinct.

In Scotland was a breed of small horses, called Galloways. They were from thirteen to fourteen hands high; of a bright bay, or brown, with black legs, small head and neck, and peculiarly deep and clean legs, Their properties were speed, stoutness, and sure-footedness over a rugged and mountainous country. Some remains of them still exist; but they are now much neglected and are fast degenerating from admixture with inferior breeds. Dr. Anderson says they much resembled a breed of ponies in Iceland and Sweden. He says that one he possessed when a boy, was in elegance of shape a perfect picture; and in disposition was gentle and compliant. It moved almost involuntarily and never seemed to tire. He rode it, as he had occasion, during twenty-five years, and twice in that time one hundred and fifty miles at a stretch, without stopping except to bait, and that for above an hour at a time. It was wont to come in at the last stage with as much ease and alacrity as it had travelled the first. He said that when in its prime it might have travelled sixty miles a day for a twelve-month without

THE PONY.

extraordinary exertion. In 1754 a Mr. Corker had a Galloway that went one hundred miles a day for three successive days, and without the slightest distress. A Galloway of a Mr. Sinclair performed the extraordinary feat of travelling one thousand miles in a thousand hours.

The Shetland pony, called in Scotland *Sheltie*, an inhabitant of the extremest northern Scottish Isles, is a very diminutive animal, sometimes not more than seven or eight hands high and rarely exceeding nine. He is exceedingly beautiful, with a small head, good tempered countenance, a short neck, and shoulders thick and low. These ponies possess immense strength for their size; will fatten upon any thing; and are perfectly docile. One of them, nine hands or three feet high, carried a man forty miles in a day.

PROGRESS OF AGRICULTURE.

Agriculture is the science which explains the means of making the earth produce, in plenty and perfection, those vegetables which are necessary to the subsistence or convenience of man. Its practice demands a considerable knowledge of the relations subsisting between the most important objects of nature. It is eminently conducive to the advantage of those actively engaged in it, by its tendency to promote health, and to cherish in them a manly and ingenuous character; and every improvement made in the art must be considered as of high utility, as it facilitates the subsistence of a greater proportion of rational and moral agents; or, if we suppose the number to be unincreased furnishes them with greater opportunities than could be possessed before, of obtaining that intellectual and moral enjoyment which is the most honorable characteristic of their nature. The strength of nations is in proportion to their skilful cultivation of the soil; and their independence is secured, and their patriotism animated, by obtaining from their native spot all the requisites for easy and vigorous subsistence. And not only to raise vegetables for the use of man, but those animals also which are used for food, is obviously, therefore, part of the occupation of the husbandman; and to assist him in his operations, other animals are to be reared and fed by him, to relieve his labors by their strength and endurance of exertion. In cold, and comparatively infertile climates, the services of these creatures are particularly important, if not absolutely indispensable, and their health and multiplication become, consequently, objects of great and unremitted attention.

Agriculture may be considered the most ancient, and is certainly, no one can deny, the most important of all arts. It even forms the basis of society, and constitutes the grand characteristics between savage and civilized life. In the necessity of cultivating the earth for subsistence, social order commenced. The wandering life of a nation of hunters admits of little or no improvement. Agriculture has the credit of having reclaimed mankind from this hopeless state; by drawing them together in communities, and imposing on them the necessity of a fixed habitation. Hence, the ancient nations, amongst which this act originated, held it in the greatest veneration. The Egyptians considered it as a gift from the gods, and even paid divine honors to the ox on account of his usefulnesss in agricultural labors. The ancient Romans venerated the plough, and in the earliest and purest times of the republic, the greatest praise of an illustrious citizen, was to be called an industrious and skilful husbandman.

It is most unfortunate for the interests of society that the same views are not now more generally cherished on this subject. It is a sad fact that vast numbers of our young men, especially, place a low estimate on agriculture, reckoning it a less honorable vocation than other pursuits, and also, though most toilsome, yet the least likely to yield adequate remuneration. Too many of those born in the country have been led to the belief that fortune, success, and happiness are found only in a city life. Every thing, in their estimation, seems to glorify the crowded marts of trade, or gaudily to paint the pleasures there supposed to bud and blossom. Stories of ready made fortunes, of wonderful success in business, and of remarkably rapid advances in social elevation, serve to feed and increase this diseased appetite; and in the end, too often it happens, that sure prospects and certainties of moderate success are sacrificed for mere golden illusions. They walk through the streets of a commercial emporium, and see its magnificent stores and princely residences; and upon looking at large signs in fantastic letters, said to belong to men of wealth, they fancy that all the stories often told them of some who rose from poverty to wealth; of some who commenced in rags and are now rolling in luxury, are matters of genuine truth. But have these young men heard or read of the history of such as failed in their attempts to become rich? They see the names of the successful ones at the corners of the streets, but where will they find a list of those who sunk in the ocean of trade? The history of such is rarely written.

The wave of oblivion passes over them, and they are soon forgotten by their own associates and friends.

Or, if our young men of the country desire the fame arising from professional or literary distinction, it does not follow that they must abandon the farm in order to attain the object of their aspirations. Have they ever reflected that all or most of those who have risen to the first office in the American Union were bred and reared in the country? How is it that almost every individual—our statesmen, poets, authors and men of science, emerged from the walks of rural life? Such facts plead volumes against the idea that the city alone is the nursery for literary distinctions! Supposing they have opportunity for pursuing agriculture, as it may be pursued by men of liberal endowments, does it occur to them how profitable would be to them a thorough scientific education. Scientific men, men with a knowledge of botany, chemistry, geology, and other sciences appertaining to the cultivation of the soil, will, by the application of what they know, make an acre of ground yield two or three times as much as obtained from ordinary culture. The cultivation of the soil may be made a learned profession. Farmers have leisure time which might always be devoted to study. Books are cheap, and a few dollars annually spent in procuring them, and then a little time in perusing them, would enable agriculturists to reach a most desirable reputation. And, it is believed that the same capital, and talents, and labor applied to farming that is devoted to merchandize will, on an average, lead to better results, so far as remuneration is concerned. In agriculture, well conducted, all may be assured of a good living, and moderate accumulations. In no other occupation do all realize as much. The good old Roman doctrine on this subject is the true doctrine. If one man is more honorable than another, other things being equal, it is the farmer.

In the season for carrying on the processes of tillage—of ploughing, sowing, and harvesting, there is much calculated to lead the mind to the most inquisitive habits of philosophic investigation—there is much to produce that excitement so congenial to active temperaments; and, indeed it is impossible to conceive of any employment more favorable to mental improvement than that of the farmer. And, in winter there is nothing disagreeable in the business of the farm. On the contrary, it is full of interest, inasmuch as the welfare of living animals is presented to the attention more forcibly than during the cultivation of the soil. The well marked individual character of different animals engage the sympathy; and the more so, that animals seem more domesticated under confinement than when at liberty to roam in quest of food and shelter. In the evening, in winter, the hospitality of the social board is peculiarly pleasing at the farmer's fireside. Neighbors interchange visits in that season, when topics of conversation common to all classes of persons are varied by remarks on what is peculiar to scientific agriculture. Thus, the business of summer is brought under the most beneficial review—its programme and its results carefully compared; its errors become detected and its deficiencies made manifest, so that in the coming year, the wise and prudent agriculturist is enabled to return to his labor under the most favorable auspices. And, should society in this way furnish few or no charms, the quieter companion of books, or the severer task of study is always at command, so that if the farmer and his family pass the winter without pleasure and profit, it denotes an inaptitude for the most essential purposes of life, if not of downright mental paralysis.

The early and the high origin of agriculture, with the philosophical and Christian mind, is a subject of deep and impressive interest and cannot fail to create a fondness for rural life, if all other motives were to be unavailing. Allusion has already been made to it; but, a more formal reference to it is proper. We learn from the writings of Moses, in an explicit manner that the cultivation of the ground was the primitive employment of man. The earth no longer yielded her productions spontaneously after the defection to the Divine government in Paradise. It has been cursed with barrenness for Adam's disobedience; and under the new constitution of things could be made to minister to his wants only by patient toil, and careful and assiduous tillage. Adam was therefore removed from the garden of Eden in which he had been placed for sustenance without labor and for a high state of meditative enjoyment, to till the ground; and the presumption is a rational one that he and his immediate descendants were instructed by God himself in the mysteries of this art, for it can scarcely be imagined, that in an art depending upon the application of some of the most curious principles of natural philosophy, unassisted reason would have been sufficient for the purpose.

In the early ages of the world, before mankind had become numerous, and whilst every tribe or family could range over a large extent of country, their principal wealth consisted in flocks and herds, and their chief employment in the care of them. This continues to be the condition of the Nomade nations of Northern Asia to the present day; and under such circumstances agriculture is but little attended to. The Egyptians were undoubtedly the first people, who applied themselves successfully to the cultivation of the earth; and they were invited to it by the extraordinary fertility of their soil, occasioned by the annual overflowings of the Nile. The wealth and power which they acquired from this source, and their extraordinary advances in knowledge and the arts, are fully attested by those wonderful monuments still remaining of their former greatness. The Greeks probably borrowed their agriculture, as they did their arts and early principles of science, from the Egyptians. The Chaldeans and Phœnicians held husbandry in the highest estimation. The Carthagenians descended from the latter, carried it to great perfection. The Romans devoted themselves, as above stated, to agriculture, with extraordinary success, and several of their treatises on the subject are still extant. In fact all the celebrated states of antiquity rivalled each other in promoting and improving this art.

The period of the introduction of agriculture into Great Britain is unknown. Pliny observes that at the time of the Roman invasion, the inhabitants were acquainted with certain manners, particularly rural. During the possession of the island by the Romans, great quantities of grain were exported from it, and it cannot be doubted that, as in various other respects, the rude inhabitants derived advantage from their enlightened conquerors; they were eminently benefited by their agricultural experience. Amidst the series of contests and confusions which followed the final abandonment of that island by the Romans, the art and practice of husbandry, in all probability, must have been retrograde. From the Norman conquest, however, it derived fresh vigor, as a considerable number of Flemish farmers, by that revolution, became proprietors of British estates, and introduced that knowledge of the means of cultivation for which their own country had been long distinguished. Prior to the sixteenth century, few data are furnished with respect to the details of agricultural practice in that country. Subsequently to the commencement of it, several gentlemen of talents published treatises on the subject, among whom were Fitzherbert, Sir Hugh Platt, Gabriel Plattes, Blyth, Hartlib, Evelyn, and Jethro Tull. The last two were of a date somewhat later than the others, and of course are more known to the readers of the present day than the other five. Evelyn published a work on plants, and Tull recommended drill husbandry, which contributed essentially to improve the taste and moral cast of the people in relation to this subject.

Since the successful and ingenious efforts of those gentlemen, a series of valuable experimentalists and writers have performed to their country very essential service, by communicating the most useful information, and exciting a spirit of acute research and unwearied exertion. From a happy combination of circumstances, the exertions of individuals, societies, and government, have been directed within the last three quarters of a century, to the subject under consideration, with more energy and effect than have been displayed in other parts of Europe. The gentry and nobility have liberally patronized, and many of them judiciously and successfully practiced it. The Royal Society, the Society of Arts, and various others, have been of distinguished service in collecting and diffusing information, and in promoting a spirit of emulation, with respect to the management and productions of their native soil. The names of Kaims and Hunter, of Anderson and Marshall, of Sinclair and Young, and of others more recent, are celebrated by publications, exhibiting a union of philosophical sagacity and patient experiment; the results of which have been of incalculable advantage; and to these and other individuals, it may be ascribed that a board of agriculture was established by the government in 1793, whose exertions in procuring and publishing intelligence on the objects of its establishment, have entitled it to the highest credit. By its agricultural surveys, by its diffusion of rewards for important discoveries, and of premiums for valuable treatises, and by its exertions at critical periods of scarcity, its utility and merit may be pronounced not only as decided, but distinguished. It has the power of directing public attention to any topics particularly requiring practical research or illustration, and possesses the means of most advantageously diffusing its collections, circumstances of high importance to the utility of the establishment; and, not less in calling the attention of the national legislature to any obstacles that may exist, or new facilities required in the accomplishment of its original design.

The pursuits of agriculture have always been and still are held in high estimation by the Chinese, who commence the agricultural year with a grand festival in honor of spring. On this occasion the emperor, in imitation of his ancient predecessor, performs the operation of ploughing and sowing seed in a field set apart for that purpose, a custom that has very seldom been neglected by the sovereigns of the Chinese Empire, who have thus by their own example, stimulated their subjects to the performance of these useful and necessary labors, and maintained the honorable position and character of husbandman, who even now holds a rank in society above that of the soldier or merchant, however wealthy the latter may be.

Among the ancients, particularly the Egyptians, Persians, and Grecians, it was a common practice to hold games and festivals, mingled with religious ceremonies, at that season when the earth is ready to receive the seed, thus showing the cheerfulness with which the farmers returned to their rustic toils, and the reliance they placed on the Supreme Being to reward them with an abundant harvest. The old festival of Plough Monday in England, was probably derived from these customs of the ancients, and was formerly celebrated in all of the rural districts with great merry makings; some of the rites observed being not unlike those among the Chinese, as an instance of which the plough-light was set up before the image of some patron saint in the village church; an usage somewhat similar to that observed among the Chinese, who place lighted candles opposite certain images in their temples.

The plough, the harrow, and the hoe, all of the rudest construction, are the chief implements used by the Chinese farmer, the spade being only seen occasionally. The plough is usually drawn by buffaloes, but sometimes that labor is performed by men, and even by women, among the lowest class of farmers. The great object of cultivation is rice, the staple food of all classes, from the prince to the peasant. Most of the plains present an endless succession of rice or paddy fields, which, in the early stages of the crops, exhibit a vast surface of bright green, but turn yellow as the grain ripens. The seed is first sown in small patches, flooded with a particular preparation of liquid manure, which promotes its immediate development, so that in a few days the shoots are five or six inches in height, when they are transplanted to the fields, some of the laborers being employed in taking them up, others in making holes to receive them, and a third party in dropping them into the holes, about six together. All these men stand up to their ankles in water, for it is requisite that the rice should be kept constantly wet, or it would be spoiled; but when the rice is ripe, the fields are drained, so that the reapers, whose labors commence about midsummer, work on dry land.

In France, the political expedience of guarding against that scarcity which, in time of war, either necessitated the yielding to harsh terms from the enemy, or exposed to the miseries and horrors of famine, by continued hostilities, induced the government, when in Great Britian the subject was early agitated, to bestow on agriculture considerable attention, and to hold out numerous encouragements to it. The court was present at various experiments in husbandry. Prize questions were proposed at Lyons, Bourdeaux, and Amiens, for its promotion, and no less than fifteen societies for the express purpose of advancing agriculture were established with the approbation, probably at the suggestions of the governing powers. But, notwithstanding all these efforts, which, however, can by no means be presumed to have been totally useless, French husbandry continued in a very deplorable state, ascribable, it may be supposed, in a great degree to that tenure of lands by which through the greater part of the kingdom the landlord contributed the stock, and the occupier the labor; dividing the profits in certain proportioned shares. This circumstance, with several others, operated to keep the cultivation of that country in an extremely low state. But the revolution of France, changing everything, in part swept away landed usages that were so unpropitious to agricultural improvement. Bonaparte established many new agricultural societies and professorships, botanical and economical gardens, for the exhibition of different modes of culture, and the dissemination of plants. He also enlarged and enriched that extensive institution, the National Garden. The lands of France are not generally enclosed and subdivided by hedges as in England, or fences as in the United States. Some fences occur near towns, but, in general, the whole country is open, the boundaries of estates being marked by slight ditches or ridges, with occasional stones or heaps of earth, trees in rows or thinly scattered. Among the objects of French husbandry sheep are prominent. The vine is cultivated there in fields and on terraced hills, like Indian corn, is kept low, and is managed much like a plantation of raspberries. The white mulberry, to furnish food for the silk worm, is likewise an object of attention. And the olive, the fig, and the almond, with several other kinds of fruit, are extensively cultivated in France.

At a comparatively early date, in the various states of Germany, lectures have been delivered on agriculture, and other means have been used to accelerate its progress. Some of our best books on the subject are from that country. Among the German writers on agricultural science who have become extensively known, and who have done much to advance it in their own country, as well as in others, Thaer and Liebig stand pre-eminent. No one can fully estimate the value of their labors. The impulse received from them, on numerous individuals, if not upon the masses, has been like galvanic action. As little inclined as are the thousands, and the hundreds of thousands, of our farmers for scientific inquiry, the treatises of these gentlemen have passed through successive editions, and they may be found in our best libraries. Germany includes the varieties of soil, surface, climate and culture adapted to a great number of the staple vegetable products; and, although the greater portion of them is consumed within her own limits, by her own citizens, excellent wines are exported from Hungary and the Rhine, together with flax, hams, geese, silks and other articles. The culture of the mulberry, and the rearing of the silkworm, are carried on as far north as Berlin. Theoretical agriculturists are well acquainted with all the improved implements of Great Britain, and some of them have been introduced especially into Holstein, Hanover, and Westphalia. In parts of Germany, particularly in Prussia, fish are carefully bred and fattened; and poultry is every where attended to, particularly in the neighborhood of Vienna. The culture of forests, likewise, receives particular attention in that country as well as in France. The common agriculture of Germany is every where in a progressive state. Government, as well as individuals, have formed institutions for the instruction of youth in its principles. The Imperial Society of Vienna, the Geological Institution of Presburg, and that of Professor Thaer in Prussia, may be numbered among the institutions of this description.

The circumstances that have given tone to American agriculture have been of a peculiar character. On the first settlement of the country, the inhabitants were so few and the land so abundant, that no attempts were made to preserve it in good condition. Everything was to be done, and labor was scarce, and of high price. Not only a subsistence was to be obtained, but houses and barns were to be built, mills were to be constructed, and every thing seemingly necessary in civilized life was to be made in the best and cheapest manner possible. To us, at this distance of time, it seems impossible that they should, as well as they did, have sustained themselves, and obtained for themselves so many of the comforts in domestic economy. True, they were often straitened; to suffer privation, with them, was habitual; and, at best, they were only able to obtain the most common fare, and were taught to be content with a small number of auxiliary helps to ease and adornment. This was the more trying, as the bulk of the first settlers in this country were persons who had never been accustomed to the severities of the lower classes in society; they had belonged to the better classes; had enjoyed plenty, and as many of the elegances and the superfluities of life as were generally known at that day. To plunge, therefore, into the depths of forests, never before penetrated by civilized man; to create for themselves on a desolate shore, separated from their native land and from all friends, by an ocean of four thousand miles in extent, at that time not to be crossed without perilous imaginings, required a moral and physical energy, seldom at this age of the world found requisite. Such was the condition of our first American ancestors.

The land was so abundant, it could be had at pleasure, or at only a nominal valuation. Thus, being able to cultivate as much as they pleased, they had no motive of interest to keep it in good condition, to enrich it, or to render it capable of future fertility, but simply to get out of it all they possibly could, with the least possible labor and expenditure. Accordingly, each one would take and cultivate such a piece of land as would yield for his family a support; this was cultivated year after year till it became exhausted, and would not pay for culture; then it was abandoned and another similar piece substituted in its place, and in the same manner exhausted ; then a third, and a fourth, and a fifth, till whole townships or districts were rendered infertile and were hence abandoned. It is known that a vast portion of the land at first used by settlers on the American continent were thus rendered of little value. With such a course of husbandry, the art of sustaining the soil became, as it were, lost, if it had been known. However, individuals became alarmed at the impending consequences, a spirit of inquiry was called into action, and here and there a resort to science for restoring this worn out land, has been attended with successful results. Agricultural Societies, more or less of them, sprung up in different sections of the country, and by their annual fairs, have already done something, and are now doing much, in placing agriculture in its proper position ; and not only in causing it to be duly respected, but in the restoration afforded by them to thousands of these worn out farms.

Agriculture in our country is on the ascendant ; it is every year becoming more and more an object of general interest ; wherever the influence of agricultural societies is duly felt, crops are in a process of augmentation, the breeds of farm animals are undergoing improvement, and the general business of husbandry is not only viewed with increasing favor, but is evidently receiving a more adequate remuneration. Between the laboring yeomanry and those of other departments of industry, there is less of estrangement, and more of a feeling of good will and common brotherhood than formerly. Our agricultural fairs are becoming the most attractive holidays of the country, and are hence annually bringing together the best talent of the nation. The highest in office deem it no degradation, on these occasions, to deliver addresses, and to mingle congratulations with the working farmer. Where do we see collected more talent than at the fair of the State Agricultural Society of New-York ? Where else is there such a rush of mechanical genius, of rural enterprise, and even of female beauty, for a succession of weeks, as at the fair of the American Institute ? Here, too, the products of the plough, the loom, and the anvil, are displayed in juxtaposition; and the factor and the merchant lend their influence to the joyfulness of the occasion. In the vicinity of Boston, also, on these anniversaries, there is an exhibition of princely hearts—of nature's true nobility—as well as of agricultural products, that would do honor to the best fair in Great Britain.

Among the most encouraging tokens of agricultural progress in our country, is the disposition frequently manifested among merchants to have a country summer residence, where their feelings become assimilated with those of a rural population, and where they can participate in the invigorating influence of a pure air, fresh vegetable products, and the pastimes of rural society. Here their social affinities will soon begin to manifest themselves; here they will begin to make permanent investments; and, here, on relinquishment of business in the city, they will resolve to make a fixed residence, where in peaceful retirement they can finish their earthly pilgrimage. Many professional gentlemen, on finding themselves able to give up professional toil in the city, are here constantly bending their way. And it is a rare occurrence, that an exhausted or spent politician will think of spending the remnant of his life amid the excitements of a dense population. Such an adhesion to influences, which are, even in the mid-day of life, unfavorable to physical energy, would also be in bad taste as well as prejudicial to personal and domestic enjoyment. When the retired merchant, the retired professional man, and the retired politician, thus plant themselves on farms in the country, their step becomes more elastic, their minds more buoyant and cheerful, and there will be a freshness in their old age, to which they would have been strangers amidst the crowds and the impure exhalations of a commercial emporium. And, they will, moreover, by their wealth and intelligence, give a new tone to the social character of country life, and also an impetus to the enterprise and industry which belong there. The benefit between the two classes is mutual. The one is rendered more polished in manners, and more systematic and successful in labor, and the other is rendered more healthful, and complaisant, and useful.

The most recent movement made in this country to advance the interests of agriculture, is the formation of a National Agricultural Society, its central point of action being at the seat of the National Government. It was organized by delegates from the several state societies, with a president and other officers. These officers so far as known to us, are gentlemen of the first intelligence and respectability, and will inspire the fullest confidence in the efficiency of the institution ; and the president especially, a distinguished merchant, horticulturist and farmer, has yet by his great industry, become well known as a good writer and orator, no one to our knowledge having rendered the cause more service by public addresses. Should this institution receive the patronage of the National Government, its sphere of operation will become large, and its utility of the utmost importance to the community. It is matter of inexplicable mystery, that the American government, which emanates from the people, and is of course responsible to the people—a people, moreover, four-fifths at least farmers, has refused or neglected to render any material service to this portion of their constituents. Congress has been accustomed to vote away money as freely as though it were water, for almost any quixotic purpose, where political capital would arise to themselves, or pecuniary advantage to their personal friends ; but if money were asked to found and sustain an agricultural bureau in connection with the government, a deaf ear was turned to the subject. It is hoped that a more auspicious conjunction of circumstances has arrived ; and that the recently organised National Agricultural Society of the American Union, will receive from the national treasury that encouragement it deserves.

The president of the institution having on unanimous vote, been elected to that office, rose and made the following address, so appropriate, that we annex it to the present article :

" *Gentlemen of the Convention :*—I will not interrupt the proceedings of this body by any extended remarks from the chair ; but I cannot forbear to tender to you my heartfelt gratitude for the honor you have conferred upon me in selecting me to preside over your deliberations—an honor which is connected with a pursuit which has ever laid near my heart.

"Permit me also to express my great gratification that there are present so many members representing the agricultural interest of this great republic—some gentlemen coming from different and distant parts of the Union, at great personal sacrifice; but whether from the North or the South, the East or the West, I extend to each of you the hand of fellowship, and I greet you as brothers in a common cause.

"Gentlemen, we come here with no sinister motives; we have no political arguments to advance; we have no sectional or party purposes to promulgate, but we are here for more important purposes. We are here to advance an art coeval with the existence of the human race—an art which employs eighteen millions of our population, and four-fifths of all the capital in our fair land—an art which lies at the very foundation of national and individual prosperity and wealth, the basis of commerce, of manufactures, and of industrial pursuits. We are an agricultural people; our habits, our dispositions are rural. I rejoice that it is so, and I pray that it may ever continue to be so. Our country embraces every variety of soil, and is capable of producing most of the products of the torrid and temperate zones; and with a suitable application of science to this art, and a wise division of labor, with proper governmental aid, there is no reason why American agriculture may not sustain competition with that of any other nation of the civilized globe.

"The progress of agriculture, as you all know, gentlemen, has been slow in the United States, but a new era has now commenced. The old worn-out systems of cultivation which have been followed by father and son, and from generation to generation, are now to be swept away, and science is to take its place in aid of honest industry. I rejoice, gentlemen, that we live at this day; I rejoice that the seed planted by the immortal Washington, and which has been watered by thousands of other eminent agriculturists, is now taking root, and that we live in our day to realize some of the proud results of their hopes.

"Much of the progress which has been attained in our country is the result of individual enterprise, aided by the agricultural press; but the great motive power is confederated action, is associated effort. Gentlemen, we have met on this occasion to avail ourselves of this powerful impetus. At no period in the history of our country has there been such an assembly collected for the purpose of considering those objects for which we are brought together, and there has been no opportunity which is so favorable to the interests of the farmer."

ARTIFICES OF ANIMALS.

Every person who has lived on a farm is familiar with the artifices of domestic and farm animals. With what shrewdness will the old cat watch for an opportunity to steal into the pantry, or an inner cellar, at the heels of some one having occasion to go thither, and to be left alone, that she may have full liberty to lubricate her parched lips and tongue with a plate of fresh butter, or a pan of rich cream—or to put into requisition the gastric juices of the stomach upon the remains of a recent dinner, or beef-steak designed for the coming breakfast. This done, with equal adroitness she watches the first opening of the door to make her egress unnoticed, and then reposes in a retired corner, as if not to be suspected, looking as demure as a saint, and no one suspecting the mischief till surveying the premises of the missing article, and even then oftentimes doubting what has become of them. And with what cunning will this old feline pet of the family feign herself asleep, behind the stove, while the inmates are at their meals, as if the crash of a thunder clap would scarcely disturb her somniferous appearances; yet, when they all retire, and she alone having life, remains in the room, with the quickness of lightning she darts upon the table, regales herself without ceremony, upon the best to be found, and then, frequently, undiscovered, resumes her previously reposing posture ! And the family dog, although possessed of certain kinds of business-like honor and honesty, when left in watch of property, understands quite as well as Mrs. Puss, how to take care of himself, especially if he is a lover of uncooked mutton and lamb, and is kindly disposed to relieve the butcher from some of his official toils.

These are but trivial specimens of their kind of what may be seen upon the premises of the ruralist. There is old Brindle, the very personification of simplicity and artlessness, will stand or lie for hours, chewing her cud, as if all were perfectly safe under her jurisdiction, yet if her powers of self-support have been fully developed, and she is left alone in the midst of all she surveys, it may be relied upon that she will begin to study the code of ways and means. If she have a calf shut up in close pen, and suspects that its owner allowed it but a scanty breakfast, the probability is that with the tip of her horn she will raise the iron hook that fastens the door and allow her darling to partake of an early lunch. Or, if her ladyship espies a tempting field of waving maize, the chance is she will become dissatisfied with the short grass in the highway, and the closely fed pasture, as many a one in human form is dissatisfied with what belongs to him, and is envious of what belongs to others, and with the tip of her horns will remove a rail, that she may luxuriate almost to self-destruction on the rich herbage. And she has been known, when honest people, and honest cows, too, should be asleep, to recollect the fruitful cabbage-plat in the kitchen garden, and also the long wooden peg that confines the gate leading thither; and as soon as the image of such good things, and the means of reaching them is fairly in her mind's eye, she seizes between her teeth the stopper to the gate, opens it, and then gives a demonstration of fondness for this wide-leaved esculent, equal to that of the most voracious lover of sour-krout in Christendom. But, who can help loving old Brindle, if she, twice a day, fills a ten quart pail with rich milk?

The horse, too, has his artifices. He will learn to open a

gate and a door with as much complacency as old Brindle ; to shake a disagreeable rider from his back ; and, if a little malicious, to give the groom a clandestine grip with his teeth, or an involuntary somerset from the unperceived application of his heels. Nor is he at a loss for means to arouse his sleepy attendant, who forgets at the proper hour to furnish him with his regular quantum of provender. The old pet ram, also, has his artifices. They may not always be very scientific. They may not always denote profound premeditation ; yet even he is able, when having the free use of his legs, to go where he pleases, and to determine, as well as his owner, the places where the best feed is to be had, and where his female associates can be reached. And the stupid hog, likewise,—he is often called stupid—is by no means all passive simplicity. If kept to the age of many other animals, and the same pains were taken to train him that are often taken with other animals, it would be found that he is not to be despised for his want of intelligence. We once heard a Commencement Oration on the intellect of swine ; and for its wit, and sarcasm, and fun, acquired more applause, even from the savans of literature than all the other performances. The author of it was a young man of genius and originality, as proved from the fact that he afterwards became an eminent divine. In previous articles, we have alluded to this subject, saying that turkeys in stealing away their nests, and the males of the barnyard gentry, in picking quarrels with their comrades, as well as in divers kinds of mischief, will bear comparison in this respect with bipeds of the larger growth in the animal kingdom.

These artifices of farm and domestic animals are in nature, much the same of what is called cunning and treachery in the human species. The chief difference is in the degree in which they are possessed ; and, there is a difficulty in calling the former instinct and the latter reason. Indeed, there are numerous cases where the intelligence of the brute kingdom approximates so near to what we call reason in man, that we do not well know how to give it any other name. Nor is it consistent to be very harsh in our condemnation of the artifices spoken of, or of kindred ones found among brutes, when many of us practice what is similar upon each other, an hundred times as much. That is, in condemning the stealthiness of the cat, certainly one of the most provoking things in the culinary department of the farmer's house, many persons condemn themselves. Does not the knave swindle his neighbor by false pretences and false appearances? Does he not overreach and defraud him of his property by deceiving him in some form or other—by sign or by word—by trick or ledgerdemain? Does not the assassin spill the blood of his brother, by lying in wait for him when unperceived? And does not the libertine destroy his victim by stealth or treachery?

The study of the artifices of animals generally has ever been peculiarly interesting to us; we know not how it can be otherwise than interesting to the common reader as well as to the philosopher. These artifices are usually attributed by writers on the subject, either to instinct, or to habit acquired from experience, a species of reason. We have a paper before us, prepared many years since, containing examples not so generally known as the ones already named, which must tend to amuse and to enlighten the reader, causing some gleams of pleasure about the fireside on a winter evening. Probably they were obtained from the philosophy of natural history; and the first one given is that of cattle when attacked by a bear or any ravenous animal. They will without hesitation unite and form a phalanx for mutual defence, like a company of soldiers Horses in the same circumstances fall into regular lines and beat off the enemy with their heels. Of this we have spoken in the history of the horse. The Norwegian horses, we are informed by Pontoppidan, when attacked by bears, instead of striking with their hind legs, rear, and, by quick and repeated strokes with their fore feet, either kill the enemy, or oblige him to retire. This curious, and generally successful defence, is frequently performed in the woods, while a traveller is sitting on the horse's back. It has often been remarked, that troops of wild horses, when sleeping either in plains or in the forest, have always one of their number awake, who acts as a sentinel, and gives notice of any approaching danger.

There is a curious illustration of our subject in the monkeys of Brazil, according to the account of Margraaf. It is this. No captain of a fort or a reposing army could better provide for its safety. When sleeping on trees, these monkeys have uniformly a sentinel to warn them of the approach of the tiger or other rapacious animals ; and that, if ever this sentinel is found sleeping, his companions instantly tear him in pieces for his neglect of duty. Could the venerable General Jackson in his best days, have been a better disciplinarian ? For the same purpose, when a troop of monkeys are committing depredations on the fruits of a garden, a sentinel is placed on an eminence, who, when any person appears, makes a certain chattering noise, which the rest understand to be a signal of retreat, and immediately fly off and make their escape. Did the old buccaneers, or do the modern West India pirates, or do our own marauders on ship-wrecked property, manage more adroitly, when committing unlawful depredations? We think not. They may have learnt the use of precautions from the monkeys.

The deer kind are remarkable for the arts they employ in order to deceive dogs. With this view the stag often returns twice or thrice upon his former steps. He endeavors to raise hinds, or younger stags to follow him, and to draw off the dogs from the immediate object of their pursuit. If he succeeds in this attempt, he then flies off with redoubled speed, or springs off at a side, and lies down on his belly to conceal himself. When in this situation, if by any means his foot is discovered by the dogs, they pursue him with more advantage, because he is now considerably fatigued. Their ardor increases in proportion to his feebleness, and the scent becomes stronger as he grows warm. From these circumstances the dogs augment their cries and their speed ; and, though the stag employs more arts of escape than formerly, as his swiftness is diminished, his doublings and artifices become gradually less effectual. No other resource is now left him but to fly from the earth which he treads, and to go into the water, in order to cut off the scent from the dogs, when the huntsmen again put them on the track of his foot. After taking to the water, the stag is so much exhausted, that he is incapable of running farther, and is soon at bay, or in other words, tnrns and defends himself against the hounds. In this situation he often wounds the dogs, and even the huntsmen, by blows with his horns, till one of them cuts his hams to make him fall, and then puts a period to his life.

The fallow-deer is more delicate, and less savage, and approaches nearer to the domestic state than the stag. They associate in herds, which generally keep together. When great numbers are assembled in one park, they commonly form themselves into two distinct troops, which soon become hostile, because they are both ambitious of possessing the same part of the enclosure. Each of these troops has its own chief or leader, who always marches foremost, and he is uniformly the oldest and strongest of the flock. The others follow him; and the whole draw up in order of battle, to force the other troop, who observe the same conduct, from the best pasture. The regularity with which these combats are conducted is most curious. They make regular attacks, fight with courage, and never think themselves vanquished by one check, for the battle is daily renewed, till the weaker are completely defeated, and obliged to remain in the worst pasture. They love elevated and hilly countries. When hunted they run not straight out, like the stag, but double, and endeavor to conceal themselves from the dogs by various artifices, and by substituting other animals in their place. When fatigued and heated, however, they take the water; but never attempt to cross such large rivers as the stag. Thus, between the chase of the fallow-deer and of the stag there is no material difference. Their sagacity and instincts, their shifts and doublings, are the same, only they are more frequently practiced by the fallow-deer.

The roe-deer is inferior to the stag and fallow-deer, both in strength and stature; but he is endowed with more gracefulness, courage, and vivacity. His eyes are more brilliant and animated. His limbs are more nimble ; his movements are quicker, and he bounds with equal vigor and agility. He is likewise more crafty, conceals himself with greater address, and derives superior resources from his instincts. Though he leaves behind him a stronger scent than the stag, which increases the ardor of the dogs, he knows how to evade their pursuit, by the rapidity with which he commences his flight, and by numerous doublings. He delays not his arts of defence till his strength begins to fail him ; for he no sooner perceives that the efforts of a rapid flight have been unsuccessful, than he repeatedly returns upon his former steps ; and, after confounding, by these opposite motions, the direction he has taken, after intermixing the present with the past emanations of his body, he, by a great bound, rises from the earth, and retiring to a side, lies down flat upon his belly. In this immovable situation, he often allows the whole pack of deceived enemies to pass very near him.

The roe-deer differs from the stag in disposition, manners, and in almost every natural habit. Instead of associating in herds, they live in separate families. The two parents and the young go together, and never mingle with strangers. The female commonly produce two fawns, the one a male and the other a female. These young animals, who are brought and nourished together, acquire a mutual affection so strong, that they never depart from each other. In a week or two after birth, the fawns are able to follow their mother. When threatened with danger, she hides them in a close thicket ; and so strong is her affection, that, in order to preserve her offspring from destruction, she allows herself to be chased. It is said by hunters, that the American deer evinces a very strong degree of animosity towards serpents, and especially to the rattle-snake, of which it has an instinctive horror. In order to destroy these creatures, the deer makes a bound into the air, and alights upon the snake with all four of its feet brought together in a square, and these violent blows are repeated till the hated reptile is destroyed ; apparently comprehending the effect of such concentrated force, as well as those who use a battering ram in demolishing the walls of a besieged city.

Hares possess not, like rabbits, the art of digging retreats in the earth. But they neither want instinct sufficient for their preservation, nor sagacity for escaping their enemies. They form seats or nests on the surface of the ground, where they watch, with the most vigilant attention, the approach of any danger. And what is more curious and wonderful than every thing else about them is this, in order to deceive, they conceal themselves between clods of the same color with that of their own hair. Chickens and young turkies of a brown color, resembling the ground, are not so likely to be seen and taken by hawks as those which are of a different color. This, hares understand; or which is the same thing, they know that they are not so likely to be discovered when in juxtaposition with objects of their own color, as when near to those of a different aspect.

The fox has, in all ages and nations been celebrated for craftiness and address. Acute and circumspect, sagacious and prudent, he diversifies his conduct, and always reserves some art for unforeseen accidents. Though nimbler than the wolf, he trusts not entirely to the swiftness of his course. He knows how to ensure safety, by providing himself with an asylum, to which he retires when danger appears. He is not a vagabond, but lives in a settled habitation and in a domestic state. The choice of situation, the art of making and rendering a house commodious, and of concealing the avenues which lead to it, imply a superior degree of sentiment and reflection. The fox possesses these qualities, and employs them with dexterity and advantage. He takes up his abode on the border of a wood, and in the neighborhood of cottages. Here he listens to the crowing of the cocks and the noise of the poultry. He scents them at a distance. He chooses his time with great judgment and discretion. He conceals both his route and his design. He moves forward with caution, sometimes even trailing his body, and seldom makes a fruitless expedition. When he leaps the wall, or gets in underneath it, he ravages the court-yard, puts all the fowls to death, and then retires quietly with his prey, which he either conceals under the herbage, or carries off to his kennel. In a short time he returns for another, which he carries off in the same manner, but to a different place. In this manner he proceeds, till the light of the sun, or some movements perceived in the house, admonish him that it is time to retire to his den.

The fox practices similar sagacity and cunning when he finds birds entangled in springs or snares laid for them by the fowler; with whom the fox taking care to be beforehand, very expertly snatches the birds from their entanglement, conceals them in different places, leaves them there sometimes two or three days, and is never at a loss to recover his hidden treasure. It is known that the hedge hog will roll himself up like a ball, so that when in this condition the fox cannot encounter him on account of his quills. Nevertheless, he teases the poor animal till forgetful of danger or exhausted, he is obliged as it were to stretch himself out, when the fox seizes and devours it. He will also attack the wild honey bee. The little creatures will fly upon him; but he lies down on the ground and rolls over, by which means he kills them, when he will make sure of their treasure—comb, wax, and honey.

In Kamtschatka, the animals called gluttons employ a singular stratagem for killing the fallow-deer. The animal is voracious, but at the same time slow and heavy in its movements, though it is remarkably acute in its sight and hearing. It is amazingly powerful, and an overmatch for any animal of its own size. Travellers say they will climb up a tree, and carry with them a quantity of that species of moss of which the deer are very fond. When a deer approaches near the tree, the glutton throws down the moss. If the deer stop to eat the moss, the glutton darts down instantly upon his back, and fixing himself firmly between the horns, so that it can suck out the blood of the suffering creature, which will quickly fall to the ground from exhaustion, or, else will tear out his eyes, so severely tormenting him, that he will sometimes strike his head against the tree with such violence as to terminate his own life. The glutton divides the flesh of the deer into convenient portions, and conceals them in the earth to serve for future provisions. The gluttons on the river Lena kill horses in the same manner.

With regard to birds, their artifices are not less numerous nor less surprising than those of quadrupeds. The eagle and the hawk kinds are remarkable for the sharpness of their sight, and the arts they employ in catching their prey. Their movements are rapid or slow, according to their intentions, and the situation of the animals they wish to devour. Rapacious birds uniformly endeavor to rise higher in the air than their prey, that they may have an opportunity of darting down forcibly upon it with their pounces. To counteract these artifices, nature has endowed the smaller and more innocent species of birds with many arts of defence. When a hawk appears the small birds, if they find it convenient, conceal themselves in the hedges of brushwood. When deprived of this opportunity they often, in great numbers, seem to follow the hawk, and to expose themselves unnecessarily to danger, while in fact, by their numbers, their perpetual changes of direction, and their uniform endeavors to rise above him, they perplex the hawk to such a degree, that he is unable to fix upon a single object; and after exerting all his art and address, he is frequently obliged to relinquish the pursuit. When in the extremity of danger, and after employing every other artifice in vain, small birds have been often known to fly to men for protection. This is a plain indication that these animals, though they in general avoid the human race, are by no means so much afraid of man as of rapacious birds.

The ravens often frequent the sea shores in quest of food. When they find their inability to break the shells of muscles and other fish of that class, to accomplish their purpose they resort to a very ingenious stratagem. They carry a muscle, or other shell fish, high up in the air and then dash it down upon the rock, by which means the shell is broken, and they obtain the end they had in view. Some anecdotes are told of the secretary vulture, illustrative of the present subject. This bird is styled the serpent-eater, from the avidity with which it catches and devours those noxious reptiles. The manner in which it seizes them, displays great intelligence. On approaching them, it carries forward the point of one of its wings, in order to parry their venemous bites, and waits till it finds an opportunity of spurning or treading on its adversary, or taking him on his pinions, and throwing him into the air. When he has at last wearied him out, he kills and devours him at leisure. Vaillant witnessed one of those combats. Finding itself inferior in strength, the serpent endeavored to find its hole, but the falcon, by a single leap, got before him and cut off his retreat. On whatever side the serpent strove to escape, the enemy still faced him. The serpent then erected himself to intimidate the bird, and hissing dreadfully displayed his menacing throat, inflamed eyes, and a head swollen with rage and venom. Sometimes this produced a momentary suspension of hostilites; but the bird soon returned to the charge, and, covering her body with one of her wings as a buckler, struck her enemy with the bony protuberance of the other. The serpent at last dropped, and the bird laid open his skull with one stroke of its beak.

The flight of the fish hawk, his manœuvres while in search of fish, and his manner of seizing his prey, are deserving a particular notice. In leaving the nest, he usually flies direct till he comes to the sea, then sails round in easy curving lines, turning sometimes in the air as on a pivot, apparently without the least exertion, rarely moving his wings; his legs extended in a straight line behind, and his remarkable length and curvature or bend of wing, distinguishing him from all other hawks. The height at which he thus elegantly glides is various, from one to two hundred feet, sometimes much higher, all the while calmly reconnoitering the deep below. Suddenly he is seen to check his course, as if struck by a particular object, which he seems to survey for a few minutes with such steadiness, that he appears fixed in air, flapping his wings. This object, however, he abandons, and is again seen sailing around as before. Now his attention is again arrested, and he descends with great rapidity;

but ere he reaches the surface, shoots off on another course, as if ashamed that a second victim had escaped him. He now sails at a short height above the surface, and by a zigzag descent; and without seeming to dip his feet in the water, seizes a fish, which, after carrying a short distance, it may be he drops or yields it up to the bald eagle, and again ascends by easy spiral circles, to the higher regions of the air, where he glides about in all the ease and majesty of his species.

At once from this sublime ærial flight he descends like a perpendicular torrent, plunging into the sea with a loud rushing sound, and with the certainty of a rifle. In a few moments he emerges, bearing in his claws his struggling prey, which he always carries head foremost; and having risen a few feet above the surface, shakes himself as a water spaniel would do, and directs his heavy, laborious course directly for land. A shad was once taken from a fish hawk, near Great Egg Harbor, on which he had begun to regale himself, the remainder of which weighed six pounds. Another hawk, at the same place, was seen with a flounder in his grasp, which struggled and shook him so that he dropped it on the shore. The flounder was picked up, and served a whole family for dinner. It is singular that the fish hawk never descends to pick up a fish which he happens to drop either on the land or on the water.

The white or bald eagle is very fond of fish, and hence seeks a location near the sea shore. In procuring this favorite prey, he displays in a very singular manner, the genius and energy of his character, which is fierce, contemplative, daring, and tyrannical; attributes not exerted but on particular occasions; but when put forth, overpowering all opposition. Elevated on the high dead limb of some gigantic tree, that commands a wide view of the neighboring shore and ocean, he seems calmly to contemplate the motions of the various feathered tribes that pursue their busy avocations below; the snow white gulls slowly winnowing the air; trains of ducks streaming over the surface; silent and watchful cranes intent on wading; clamorous crows, and all the winged multitudes that subsist by the bounty of this vast liquid magazine of nature. High over all these hovers one whose action instantly arrests his attention. By his wide curvature of wing, and sudden suspension in air, he knows him to be the fish hawk, settling over some devoted victim of the deep. His eye kindles at the sight, and balancing himself with half-opened wings on the branch, he watches the result. Down, rapid as an arrow from heaven, descends the distant object of his attention, the roar of its wings reaching the ear as it disappears in the deep, making the surges foam around! At this moment the eager looks of the eagle are all ardor, and levelling his neck for flight, he sees the fish hawk once more emerge, struggling with his prey, and mounting in the air with screams of exultation. These are the signals for our hero, who, launching into the air, instantly gives chase, soon gains on the fish hawk, each exerts himself to the utmost to mount above the other, displaying in these rencontres the most elegant and sublime evolutions. The unencumbered eagle rapidly advances, and is just on the point of reaching his opponent, when, with a sudden scream, probably of despair and honest execration, the latter drops his fish; the eagle, poising himself for a moment, as if to take a more certain aim, descends like a whirlwind, snatches it in his grasp ere it reaches the water, and bears his ill-gotten booty silently away to the woods.

The predatory attacks and defensive manœuvres of the eagle and the fish-hawk, are matters of daily observation along the whole of our seaboard, from Georgia to New England, and frequently excite great interest in the spectators. Sympathy, however, on this, as on most other occasions, generally sides with the honest and laborious sufferer, in opposition to the attacks of power, injustice, and rapacity; qualities for which our hero is so generally notorious, and which in his superior, man, are certainly detestable. As for the feelings of the poor fish, they seem altogether out of the question. When driven, as he sometimes is, by the combined courage and perseverance of the fish-hawks, from their neighborhoods, and forced to hunt for himself, he retires more inland, in search of young pigs, of which he destroys great numbers. In the lower parts of Virginia and North-Carolina, where the inhabitants raise vast herds of these animals, complaints of this kind are very general against him. He also destroys young lambs, in the early part of Spring; and will sometime attack old sickly sheep, aiming furiously at their eyes.

The woodpecker is furnished with a very long and flexible tongue. It feeds upon ants and other small insects. Nature has endowed this bird with a singular instinct. It knows how to procure food without seeing its prey. It attaches itself to the trunks or branches of decayed trees; and, wherever it perceives a hole or crevice, it darts in its long tongue, and brings it out loaded with insects of different kinds. This operation is certainly instinctive; but the instinct is assisted by the instruction of the parents; for the young are no sooner able to fly, than the parents, by the force of example, teach them to resort to trees, and to insert their tongues indiscriminately into every hole or fissure.

There are some peculiarities in the American sparrow hawk deserving notice. It approaches the farm-house, particularly in the morning, skulking about the barn yard for mice or young chickens, apparently well aware that at a later hour its visits would be attended with hazard to itself. It frequently plunges into a thicket, after small birds, as if at random, but always with a particular, and generally a fatal aim. Though small snakes, mice, and lizards, be favorite morsels with this active bird, yet we are not sure it is altogether destitute of delicacy in feeding. It will seldom or never eat of anything not killed by itself; and even that, if not in good eating order, is sometimes rejected. A very respectable friend informs me, says Wilson, that one morning he observed one of these hawks dart down on the ground, and seize a mouse, which he carried to a fence post; where, after examining it for some time, he left it; and a little while after pounced upon another mouse, which he instantly carried off to his nest, in the hollow of a tree in the neighborhood. The gentleman, desirous of knowing why he had rejected the first mouse, went up to it, and found it to be almost covered with lice, and greatly emaciated! Here was not only delicacy of taste, but sound and prudent reasoning. "If I carry this to my nest," the hawk might have thought, "it will cover my nestlings with vermin; and besides, it is so poor as scarcely to be worth eating."

It is well known that the owl kind are incapable of supporting the light of day, or at least, of their seeing and readily avoiding danger. Hence, in the day time they are usually in concealment, in the cavern of a rock, the darkest part of a hollow tree, or in some obscure recess of the farmer's outhouse, where they remain till evening, when they emerge in quest of prey. If by accident one is discovered, not thus protected, by the other birds of the place, for nearly all of them are the inveterate enemies of the owl, sad will be the treatment he receives. The blackbird, the thrush, the jay, and the redbreast, all come in full force, and employ their little arts of insult and abuse. The smallest, the feeblest, and the most contemptible of this unfortunate bird's enemies, are then the foremost to injure and torment him. They increase their cries and turbulence round him, flap him with their wings, and are ready to show their courage to be great, as they are sensible that their own danger is but small. The unfortunate owl, not knowing where to attack, or whither to fly, patiently sits and suffers their insults. Astonished and dizzy, he only replies to their mockeries by awkward and ridiculous gestures, by turning his head, and rolling his eyes with an air of stupidity. It is enough that an owl appears by day to set the whole grove into a kind of an uproar. Either the aversion all the birds have to this animal, or the consciousness of their own security, makes them pursue him without ceasing, while they encourage each other by their mutual cries, to lend assistance in their supposed laudable undertaking.

THE MODEL COUNTRY FIRESIDE.

NEW-ENGLAND FARMHOUSE.

In the year 1829, by the almost simultaneous death of his aged father and mother, George Boardman was left in the homestead a solitary bachelor; not what is called an old bachelor—a Cœlebs of forty-five; but just of an age to appreciate domestic society, and to regulate in the best manner the internal elements of the household, as well as the business that is to sustain the household. It is quite problematical whether what are called early marriages are generally to be recommended. Occasionally they are followed by the best consequences; but where this happens once, in numerous cases the results are different. It is a prevalent opinion among those the most competent to judge, that the age of twenty-one years does not necessarily bring with it full maturity of intellect and the sound discretion required for the best success in business or professional life. Money is rarely made and retained from business, or high reputation from professional skill, by persons before the age of thirty. When it occurs, the event is rather an exception to the order of nature and human development, than one from which a general inference can be drawn.

It might indeed be exciting in this age of rapid progress to see girls and boys with one long stride pass from childhood to manhood and womanhood; and from the latter to the highest grades of human attainment, without the patient labor, hitherto deemed indispensable for it. On the same principle, we might think it a defect in the economy of human existence that human beings do not, like mushrooms, in a few hours start into life in full stature. Such, however, was not the purpose of the Almighty in giving birth to our species. That purpose is well known. The opening of our destiny is like the first blade of vegetable life, in size most diminutive, and so delicate and feeble as scarcely to bear the slightest touch. Infancy is interesting from its helplessness; childhood from its laughter and its frolicksome glee; and not less the buds and the opening flowers of youth which denote the fruits of matured life. Who can deny that the extinction or removal of either of these seasons would render our species less interesting and less lovely? Who can affirm that either one of them is not needful as a preparation for that which now succeeds it? If that one were not to exist which seems to be of the least use, and is the greatest burden, for instance, the period of infancy, what a blank would be created in the life of the mother? Does woman ever appear so lovely as when nursing and fondling her tender offspring? Where else could she bestow such affectionate assiduity? If it were stricken from the record of her life, what a charm would be removed from the domestic precincts!

It is not only a matter of fact, but it is one of peculiar fitness and wisdom, in the Divine appointment, that there should be different grades of maturity in the human species subsequent to the age of twenty-one, as well as prior to it. If all on reaching that age, were to reach the perfection of their nature, and at one bound were to aim at the front ranks of labor and enterprise and responsibility, would the condition of human society be improved? If the wisdom of long experience were rendered unnecessary, would the world on the whole be more desirable than it now is? According to existing organizations, the sobriety and the moderation of old age and the impetuosity of young manhood, may work together and in harmony; neither is in the way of the other; each is needful to the other. It appears to us, that a matured intellect is as necessary in entering upon the management in domestic economy, as in entering upon the great business of life. Fancied and untried theory in the former is often attended with peril as in the latter. Is there no difficulty in arranging the family equipments upon a basis of sound wisdom? Is there no difficulty in instituting and maintaining the best mode of family discipline, especially in rearing a family of children? Few understand this difficulty till called to encounter it.

There is another consideration that would render us cautious in recommending what may be called early marriages. Physiologists have said that the offspring of young parents, cannot be expected, as a general thing, to have that mental and physical vigor, all other things being equal, to be anticipated from those born of parents more advanced in life. They say that deterioration in families may be the natural consequence of too early marriages; and that most of our greatly distinguished men have been born of parents past the age of thirty, and when in the perfection of their nature. If so, the fact is a most interesting one, a most important one. It would be easy to specify illustrations of it; but, this our readers can do for themselves. We request them to do it. Perhaps in the circle of their own acquaintance they can refer to cases where the younger born of the same parents have more talents than the first born. To make such examples appropriate, the latter must have been born of young parents; and, the former of parents who had attained subsequent to the birth of their early offspring, a manifest progress in intellectual strength. There is surely nothing fanciful or unphilosophical in this hypothesis. Is it not the same principle recognised by those who labor to improve their domestic animals. Do amateur stock breeders pay no attention to the age and perfection of the animals from which they seek a prime progeny? All this is well known. Why then does not the same law prevail in reference to the human species? It is proverbial that physical defects—for example, pulmonary complaints, descend from parents to children; also a tendency to insanity. Why then may not the mental vigor or imbecility of the parent be impressed on the mind of the child? If so, the impression of the mental feature is made according to the original as existing at the time, and not as it may become ten years afterwards.

It has been stated that George Boardman, at the time of which we are speaking, had arrived at the age, according to our theory, for entering advantageously upon the marriage state. He was about the age of twenty-eight. And we advise, as a general thing, that men make their plans to marry somewhere during the age from twenty-five to thirty; and that girls lose no good opportunity for doing it after arriving at the age of twenty-one. In this no invariable rule is applicable. Circumstances may render that expedient with some that would not be advisable with others. Indeed, some persons of either sex may have acquired a maturity and vigor not acquired by others ten years older than themselves, and which others perhaps never will acquire. All in a case so to be modified by circumstances must exercise a sound discretion of their own. They should be well advised as to the principle to be observed, and then make as little departure from it, as their own situation will admit. Matrimonial alliances are usually to appearance the result of accident, or impulse, or bewildered fancy; whereas, they should be under the guidance of reason. If there is any one occurrence in life above all others, deserving the most deliberate and rational consideration, it is the adoption of the marriage vows. From the state consequent to these vows, every thing seemingly is to proceed which is to render life a blessing or a curse; which is to make the fire-side radiant with smiles, or sombre with the shadows of that forever present evil genius, which there presides over the discordant elements of ill-matched and discordant spirits. If an offspring is to arise blessed with health, and talents, and social virtue, the years of the fond parents will roll on to the ocean of time, like the crystal streamlet, gently pursuing its course, till buried and lost in its own kindred abyss. If this offspring inherits disease and mental weakness, or it becomes loathsome from moral pollution, the otherwise happy parents will spend their days in weeping and anguish, and at last in wretchedness go down into the grave.

George Boardman would have chosen the bands of matrimony ere this period of his life, but like a thoughtful and affectionate son, he had resolved not to marry so long as his parents might live, knowing, that according to the course of nature, their last sands were mostly fallen. His mother oftentimes, when oppressed with infirmities and the responsibilities devolving on every female head of a family, would fervently urge him to get married. He appreciated her kindness; but prudently aware that there might spring up some noxious weed to mar the fragrance of the matrimonial flower-bed, which had found place in her imagination, he still persevered in deferring all thoughts of any such change. He knew that no one else could love an aged father and mother, and bear in them the infirmities of a second childhood, like an own child. To him it was no hardship to wait upon them, to cheer them, and to gratify them in every desire they might indulge. To another one, even the wife of a son so devoted, there might not be this ready sacrifice of her own individual preference, and self-devotion to those with whom there had not long been that identity of purpose felt by himself. He knew also, that however self-sacrificing such an one might be, on his account, if not otherwise, and however affectionate in manner she might be, they could not feel towards her as towards him. He was their own child, and they fancied him almost perfect. With her the case was widely different. Things done by him would be esteemed unexceptionable; but if done by her would be judged faulty. From such causes alienations might arise even between himself and his companion, that would survive the occasion in which they originated. In such a dilemma it is no unusual thing for a man to become estranged in affection from one party or the other. George Boardman had that kind of native good sense to perceive all this; and in perceiving it, to resolve on avoiding it. He did avoid it, living as already stated, a bachelor, till he performed every act of assiduity to his venerable parents, till called to lay them in the cold grave.

If there is is anything amidst the imperfections of our world truly lovely; if there is anything that exalts human nature above its ordinary level; if there is anything that casts over the domestic altar an unfading halo, it is such an union of hearts and such a concentration of interests as had thus existed in the Boardman homestead to the termination of the generation now closed; and, what had thus characterised the family annals of these venerated saints, was destined to continue, even in augmented loveliness, during the lifetime of this excellent son, who survived them. On witnessing such exhibitions of domestic union and love, we almost forget that our nature has lost its original perfection; and, that there is for the pure in heart even, a better life beyond the grave.

The death of the parents of George Boardman occurred in the latter part of October, just after the close of a plentiful harvest. The soil of the homestead, as well as the minds of its inmates, had ever received an enlightened and substantial culture, so that the barns, and the granaries, and dairy rooms were all filled to overflowing. It was not an uninteresting circumstance, that this time-honored pair, at the same season of the year, were taken to the great moral storehouse of the world. The analogy between an earthly and a heavenly harvest is most striking. The luscious fruits of the orchard and the heavy sheaves and the long ears of the corn field, all fully ripe and ready for use, fitly represent those plants of celestial origin bending under the matured weight of a rich moral discipline. While a fertile soil, and the genial influences of the sun and the descending rain, had caused the one to reach the fulness of its destiny, the showers of divine grace had caused the other to expand till made meet for heaven. Upon the Boardman homestead both grew and ripened simultaneously; and both, as it were, simultaneously, were transferred to their respective garners.

The surviving son, alone, as it were, in one of the largest mansions of that day, must have been desolate. In addition to the loss he had experienced, the fading of nature all around him, the long nights of frost, and the unwonted stillness, every where to his imagination, excepting the loud and well measured ticking of the family clock, reigning supreme, seemed to impress him with an unknown sadness, and admonish him that man was not made to live alone. During his dreams, in midnight solitude, the retrospect would rise up before him in full freshness. At one moment he would imagine himself in assidious ministrations to his languishing parents, or listening to their fervent ejaculations of prayer; at another he would seem to be following their lifeles remains, in the long procession, to the cold grave; but, at the next moment, the vision would vanish, and he would find himself, as before, a mere unit in one drear blank, spreading over the whole creation. Who can say that to such an one existence is a blessing? In what mortal can the social breathings of the soul be hushed to silence? The world, with all its riches and all its glories, would be a worthless treasure to one in complete and unending solitude. The perpetually doomed recluse, if possessed of human sympathies, would be a personification of wretchedness.

George Boardman was known throughout the county in which he lived as a young man of a sound mind, of unblemished morals and as the heir of a valuable estate. His dutiful kindness to his parents had endeared him to all that knew him, and all sympathised with him in his present loneliness. Nor was this all. Perhaps fifty of the most eligible single women were named, by different kind-hearted souls, common in such cases, as fit companions for him. One advised him to marry the daughter of Esquire Peabody, another the daughter of a wealthy farmer in the adjoining town, another the daughter of a retired merchant from a distant city, living close by, another a rich heiress in the neighborhood; and, indeed had he taken to himself one-fourth part of those thus kindly recommended, he would have been obliged to enlarge the old mansion, big as it was. All this was done too before the dirt upon the fresh graves of his deceased parents had become well seared over. The poor man was annoyed at it. He knew well the peculiar exigency of his situation, and resolved to seek a remedy for it; but felt that he was competent to do so without any such aid, although he doubted not that all this volunteer service was well intended, and that each of the ladies named to him might deserve a husband better than himself.

The subject thus brought to his attention, he must have been a being without the ordinary human social attributes, had he let it go from him without careful consideration. He made it a matter of business. When the business features of the case had been adjusted in a preliminary way, he presumed, as a matter of course, that love would complete the enterprise in a most satisfactory way. All persons know that love turned loose, without the guidance of reason, is like a beautiful, young, unbroken colt, without a bridle and traces to keep it steady, playful, frolicksome, and capricious; and, oftentimes, in its antics and merriment, will founder and do irreparable mischief to those in its way. But, when this colt becomes well trained to the harness, it is one of the most beautiful and useful of animals. Love, too, when well disciplined and guided, becomes one of the fundamental elements in all sublunary bliss; it gives a delightful charm to the social circle, found no where else; it fills the family precincts with a most vivid radiance, dispelling gloom, lighting up joy on every countenance, and oftentimes converting the abode of poverty into one of cheerful contentment. Without this passion, the world is heartless as a steam engine, and cold like an iceberg. With it, a genial warmth penetrates and enlivens the whole mass brought under its influence.

So reasoned our young friend. So reason all men of sense.

He disposed of the preliminaries to his new undertaking in a most systematic manner. In the first place, negatively—that is, whom he would not marry; and, in the second place, affirmatively, whom he would like to marry. Hence, he resolved that he would not marry a lady who had a mother, and, perhaps, a grandmother, to accompany her; perhaps, under other circumstances, excellent women, but to attend a new married daughter to her new home, frequently a great nuisance. He resolved that he would not marry a person older than himself, because, in most cases, such an one would be insolent, and in all cases, because women at forty are, physically, as old as men at fifty. He resolved that he would not marry one so young, as to be incompetent for every duty devolving on the mistress of the family. He resolved that he would not marry one so handsome that she might be vain of her beauty; or so ugly in her appearance, or so illiterate, that he might be ashamed of her. And be preferred to marry one who was comparatively poor, rather than being rich, lest that on that account a feeling of independence, and a consequent disposition to extravagance, should be attended with evils more than equal to all the benefit derived from her possessions. On the other hand, he chose to marry one of respectable family—no matter if poor; of good personal appearance; of agreeable manners; of domestic habits; of religious principles, as a woman without them is, on no account, fitted to be a wife or a mother; of good talents in reference to an offspring, and to her competence for presiding with dignity over her family; and, especially, of good education, which enables a wife to be admired as a companion, and as a mother to render her own parlor the best of all school rooms for her own children and all the members of her household.

Such was the directory of George Boardman in the choice of a wife. With so specific a delineation of the points to be observed, the attainmant of his object might have been deemed problematical. He doubtless fancied it so himself. At any rate, as he afterwards admitted, with all the candidates that had been recommended to his consideration, no one stood so high in

his estimation, as to lead him to any decided action. The more he reflected on the subject, the less inclined was he to cherish any strong confidence of an auspicious termination of his wishes. Things thus stood for about four weeks, when he had occasion, on business, we believe, to collect interest on a mortgage ; to visit the shire town of a neighboring county. It happened to be at the time of the annual examination and public exhibition of the Academy—an institution that had long been noted for having educated some of the most distinguished women in that region, and for having had some of the first female teachers in the country. The occasion in such a locality was one of great interest. What can be more interesting than to see a dozen or twenty fine boys picking up their books, bidding adieu to their companions and friends, starting for the University or the place of active business? What can be more interesting than to see thirty or forty young ladies, having completed their education, and then giving such evidence of their proficiency, preparatory to their return to their respective homes, some of them perhaps at once to enter upon the most important relations of life? Mr. Boardman was induced to remain and witness the exhibition ; and, the sequel shows the close connection between some rare and unexpected concatenation of incidents and the remaining portions of one's life.

Although the prominent individual of this memoir was not accustomed to the gala of literary institutions, he was so habituated to read the current domestic literature as well as news of the country, and was withall possessed of so much native discrimination and good taste, as to be enabled to appreciate the performances of the pupils. The spectacle of sixty or seventy fine boys, with manlike demeanor and prompt intelligence, answering the questions proposed to them in all the branches of elementary education ; and, of as many females, more than half advanced to the stature of womanhood, giving evidence of matured intellect and proficiency in their studies, seemed to open to him a more beautiful picture than he had ever before seen. The pleasure he had experienced the day previous, in the recep-

tion of a few hundred dollars of interest money, did not compare with the pleasure now felt on seeing this display of youthful progress in useful knowledge. Indeed, that person must be void of all the intuitive perceptions and of all mental culture, who is unmoved when witnessing the social and rational embellishments with which such juvenile groups are being ushered upon the great drama of life. A northern climate has made us cold and phlegmatic; no common event is ever able to raise our feelings above a temperature of thirty-five or forty degrees—a little over the freezing point; perhaps we ought to be ashamed of it; but, at an exhibition of the kind described, our impulses begin to swell till we get up to boiling heat: we become enthusiastic in all matters on which depend the usefulness and the respectability of the rising generation.

Mr. Boardman was especially delighted with the performances of a class of young ladies, under the direction of Miss Jane Bickford, the preceptress, in vegetable physiology. Both teachers and pupils seemed as familiar in all the principles and processes of the growth of plants—of their elementary constituents and of the manures that nourish them—as women in the country usually are in the making of bread or in other kinds of cookery. Is it possible, that this lady of twenty-four, thought he, devoted as she is to the duties of instruction in a public institution, knows more of the formation of the vegetable tissues, and of the nature of the manures which become the food of grasses, fruit, shrubs, trees, and the various culinary esculents, than the mass of farmers? He was so delighted and astonished, that he seemed lost in a kind of a revery. Even after the close of the public performances, when returned to his lodgings, he could not drive the subject from his mind; and very soon, such an impression was made upon him, he had no wish to do it. The truth is, the agricultural chemistry, or her comely looks, or the lady-like deportment of the preceptress, had made havoc upon the social equanimity of the young bachelor. Nothing would do, but he must remain another day and make acquaintance. On introduction to her, he became satisfied she was the ery individual sketched out in his idea, of the qualifications for a ,ife. She was a poor orphan, educated, in consequence of her uperior talents, by a distant relative ; was an accomplished ady in her manners ; was proverbially amiable in her temper ; nd, was as much distinguished for the Christian graces, as for er mental endowments. Almost as a matter of course, with a usiness-like precision, before he returned to his own home, it vas agreed by the farmer and the fatherless and the motherless eacher, that they would enter wedlock on the succeeding new ear's day.

As this memoir is designed, not to gratify an idle curiosity vith those who think of nothing but silly love stories, we desire ere, to say a few words to all the young men and young women who may read this volume, on the kind of education most useful or them in the country. If every young farmer were to seek a vife, having that kind of mental culture possessed by Jane Bickford, we should in time have no occasion to complain of an lliterate yeomanry. Such wives would be as good in revolutionzing the processes of husbandry, as the lectures and prescripions of professors Mapes, Norton, and Fowler. Could the husands and their sons fail to become improved under such influnces? We would far rather have ten thousand such wives cattered over the State of Massachusetts, than all the agriculural colleges that have been dreamed of during the last ten ears in our country, and have given some of our best and wisest nen such paroxysms of the nightmare, as almost to have endangered their health and their sanity. Could we have appropriate brief treatises on animal and vegetable physiology, poitical and domestic economy, and the history of agriculture, nd compends of moral literature, in all our schools, instead of he worthless trash with which many of our youth are stuffed, here would be a speedy regeneration in rural aspects and intersts throughout the land. This might be done promptly and at mall expense, without legislative enactments or princely endowments; but, when so told, our great men interested on the subect, wishing to do or to have done some great thing, as it were despising this for its simplicity, resemble the Syrian, who went to the prophet for a cure of his leprosy, and then turned away in a passion, saying, "Are not Abana and Pharpar, rivers of Damascus, better than all the waters of Israel? May I not wash in them and be clean?" Or, if our suggestion here on agricultural education were to be generally adopted, it would be but a few years before all our young farmers, in masses, would flock to the polls to decree that we must have agricultural bureaus and agricultural colleges, and agricultural lectures, where needed for the public good.

George Boardman, on returning to the homestead, quietly commenced giving it inside and out a new coat of paint—worth double at this season of the year, what it would be, if put on in the hot weather of summer. At the same time, the old bedsteads, bureaus, and chairs, instead of being put up in the garret, or out in the wood sheds, as if ashamed of them, because they were old, are treated with the respect and veneration that had been shown to their former owners, being subjected to the scrutiny of the cabinet maker, for any slight repairs needed, and for a new coat of varnish. The old family carriage was treated in the same manner. Even the farm wagons, and ploughs, and sleds, were similarly honored. New curry-combs and brushes were furnished for the horse stables. Not a door handle was passed over untouched, if out of order. The same may be said of the gates, and the fences, and the out-buildings. In two weeks time, every thing inside and out gave evidence of renovation. The old furniture and carriages looked like old boots that had just received a visit from Day & Martin's best Japan blacking. The neighborhood was in astonishment, as to what it could all mean, not a word having been dropped, concerning the events that grew out of the academical exhibition. Here was a dilemma. It was not possible that his kind-hearted neighbors could contentedly remain ignorant of what was evidently in immediate prospect. The movements of Cœlebs, to ascertain the point of compass to which his magnetic powers tended, were watched as closely as though he had been a suspected refugee from justice. This supervision was nothing better than an odious persecution. It was an intolerable nuisance. Such is human nature! To such vexations every bachelor, on getting married, is subjected.

The first of January was soon on the docket. The day did not pass away without witnessing the vows of George Boardman and Jane Bickford to be forever one in interest and affection, and of twain they became one flesh. She was at once the mistress of the Boardman mansion, and soon became a favorite throughout the town as well as the idol in her own household, giving evidence that she was as competent to preside with dignity and to impart useful thoughts at the head of a family, as at the head of her department in the literary institution which she had served with so much fidelity to her patrons and with so much honor to herself. Her labors there had less reference to the remuneration she was to receive than to a desire of being useful to her pupils and to the public. To toil simply to secure the former is induced by a degrading and sluggish sense of one's destiny; being a submission to the penalty, entailed on man's first act of disobedience; and the idea of it is calculated to oppress the soul with a moral burden more exhausting and humiliating, than that arising from any muscular action to which we can be doomed. But, when our labors are of a nature, as in the business of education, or in any act where we become the co-workers of our Heavenly Father in ministrations for the mitigation of human woe, or the advancement of the great purposes of man's moral and intellectual exaltation, there is felt a dignity and a pleasure far outmeasuring the evil of subjection to any physical necessity. The person that has not experienced a consciousness of laboring from such high motives is but little elevated above the ox or the horse, which knows no other law than that of an unyielding master. To such an one there is no moral pleasure; no spiritual delight; no intellectual ascendancy over the material and sensual elements of a mortal existence.

Jane Boardman, as she is now to be called, came to her new sphere of action with these ennobling motives for the performance of duty. She could not, indeed, have divested herself of the consideration of securing to herself for life a most desirable home, and a friend who was to abide by her under every vicissitude, to the day of death—in sickness as well as in health—in poverty and disgrace, if God should permit them to arise, as well as in plenty, and affluence, and honor; but, she came to that home and that friend, with a determination to be an ornament to the one, and help-meet to the other. She came thither, not to spend his money, but to help him to take care of it, and to use it as a faithful steward of God's bounty. She came thither, not to make him a slave to her caprice or extravagance, but to be to him a kind-hearted and enlightened companion. She came hither, not to extort an unwilling service from the numerous laborers, male and female, on the premises, but to make them cheerful and honest in the discharge of their duties; to instruct them to toil for a better reward than silver and gold; to make them realise that there is no degradation, but that a moral dignity may attend persons in their situation, as if they were the proprietors of all they beheld. With such lofty conceptions she entered on her new relation and new sphere of duties; and to this day she has been, and is now, a pattern to all wives in the country; she would be a pattern for wives in all situations, whether in the city or the country. Her domestics, and the hired male laborers, love her as good children love a mother; indeed she is a mother as well as a mistress to them. Her husband and all delight to honor her; and throughout the country, where she has now lived twenty odd years, her name is a sweet savor, and the prayers of the destitute and the afflicted have never ceased to ascend to Heaven in her behalf. Such is the general character of Jane Boardman.

It is thought, and frequently said, that there is no release from the dull monotony, as it is called, in the labors of a farm; and, that those who perform these labors, are scarcely fit for enlightened and refined society. There has, indeed, been an appearance of truth in these assertions; but, on a careful examination of all the circumstances connected with rural life, it will be found that there is no necessity for such assertions, or rather for the evils on which they are predicated. Of this Jane Boardman was firmly persuaded, and she resolved to demonstrate the correctness of her opinion. Her husband had, as in the country is very common, become negligent in his personal appearance, if not in reality slovenly. The tendency to this is very strong, and sometimes, seemingly, irresistible. Where the situation is retired, and frequently no one for week after week is expected on the premises, save those belonging to them, the husband will go day after day without shaving, and otherwise being equally regardless of his looks, and then sitting down with his family for the evening, having on the same clothes he had worn through the day, boots and all, clammy with sweat, soiled by dirt, and impregnated with the odors of the barn-yard. The wife, too, not expecting to have company, as it is called, would make no change of dress, retaining upon her person the same she had worn in the kitchen, saturated with the juices and the odors of potluck. Thus they would sit down together, parents and children, and no one can tell how many others, as uncomely as a deck load of foreign emigrants after a month's voyage. This is no uncommon practice, and is one of the most offensive customs in country life.

Jane Boardman, the very first week in her new home, began to argue for a reform in this matter. She maintained, and with great truth, that she could not expect to see any one to be treated with more respect than her husband, and that whenever he had finished his out-door cares and was ready to sit down with her, she was disposed to dress up to appear as well to him as she would to please the Governor of the State, were he expected to make them a visit. On the other hand, it was urged that her husband should be equally complacent to herself. Accordingly, an evening suit, from the clean dickey to the fresh slippers, as well as the day suit, was ever in readiness; and as soon as her own family duties would admit of it, she would adjust her own personal fixtures with as much care as when expecting him on the evening before they were married. This being done, on his return to the house, he would be assisted in his ablutions and in making all other corresponding changes of apparel, as if he were preparing for church on Sunday. Then they could look upon each other and smile in remembrance of their first love. Was there no pleasure in this to balance, a score of times over, all the labor it cost? Was there not a mutual gushing of affectionate feeling through the evening, that would not have been witnessed had this preparation been omitted? Let a man and his wife thus study to respect and please each other ever after their matrimonial union, as they did before it took place, and there will be no danger of estrangement or lukewarmness in their social affinities. Most unhappy matches are the result of an inattention to this very fact, trifling as, to most persons, it may appear.

The practice of dressing up for each other every evening, thus begun by George and Jane Boardman, has continued to the present time, and it has done much, both in rendering perpetually fresh their first ardent affection for each other, and in rendering rural life the most delightful to them. Jane, however, did not limit to the parlor this wholesome reform. A bathing room was to be furnished, and a bathing tub was each day to be filled with pure water. Here each laborer, at the close of his work, was to undergo a thorough purification, and then to be attired with a clean suit for the evening. By this means, they were all refreshed and rested, and enabled the better to sleep, so that on the coming day, they were the better prepared for their respective duties. By this means, they were relieved from liabilities to disease, otherwise experienced. By this means, the kitchen and the halls, and the lodging rooms, were exempt from the nauseating odors of the cattle, the piggery, and the horse stable. No one who has once tried the benefit of these observances will ever afterwards tolerate them, unless his taste and his olfactory nerves are in unison with those of the brute.

Cleanliness and neatness should be ranked among the cardinal virtues. It is difficult to imagine how a pure mind can inhabit a body that is defiled with filth. There can be no congeniality between them. Each, from its own nature, repels the other. On the other hand, an impure mind delights in, and draws to itself kindred objects ; a mind that is debased will attract the surrounding pollutions, both moral and material, as the magnet attracts to itself whatever partakes of its own essence, though it be the smallest dust, till the whole is consolidated in one compact mass. It is contrary to general experience to witness exceptions to this rule. The individual who does expect it, is sadly deficient in his knowledge of the world, and of the laws which govern human conduct. Nor is there any necessity for persons to be uncleanly because they are poor. If they have but little, the less time is required for purification. Neatness is within the reach of all ; and to those in poverty, there is the more occasion to enjoy every thing within their control without alloy. Moreover, the more liable persons are from their occupation to become filthy, the greater should be their efforts to counteract this liability ; otherwise the man who digs in muck or the manure heap, soon becomes unfit for any place but the cattle stalls and the piggery, and sinks to the level of whatever dwells in them. The woman, too, who casts not off the impurities that adhere to her person, and to her clothes, when engaged over the wash tub, in scrubbing the floors, and about the culinary apparatus, will soon become unfit for any other place, and in mind, for any refined society.

But Jane Boardman resolved on an effort for a still higher attainment ; while it was the more immediate province of her husband to direct all measures for the improvement of the soil and its varied products, she felt disposed to exercise a kind of supervision over measures calculated to improve the mind. Instead of acquiescing in the too frequent slander, that persons in the country, particularly on the farm, are necessarily ignorant, she maintained that there is no other situation so favorable for the acquisition of many kinds of useful knowledge ; that persons in no other occupation have so much leisure for reading ; and as there are but few interruptions from company and public amusement, all disposed have the best opportunity for reflection, and thus invigorating their own minds. Accordingly, it was her determination that the house should be amply supplied with valuable books, and a portion of each evening should be spent in reading; the rest in conversation on what might have been read, or other topics of a useful character. Few books, at that time, had been published in this country, compared with what are now published, particularly on agriculture and rural economy. Hence, the best imported volumes were procured; oftentimes at a high price, but, as the event proved, they were worth to the family all, and far more than the cost of them. Who would have anticipated the result? From that time to the present, there has been read carefully, and frequently over and over again, more than a thousand volumes, in time that would otherwise have been lost.

All farmers may do the same. In this way their fire-side may become the centre of a high intellectual and social elevation. Mechanics and traders have not the time to do it. Their evenings are devoted to their regular occupations till too fatigued for mental effort. Even the professional man has no more leisure to do it than the husbandman. This is proved by the experience in Mr. Boardman's family. The whole of that family has acquired a familiarity with general literature that would do credit to the best educated circles in the community; and, in whatever relates to the general principles of agricultural chemistry—animal and vegetable physiology—successful tillage—horticulture—the kitchen garden—the products of the dairy—the economy of the poultry yard,—the wife as well as the husband possesses an accuracy of information that would do honor to an amateur of the first class. Although she has no occasion or desire to interfere in the application of this varied knowledge to the labors on the farm, yet when alluded to, it is to her like household words; she is as familiar with it as with the processes of making pastry or sweetmeats. And, while she would make a visit to the swine and dairy yard, and specify the fine points of each animal, like a professed stock breeder, like the first class of English ladies as described by the Rev. Henry Coleman, her manners are polished and dignified, and seen in the family is deemed a fit companion for a princess.

With such high purposes and rational occupations, the winter quickly passed away; and, when it was gone, the remembrance of it was like a dream or a revery of the imagination. So sweetly had it sped from the happy pair, spring came upon them with all its enlivening influences—the singing of the birds, the opening of buds, and the balmy fragrance of flowers, before they appeared conscious of its approach. At such a season the wide range of nature for renovated existence is in travail. Man and beast partake in the excitement. The ploughman and the seedsman are on the alert. Anon the meadows are covered with verdure and the fields are waving high with their cereals. The year rolls round; the summer with sultry suns, its drenching showers, and its hay time toils, treads rapidly in the path of its vernal sister. In equal speed autumn succeeds summer laden with a bountiful harvest. Now the heart of the husbandman is wont to swell in adoration of Him who thus makes the earth fruitful, crowning it with plenty. In the spring man reposed on the Divine assurance that in due time he should be enabled to reap as well as to sow the seed. Then a living faith gave vigor to his arm and strength to his hopes. In autumn the husbandman's faith and hope are swallowed up in fruition. How beautifully typical is the plentiful harvest of the farmer, of the great harvest of the Christian, in another life, when a frail and polluted nature is to inherit a glorious immortality! When for all the toils and anxieties of a mortal existence he is to be rewarded with the bread of heaven and the waters of salvation!

As abundant as were the crops of hay and grain and fruit, in the autumn of which we speak, about the close of the harvest season George and Jane Boardman had other cause for a grateful remembrance of God's goodness to them. To them an heir was born—a son inspired hope that their most joyous days were yet in the future. Those who have experienced such occasions need not be told the emotions that now animated them. Those who have not experienced them need not be told, because if they were told, they could not appreciate the reality. The birth of a child, especially if it be the first born, gives a new era in the life of the parents ; opens to their broad vision an untried sphere of hope and fear, and solicitude and duty. Hitherto existence has been to them a kind of prelude to human destiny and responsibility ; a season of preparation for the great business of life. Now comes the development of that destiny ; the manful redemption of that responsibility or the cowardly shrinking from it. Now comes the chain that may bind us indissolubly with the future generations of our race, entitling us to share in the glory or in the ignominy that is yet to dawn upon the world.

As the seasons follow each other in uninterrupted succession, so do the years. On the commencement of the new year we lay our plans of enterprise to be achieved therein ; but, seemingly, before such an outline is complete, the entire year is gone, and another rises before us, each additional one, to our imagination, with accelerated velocity, so that the last years of our life seem no longer than months, or even weeks did, when we were in youth. The more agreeable is our life—the more numerous the flowers scattered in our path—the shorter do the years appear to us. Thus the wheels of time rolled round to George and Jane Boardman, leaving about once in the course of each alternate revolution, another pledge and token of their unfailing love, so that by the end of the tenth year of their marriage relation they had two sons and three daughters. The youngest of the number is now more than eleven years old, and the eldest twenty-one. We are often astonished at the number of events crowded into a short life, especially, between the period of our majority and forty-five. These events occur in such close proximity to each other, and we press forward with such rapid speed, that they often seem to us like the houses of a country village through which we pass, in the train of a steam locomotive, without being counted, and as if they all joined together, leaving our impressions of them vague and indistinct.

At this time, in those five children, varying in age from eleven to twenty-one, we have a fair exhibition of the results of the efforts made by their parents for bringing them up according to the most approved modes of domestic culture. Each one of them, as soon as of a suitable age for it, was kept at school as long as needful; but before reaching that age, and ever afterwards when at home, their own mother was to them the best of teachers. They learnt far more there than in the schoolrooorn; and what was of especial use to them, as soon as they could take part in the family usage of reading aloud to each other every evening, the books used were selected in reference to their respective ages. This is the best of all training for the juvenile mind. All in turn become their own teachers; all acquire an habitual feeling of self-respect, and a manly deportment rarely acquired in any other way. Children thus educated become men and women in manners, and in all the fundamental elements of character, before they attain the stature of men and women. With such influences thrown around the junior members of a household, what is usually termed family government, ceases as it were to be necessary. A woman like their mother, at the head of a young family, is better to them than a dozen college professors or boarding school dames. A woman like her, as the wife of a farmer, will do more to advance his own individual interests, and to render rural life attractive and honorable, than the best agricultural lecturer on the continent.

The male head of this family has been pretty much passed over by us for the period of twenty odd years. Since renovating the old homestead in 1829, and his consequent marriage, he has had a subordinate prominence in our hasty sketch. It would be a pleasure to us to enlarge upon his merits and the happy influence he ever exerted upon the community in which he lived. He was not inferior to his wife, excellent as she was. Each was suited to the other; each was worthy of the other; and, both together furnish the best illustration of the good that can be accomplished in such a sphere, and of the respectability that can be given to the business of agriculture. In that family it is seen, that on a farm, amidst the constant duties each one is accustomed to perform, the most substantial mental proficiency is made by old and young; and on the other hand, there is no deficiency of social culture, and especially of domestic harmony and contentment. Home is the common centre around which all delight to assemble, all being strangers to a phrenzied impatience to indulge in the fashionable amusements of the age.

It is frequently imagined by persons in the country, and especially by the sons and daughters of farmers, that they can have no influence in society. Such a thought should never be indulged. There is in rural life, the best opportunity to exercise it. In the city, people are not swayed by any individual effort, unless it be from some commanding prominence. They scarcely know each other, although living in adjoining houses. They do not make the slightest recognition of each other in the streets or public places, unless formally acquainted. Thus they frequently live for years in joy and in grief, without ever speaking or exercising any common feeling of social relationship. Thus they see each other daily, but pass along unnoticed, as though they mutually possessed no common bond of interest or affinity. An ordinary person, whether male or female, in the city, has indeed, little or no influence. In the country it is otherwise. All know each other; all recognize each other; all sympathize with each other, whether in prosperity or adversity. Nothing is done in the country but what all know it, however trifling and unimportant in itself. Especially, if one does something that is unusual, the whole population stop to gaze upon it. If the farmer adopt a new mode of tillage; if he raise double of the customary crop; if he have a cow that yields double the quantity of milk usual in the neighborhood; or if his hogs weigh as much again as customary, all know it and talk about it, for miles in every direction.

So in family matters. If his wife well understands the business within her sphere of duty and supervision; if her house is always clean and tidy; if her children are always well dressed and comely in their manners; if they are sprightly and intelligent; if the family style of living is genteel and comfortable, and still maintained at a small expense; if her butter and cheese are abundant and of excellent flavor; if she dress herself like a lady in seasons of leisure, without much outlay for clothes, and is always prompt to do what belongs to her to do; and, if they all live and appear like persons possessing abundance, while all earn their living and much besides, be assured such a domestic establishment will become proverbial throughout the town or county. Where in the city, as the female head of a household, are such facilities for acquiring enviable applause? where such facilities for being useful in the world? Does the reputation from a display of silks and laces, and jewelry in the ball room, or at the theatre, or at the party of pleasure, compare with this? No man of sense, and no woman of sense and moral principle, can hesitate in deciding which of the two is most commendable.

Here is a field of labor open alike to all, the poor as well as the rich. Here female ambition can reach an elevation seen and admired by a whole community. Here young genius may bud and blossom and bear fruit of an hundred fold. Here the nerve and the muscle, and the business talent of matured manhood, can have full scope for enterprise, and labor, and abounding success. The example above detailed, illustrates every point in our position. Look at the poor orphan girl, Jane Bickford, who, by her talents and moral worth, rises to a position in influence, in comparison of which, fashionable equipage and brainless wealth is contemptible. Look at the plain farmer's boy, George Boardman, without the aid of factitious influence, having native good sense, industry, economy, and finally, the companionship of an excellent wife, becoming a man of moderate wealth, of unblemished character, and the father of children of which a prince might be proud, and in the order of Providence destined to an honorable career rarely to be enjoyed!

SEED-TIME AND HARVEST.

If in the agricultural year there any particular periods enveloped with especial interest, and standing forth with distinctive prominence, they are the ones here selected for a brief commentary. So important is the one in the farmer's schedule for labor, as well as for the exercise of skill in his vocation, that it seems to be almost a literal personification of the elements which form his character. Without it there would be little to give buoyancy and elasticity to his energies. Without it there would be little to inspire hope in the future, or to stimulate in that career of activity designed by the Parent of nature to be commensurate, in animal and vegetable life, with the necessities incident to existence. While the one is thus constituted the door of entrance into the calender of the husbandman, the other is the fruition of his most ardent aspirations; the joyous remuneration of his labors. Indeed, without these two periods, what a paralyzing monotony would be spread over the wide creation! Were the beings existing upon it to be perpetuated by some agency now unknown to us, there would be a sluggish development of life wholly incompatible with all our present ideas of happiness. So essential were these two periods in the Divine Mind, there was given to us the perpetual assurance that they shall continue to the end of time.

The leader of an army in contemplated conquests and victories, enters not on his campaign without making a comprehensive and judicious estimate of the means requisite in overcoming obstacles and in securing the objects of the enterprise. He proportions the forces to sustain him to those which may be brought against him; man against man, and weapon against weapon, for every possible contingency. His ammunition, his cavalry, his luggage vehicles, his provisions, and all his instruments of destruction, are not only procured in full competence, but are selected with wise reference to their excellence, as well as arranged with reference to their safe preservation and their position for use upon the least expected emergency. Without such systematic and well-devised preliminaries, defeat and disgrace would be the probable results. Without them there would be no laurels—no military glory ; and the anticipations of the commander would be like the baseless fabric of a vision.

Analogous to this should be the schemes and the precautions of every tiller of the ground anterior to the sowing of his seed. His lands should be surveyed and laid out with as much deliberation as a commanding general inspects and assigns for particular use the hills and the valleys which are to be the battle-fields on which he is to triumph or suffer defeat. How much in martial tactics does success depend on this ? Not less does the success of the farmer depend on a knowledge of his soils and the adaptation of particular localities for particular crops. Should he appropriate a particular locality for a crop to which it is most inappropriate, and practice a similar indiscretion or want of skill for all his crops, his disappointment in relation to a harvest would be inevitable. A farmer can no more resist the influence of such untoward mistakes than an army in a valley or deep ravine can overcome an enemy planted on a commanding eminence. And when the grounds have been thus judiciously assigned to specific uses, there should be prompt and efficient action in the preparation of them for these uses. The occupation of a husbandman cannot be profitable, nor indeed pleasurable, unless his grounds are well prepared by tillage and manure for the reception of the seed. The best of seed might as well be cast into the highway, or upon a brick pavement, as into an exhausted and sterile soil, possibly, too, as hard as sunburnt and compact clay. It would be as useless and ineffectual as it would be in a ship of war to point her cannon to the stars instead of living men, or to throw bombshells into the adjacent ocean instead of the assaulted citadel.

But our present object is especially to impress the farmer with the necessity of more attention than is usual in the selection and use of the different kinds of seed to be applied in the processes of agriculture. It is a doctrine of vegetable physiology, founded on well known analogies, and substantiated by the experience of the most discriminating amateurs of rural science, that the quality of the seed has much influence on the harvest that is to arise from it. Poor seed is not dissimilar in its agency for agricultural wealth, to poor gunpowder in the hands of the sportsman, or for the purposes of human slaughter in martial conflict. It may be entirely destitute of vitality, and hence have no germination; and much more frequently possess only an impaired or feeble vitality, and consequently yielding only a deficient or defective product; for it is a law of physics that everything in nature creates succession in its own likeness. To this law there may be exceptions; but they are so few in number and so equivocal in character, as to furnish no sufficient reasons for neglecting the law itself. Sometimes, possibly, deteriorated seed may furnish samples of excellence in vegetable growth which had not been known to spring from it for years, being one of the transient and lingering spasms of vegetable element, once predominant in the family to which it belonged, but now having no abiding features in it. It is so in the animal kingdom. Degenerated farm animals will very rarely have an offspring of some points kindred to those which pertained to the stock years before. And now and then in the human species there may be seen one in a family neither resembling father or mother, brother or sister, but a mere fac-simile of a grand-parent, or even a great-grand-parent. We have all witnessed such cases.

At the season of harvest the best portions of the crop should be saved for seed the following spring. This should be a general rule. For Indian corn in particular the largest ears and those which are first ripe are to be selected; and where there are two, or more than two, ears on the same stalk, preference should be given to these, provided they are of a large size Similar preferences in all similar cases are to be made. And in the case of maize, not only the early and large ears are to be chosen, but the small and irregularly formed kernels at the ends of the ears are to be rejected. It is difficult to prove mathematically the beneficial effect of this caution; but it is well known that where efforts are thus made for a succession of years there will be improvement in the quality of the article, and that where such precautions are neglected there will be a corresponding degeneration in it. This is analogous to changes wrought from corresponding influences in the rearing of animals. Who would think of taking a calf, or colt, or pig, or lamb, of pigmian size, or of sickly constitution, to be reared to propagate his species? Possibly the thing may have been done when the animal was worth less for the market and the table; but if we were to see it done, the inference would be irresistible that the individual who does it is ridiculously stupid.

Scarcely less important is it that all seed to be cast into the ground should be free from impurities. A neglect of caution in this particular will occasion more trouble and vexation than the cost of seed that is pure. In the cultivation of wheat especially, and where fields are to be put into meadow, some kinds of wild seed will diminish the value of the former at least twenty per cent, and will prove ruinous to the latter, unless labor be spent to eradicate the vile weeds more than equal to the value of the crops. In traversing some portions of the country, whole farms may be seen, comparatively valueless, from being covered over with daisy, or some other worthless and ruinous vegetable nuisances. All this might have been prevented if suitable prudence had been exercised in the selection of seeds and in the application of manure. In agriculture, above all things else, there may be a parsimony which tendeth to poverty; and if there is a liberal and enlightened expenditure, it will lead to wealth. A poor farmer has the more occasion to avail himself of every contingent circumstance, on the one hand to obtain adequate remuneration for his labor, and on the other hand to be secured against liability to loss and discomfiture.

The successful husbandman will always be cognizant of the appropriate time for planting his seed, as well as of the quality of it, or the due preparation of the soil which is to receive it. In our climate especially the appropriate time for planting is to be watched with scrupulous vigilance. Too much precipitation or too much delay will always be injurious, and sometimes fatal to the crop, The laws of vegetable development depend on the peculiar atmospherical and meteorological influences arising in the revolving year. Were we regardless of these laws and these influences by planting the seed before the return of spring, it might yield no germs—or, if there were germs, they would be so chilled and stinted as to receive only a feeble growth; or, by waiting till midsummer before planting it, the hot suns and parched soil would burn it up—even if it did germinate, the tender plants could not survive the withering breezes and the mildews of that oppressive season; or, by waiting till autumn before planting it, the frosts would assuredly prevent maturity. If the farmer would secure a harvest, he must sow his seed in the spring, and the different seeds successively in the different periods of spring, designed by the Author of nature to render them the vigorous germs of vegetable life and of abundant harvests. The Christian is furnished with certain means of grace. If he use them as it is intended they shall be used, he has no occasion to doubt their efficacy. So is it with the farmer. He is furnished with the elements of the material creation with which to unite his own labors. If he do it, a harvest is to be expected. But if he exercise skepticism or sloth in regard to such co-operation; or self-conceit, endeavoring to improve upon the laws of nature, instead of an autumnal fruition, he may actually die of starvation. With the farmer, as with the Christian, faith and works must go hand in hand, harmoniously supporting each other. Without the latter the former will be but a dead letter, and the works to accompany it, or spring from it, must be prompt upon every emergency, and abundant as they are prompt, like the gushing fountain upon the hill side that fertilizes and renders verdant the whole plain which lies beneath it.

There is a delightful harmony between the world of spirit and the world of matter. To the eye of the husbandman especially each throws its own shadow upon the other; each warms the other; each gives life and animation to the other. The faith of the Christian imparts to the laborer in agriculture an equanimity, a steadiness of progress, and a self-balancing hope of reward, found in no one pursuing his toils like the ox that knows not God ; and again, the enlightened and meditative agriculturist finds mental food in his occupation favorable to the growth of the sentiments of piety and the moral virtues. Here is a species of reciprocal influences, less obvious and less efficient in all other relations of industry. With what fervor and rationality may the farmer, on casting his seed into the ground, and having well performed every other incidental duty, pray to the Father of Mercies that the early and the latter rain—that the solar rays by day, and the abundant dews by night, may cause his barns, his granaries, and his storehouses to be amply filled at autumn ! Here is a foundation for growth in spiritual excellence, as well as for accumulation in the products of the earth, found not, according to our apprehension, in any other department of human labor. Can the merchant so pray for an increased valuation of goods on hand, and for diminished prices on those to be obtained, upon which depend his profits and his wealth ?

Verily to the husbandman there is a harvest as well as a time for planting seed. To him the autumnal harvest is an annual jubilee. To him it is an occasion of triumph and joyfulness, like rent day to the landlord, stock dividends to the capitalists, and the return ship to the merchant. To them, however, there are contingencies and disappointments seldom experienced by him. To one there may have been loss from conflagration—to another from commercial revulsion—and to the other from shipwrecks. Seldom is the farmer oppressed by such casualties. Seldom does the earth fail to yield a plentiful supply of its productions. If there is a deficiency of one, there is an overstock of the others. Rarely does it happen to him that he is unable to unite in the general acclamations of praise from nature ; rarely do the mountains and the hills break forth into singing, or the trees and the fields clap their hands while he is a mute or desponding spectator. Such is not the wonted destiny of the skilful and industrious culturer of the soil. If his face is sunburnt, and his arms and his hands are brawny, his soul is buoyant and expansive. Though he toil early and late, seldom does life become a burden to him; seldom is he overwhelmed with losses and perplexities, seldom is he deficient in the staple comforts that cause existence to be desirable.

Nor should it be forgotten, that, in the spirit-world, as well as in the realms of materialism, there is a seed-time and a harvest. In the one as well as in the other, there is to be a process of culture. The mind, as well as the soil, is to be made rich and vigorous by a supply of aliment congenial to its own nature. Without this process of culture, and this supply of aliment, both will be sterile and unproductive. There is this difference between them: in one case harvest succeeds seed-time after a few months; in the other case, the period between the two is the life-time of the individual, whether it be long or short. Hence childhood and youth are the spring season or the seed-time of the soul, and the harvest is the end of life. Thus the solar year is analogous to and is emblematical of a human generation. One has a calendar of months; the other may have a calendar of years. In both cases the periods assigned are proportioned to the nature of the fruits to be reared. As the solar year is none too long for the entire process of the great staples of vegetable life; so the life of man, though it be three score and ten or four score years, is none too long for the process of a great spiritual harvest.

Man comes into the world helpless, diminutive in size, and fragile in organisation, almost like the most delicate of our garden plants. A little too much sun or too much cold, too much moisture or too much drought, would, in many instances, be fatal to each. To either there must be vigilant nurture. If neglected, they will perish; or if they survive, they will be without value —like worthless weeds. Let a human being pass from the cradle to the grave without receiving any stimulants for intellectual expansion, and he will more resemble the automaton than the offspring of genius. Let him pass from the cradle to the grave without receiving any moral culture, and he will more resemble the prince of darkness than the candidate for celestial beatitude! But in the season of childhood and youth, furnish him with the seeds of knowledge; furnish him with the elements for mental vitality, and speedily he may be seen to rise above the world that sustains him, to grasp at the hidden mysteries of science, and to soar in imagination among the works of God far beyond any limits revealed to human ken. In such a field of culture, one like Newton may analyze the laws of gravitation; another, like Herschel, may number the stars and call them by name; another, like Hervey, may unfold the ceaseless action of animal mechanism; another, like Franklin, may disarm the thunderbolt of its power; and another, like Fulton, may teach his fellow-men, as it were to annihilate space and bring society into new combinations. It is by the use of such instrumentalities, in the seed-time of human life, that the world is blessed with those master spirits which give dignity to our nature and glory to the world. If men would rise to renown and become distinguished among their fellow-men, they must not, like the slothful or the unwise husbandman who neglects to sow his seed, allow their minds in the spring season of life to lie fallow. If they would see their sons and their daughters become good and great in the world, the foundation for it must be laid in youth. It is in the buoyancy of youth that the germs of mental distinction first display themselves. Then the soul is plastic and readily receives impressions. Then the elements of thought quickly become assimilated, and tend with progressive velocity, to the developments of genius in all succeeding time.

Nor is it less important that the spring-time of life be carefully observed in reference to a moral harvest. Why is it that vice has such luxuriant growth in the world? Why is it that crime so often becomes rampant, and with lawless strides passes over the land, leaving footsteps stained with blood or marked by desolation? Why is it that society is so filled with drunkards, the neglectors of religion, and the violators of the rights of their fellow-men? Why is it that so many mothers die of broken hearts, and so many fathers go down to the grave in sorrow? Why is it that our criminal courts are pressed with the most painful responsibilities, and that our penitentiaries are crowded with the outcasts of society? It is because the moral seed-time of the soul has been suffered to pass away in neglect! It is because no plants of heaven have been caused to take root in the young bosom! It has been overgrown with a spurious progeny, as the neglected garden is overgrown with worthless weeds. Its flowers have been choked by pernicious tares; its fruit has perished from the mildew of sin; and even the verdure of its leaves has become sadly faded and seared by some hidden malady destroying their vitality.

The sentimental ruralist is always surrounded with objects of absorbing interest. Among these objects the processes of vegetation are prominent. As soon as the seed is cast into the soil he daily meditates on the swelling kernel and the opening shell. No sooner does the slender blade rise above the surface than he watches its daily growth. With what pleasure in the morning does he behold the full orbed dew-drop upon it! The refreshing shower and the genial rays of the sun, as they unfold its vital energies, are observed with ceaseless assiduity. If it be one of our rich annual garden fruits, how soon does the ripe, luscious berry succeed the delicate flower! If it be one of our staple cereals, maize or wheat, who can fail to admire the well-filled ear? If it be a fruit-bearing tree, it is cherished year after year, till its branches becomes pendant with the load that hangs upon them, and, as it were, invite us to partake of their delicious treasure. If it be the oak, we fail not to make our pilgrimages to it, if life be continued, from the day when the acorn that gave it birth was cast into the ground, till its branches wave in beauty, and its stately trunk defies the tempest; and, indeed, till its enduring wood is moulded into the majestic ship that is to ride upon the ocean's mountain wave. Is there anything plodding and dull in watchings like these? Can the philosopher find a counterpart to them for chastened reveries and pious meditation? We think not! If, in the great temple of nature, one cannot find, in his own bosom, the kindling of rapture, where can he do it?

Yet there is not less pleasure in watching over the developments of the intellectual and moral seed cast upon the human soul. How soon does it take root and rise upward? How soon does the discerning mother and the attentive father rejoice to behold the sparkling infant eye and the symbols of infant thought in the motion of its tiny hands; and, anon, in its lively prattle and its laughing glee? Here are the first dawnings of character that may animate the family circle, and cast a halo over the dark places of the earth! Then the faithful schoolmaster records his prognostications of some casual tokens of advancing greatness; and not less does the attentive watchman upon the ramparts of Zion scent the odorous incense that begins to ascend from the young heart. The progress is onward. Each revolving year furnishing new evidence of a glorious intellectual and moral harvest. The seed was carefully spread upon a rich soil. The culture was without relaxation. All noxious weeds were removed. Thus a vigorous maturity is in prospect. The fathers of state quickly discern those on whom their own mantle is to descend; the ermine of the bench is enabled to salute with certainty those who are to perpetuate its own judicial purity; and not less readily does the Church open her arms and receive to her embraces those who are to minister at her altars and to share in her highest honors.

Obtuse indeed must be those sensibilities which can apprehend no occasion in such exhibitions for pleasure. The heart that can witness them without emotion must be destitute of every social attribute, and frigid as the frosts of Lapland. Such a heart would be a burlesque on manhood? If there is aught amidst the wreck of virtue, whose ruins have been scattered broadcast upon the world for nearly six thousand years, that can justly cause enthusiasm, it is in the growing adornment of that intellectual power, and in that spiritual regeneration which may be traced backward to a due observance of the spring-day of life and the seed-time of the soul. If we have one aspiration rising above all others, it is, that through the remainder of life, our own cup of blessing may be filled with the dews of heaven descending upon the soil and the soul, in their vernal seasons, and with the profusions of that harvest, whether material or spiritual, vouchsafed to all who labor faithfully in the culture of the earth and of the mind.

BATHING, OR CLEANLINESS AND HEALTH.

The advantage to health and comfort, from a free use of water, is becoming more and more an object of attention. That pure water is among the most valuable preservative, of health, as well as curatives of certain diseases, is become well understood. Indeed, some of our most popular retreats for the restoration of health, and for the building up broken down constitutions, are based on this hypothesis. As a general thing frequent ablutions, or immersions in water, are now, we believe, universally acknowledged to be very beneficial, and are the most effectual preventives of many distressing maladies. Cleansing the skin by rubbing, washing, and bathing, is, from the very nature of the acts, a most salutary operation. Moreover, it may be pronounced next to impossible for any person to enjoy perfectly good health who lives in the constant and habitual neglect of these means.

The cold bath is mainly designed to cleanse and purify the parts exposed to the action of the water. This seems well appreciated, at least so far as the feet, face, and hands are concerned, for no one would think of neglecting this, were it only for decency. But many persons who wash their face and hands every day—perhaps several times a day—apparently think it of little value to purify their bodies in the same way. It is indeed true, that the parts constantly exposed to contact with the dust which floats in the atmosphere, and with the dirt connected with labor, are more liable to need cleansing than the parts covered by our clothing. The latter, however, are not wholly exempt from occasion for less frequent ablutions. In the summer especially, when fine dust is abundant, enough of it will ordinarily pass through the interstices of cloth, to form on the skin, in the course of a few days, a coating so complete as effectually to conceal the color of the skin, and by filling the pores of it, to prevent a free perspiration. This dust uniting with the perspiration forms a kind of paste, which intercepts the free action of the air upon the surface of the human form, as unfavorable to health, as it would be to have the body and all its members besmeared with paint. Would plants be healthy and vigorous if they were completely covered over with some oily substance that would intercept the action of the atmosphere on their external surfaces? If this action of the atmosphere were prevented upon vegetation, it would speedily languish and die. The principle is much the same, were it prevented in regard to the parts of an animal body.

A medical writer states that bathing in cold water during the warm season, is an antidote to diseases, particularly fevers, by lessening the heat of the body; that it cleanses the skin from its impure and acrid contents, thereby removing a primary source of sickness; and, that the bath braces the solids, which were before relaxed by heat, restoring and tranquilizing the irritability of the nervous system, and greatly exhilerating and cheering the spirits with an increase of strength and bodily power. The morning is the best time for bathing, or two hours before sun-set, if in a river or pond, as the water has then, from the rays of the summer sun, acquired an agreeable warmth. When the sun has disappeared, or evening begins to throw her mists over the water, it is imprudent to bathe in the open atmosphere, on account of its dampness, which is apt to produce a chill, and sometimes a fever. Ordinarily it is best to bathe when the stomach is empty; either before the regular time of meals, or sometime after having eaten. On leaving the water the skin should be well rubbed by a coarse towel, which produces a most pleasant glow or increase of heat; and this is attended with a delightful serenity and cheerfulness; from which it may be known, that the bath is producing the desired effects.

Moderate exercise should also be taken after bathing, so as to restore the circulation, and produce a reaction in the vessels and muscles.

On entering cold water for bathing, it is best, first to wet the head; or if the bather can swim, to dive head foremost, so as to make the impression uniform; for one will feel the shock less by such precautions, than by reflecting and acting slowly and timidly. The time of remaining in the bath should always be short, and must be determined by the constitution and the feelings of the persons themselves, as healthy persons may continue in it longer than those who are weakly and in bad health. It is improper and unsafe to remain in cold water longer than a quarter of an hour at most, during the hottest day in summer, as the principal object of cold bathing is purification and the influence and effect produced by the first impression made on the system. At other seasons of the year, less than five minutes should be spent in it. On aged and lean persons the cold bath acts more powerfully than on young or corpulent ones; hence a young person, or one that is fat, can remain double the time in the bath to one that is old or of delicate constitution. Much judgment is therefore to be exercised in regard to the use of the cold bath. Young persons, particularly boys, sometimes have such a passion or fondness for bathing and swimming that they become debilitated from being much in the water; and cases are known where fevers have resulted from it, if performed under improper circumstances. An instance is recorded of an individual exposed throughout the day to fatigue and the heat of the sun. In the morning he took a cold bath, and the effect was the most salutary—a sudden increase of animal spirits and vigor. But towards evening he was induced to take a second one; and, from the body being in an unfit condition for it, an attack of chills and then of fever was the consequence.

It seems to be a settled conviction with nearly all, that water ablutions are indispensable for the health of children. Infants and very young children are daily subjected to them, as regularly, each morning or evening, as adults are accustomed to wash the face and hands. Without it, they would be disgusting in appearance and sickly in constitution. Where this is habitually and thoroughly done, there may be expected a ruddy complexion, full cheeks, athletic limbs, sweetness of breath, and bright eyes not otherwise witnessed. How loathsome are children covered with filth, and feeble for the want of these regular purifications! But, although meanly clad, if kept free from dirt, and rendered sprightly, animated, and muscular by the use of these auxiliaries to health, they become objects of kind regard. The cleansing of the skin, and the removal from it of all impurities and obstacles to the free action upon it of meteorological influences, is as needful for proper developments, in children and adults, as the removal of weeds in the garden in order to produce rapid progress in vegetation. Weeds, especially on being rank, will check the growth of all tender cultivated plants, both by absorbing from the soil the nourishment intended for them, and by secluding them partially or altogether from the sun and the air. But, weeds are no more unfavorable to vegetable progress, than the obstructions we are speaking of are to animal progress. Dirt and filth are as fatal to beauty and physical soundness in the human structure, as the most rank and noxious weeds are to the useful productions of the soil. And the rubbing of the skin and the muscles of the human form, so as to produce an active circulation of the blood and the other fluids, is analogous to the stirring and loosening of the soil about the roots of a vegetable organization. In both cases they are alike necessary; and in both cases, a neglect to do it will be attended with visible detriment.

We have frequently noticed one thing relating to this subject, very absurd and inconsistent. Thousands of parents are very particular daily to wash their children; not only their limbs, but their bodies; who never think of doing the same to themselves, although from their labors and more constant exposure to all the impurities floating in the air and mixing with objects in contact with them, are infinitely more in need of it, so far as viewed in reference to purification. True, the adult, from having more muscular power, is not so likely suddenly to die from the neglect as a child; but, if not so likely to die from that cause suddenly, does, from the very laws of nature, suffer in physical and mental energy. The difference is not dissimilar to that of the germ in its first years of growth and the full grown apple tree. The former, if neglected, will become choked and die of suffocation from the weeds; whereas, if the soil about the latter is neglected, and if the moss and other excresences on the bark are not removed by scraping or scrubbing, it will become unfruitful—it may be the fruit will be knotty and will fall to the ground before ripe; and, its yellow leaves and the dead tips of its branches will indicate premature decay. It is, certainly, very paradoxical for persons in the country to scrape, and cleanse, and whitewash their orchard trees, as they sometimes do, in order that the atmospherical influences may have unobstructed action upon them, and thus be made fruitful and be endowed with the elements of long life; and yet, rarely or never exercise corresponding precautions in regard to perpetuating in full vigor their own bodily organs. The care of fruit trees here alluded to does certainly bear a resemblance to the ablutions of the human body, and the succeeding friction from being duly rubbed.

It is a law of nature, that all physical elements, in the animal and vegetable kingdoms, should be kept in due combination, and in habitual exposure to influences conducive to their vigor. In unequal proportions their legitimate functions are impaired. The same may be said of them if they cease to act upon each other, as designed to act. Running water, by its motion and contact with the air, is relieved from any deleterious properties it might have held in solution, and is enabled to impart vitality and vigor to everything within the range of its course; but, if shut up and confined, it stagnates and becomes impure. All physical elements have certain agencies to perform; but, as soon as these agencies are performed, the elements become a nuisance and are to be cast away. Thus they are designed for animal food; but, if internal obstructions take place in the alimentary canal, to prevent the passing away of what is not converted into nutriment; or if the pores of the skin are closed, so as to hinder the escape of portions designed to escape through the perspiratory functions, the entire machine becomes deranged. It is believed, therefore, that the great secret of preserving health and promoting long life lies mainly in these two things; in the first place, to prevent all internal obstructions that impair the free course of the animal functions; and, in the second place, by external purification, to prevent any obstructions through the pores of the skin, which may on the one hand prevent the exhalation of poisonous secretions, and on the other hand, receiving from the atmosphere, through the same medium, those well known principles of health with which it is impregnated.

Near akin to the water purification we here recommend, both for decency and for health, is the keeping the teeth clean. If this is not done the breath will be foul, and sometimes feverish, especially in the morning. The teeth ought therefore to be cleansed after every meal, as the refuse of the food settles about them, rapidly becomes putrid, and proves injurious to the gums. It would be well, prior to retiring to rest each night, certainly after eating the last evening bit, that the mouth be well washed out and the teeth relieved from all remains of food; and, especially this should be done in the morning before and after breakfast; the first will create appetite for the breakfast, and the second a loathing of what is eaten. The effect upon the breath where the teeth and mouth are foul from the remains of food, is similar in character to the air of a kitchen, where all the fragments of uncooked meat and the dinner table are suffered to lie round on the table and upon the floors for days and weeks, till becoming masses of putrefaction, and perhaps breeding disgusting insects. Would not such an atmosphere be the most noxious to the senses, and be likely to breed fevers? And, yet how does it differ in kind from the breath of persons who neglect to cleanse their mouths and teeth? Or suppose a butcher suffer the drippings of blood and the atoms of flesh to accumulate on his meat benches, without any ablution for days and weeks, how repulsive would be the air of the market house! It might endanger the life of the sickly to enter it! Yet, the uncleansed teeth are but a miniature of the butchers meat benches; and the mouth to which they belong is but a miniature of the market-house which contains those benches.

In the country, especially, there should be connected with every house convenient accommodations for bathing. During the hot weather, if not at other times, an ample bathing tub should be daily filled—the morning is the best time for it, the water thus becoming warm—so that at the close of the day, each person having been exposed to dust and mud—to perspiration and the odors of the barnyard—may undergo a complete purification. It is true, a little moral courage will be requisite to do this with those have been greatly fatigued with labor. But when done, all who do it, will experience a freshness and a glow of mental and physical energy before deemed impossible; sleep to them will be sweet and sound; and they will be ten per cent. better in the morning for the toils of the day, than though this had been neglected. Even the washing of one's face, hands, and feet, after a day of fatigue and exposure to the impurities connected with most kinds of manual labor, makes a vast difference to personal comfort and to the refreshment experienced from a night of sleep. The cost of accommodations for doing it would be trifling; and the time spent by each one in these habitual ablutions, would be of no value compared with the benefit derived from them. Nor should they be confined to the male laborers on a farm. All more or less need them; and to children they should be deemed indispensable. These ablutions would add much to the amount of personal comfort; to health; to the amount of labor each one performs; and probably to the aggregate of human life among those forming a rural community.

Cleanliness ought to be attended to in our houses and about our work, as well as in our persons and dress. Fevers of a malignant nature often originate among the inhabitants of close dirty houses, who breathe an unwholsome air, wear their clothes until they become filthy, and take but little exercise out of doors. Where a number of persons are collected together under one roof, cleanliness is a point of the very highest importance, as it is a well-established fact that contagions are communicated by air which has become tainted, and which soon takes place by its being breathed over and over again, and by its being further deteriorated by a want of due ventilation and proper cleanliness. In the country, unless it be in some unhealthy locality, rendered so by miasmatic tendencies, liabilities to disease from such causes are not to be apprehended if a due attention is paid to the cleanliness of the house and out buildings of a farm. Here water is abundant, and the air is pure, unless corrupted by the filth of the premises generated in a want of due purifications. Neatness ought to be classed among the cardinal virtues. To be filthy is contrary to natural instincts. Birds defile not their own nests; and swine, if left to choose their own resting places, will rarely repose in filth. For human beings, then, to disregard what may seem to be a law of nature on this subject is inexcusable. A small cabin, if neat and tidy, elicits our commendation; but the spacious mansion, if neglected till it becomes a shelter for noxious vapors and a hot-bed for disease, is an object of disgust and execration.

We are much in favor of what is called the shower-bath. The means for it are easily devised. A tin tub or box of fifteen or eighteen inches in diameter, the bottom being perforated like a grater or cullender, placed six or eight feet above the head, and then filled with water, so as to descend in a quick succession of drops upon the person for whom it is designed, makes one of the best shower-baths. The suddenness with which the water falls gives a sudden impulse to all the animal functions, not unlike the exposure of the naked body to a smart shower of rain. If the blood circulates at too slow a rate; if the pulse has become languid; and, if there is a general prostration of the animal feelings, the shower-bath is of peculiar efficacy; nature, in a measure, is at once restored to its original and legitimate pitch of action. The operation of the shower-bath may be prolonged for a longer or shorter time at pleasure, according to the inclination or health of the person receiving it. As the shower passes the head and is gently descending to the feet, it occasions little or no obstruction to breathing, and during this constant gliding of single drops downward upon the skin, there are the most delightful and thrilling sensations. The shower-bath in the morning, when rising from bed, if any undue drowsiness remains, will stimulate the whole system, and prepare it for an early breakfast or any active duty. A little closet in one corner of a chamber or outbuilding may be chosen for the purpose. The size need not be more than two by three feet; and the water may be emptied into the perforated receiver, from a revolving vessel, operated by the bather with a string.

Bathing undoubtedly took place first in rivers and in the sea, but men soon learned to enjoy this pleasure in their own houses. Even Homer mentions the use of the bath as an old custom. When Ulysses entered the palace of Circe, a bath was prepared for him; after which he was anointed with costly perfumes, and dressed in rich garments. The bath, at that period, was the first refreshment offered to a guest. In later times, rooms, both public and private, were built expressly for the purpose of bathing. The public baths of the Greeks were mostly connected with the gymnasia, because they were taken immediately after the athletic exercises. The Romans, in the period of their luxury, imitated the Greeks in this point, and built magnificent baths. For undressing, for receiving the garments, and for anointing after bathing, there were different rooms; and connected with the bath were walks, covered race grounds, courts, and gardens. These buildings, together with a number of bathing rooms, were necessary for a public bath, which was adorned with splendid furniture, and all the requisites for recreation; and resembled, in its exterior appearance, an extensive palace. Roman luxury, always in search of means for rendering sensual enjoyments more exquisite, in latter times, built particular conduits for conducting sea-water to the baths, used mountain snow, and enlarged these establishments in such a way that even their ruins excite admiration.

Among the Europeans, the Russians have peculiar establishments for bathing, which are visited by all classes of the people during the whole year. The Russian bath consists of a single hall, built of wood. In the midst of it is a powerful metal oven, covered with heated stones. Round about these are broad benches. In entering this hall, the heat is so great, one not accustomed to it can remain there but a few minutes. This heat of the Russia vapor is raised to 150 degrees, sometimes above that for melting wax, and only a little below that for boiling spirits of wine. The vapor is produced by throwing cold water upon the heated stones; a thick hot steam rises, which envelopes the bather, and heats him to such a degree, that the sweat issues from his whole body. Upon an American this would produce almost suffocation; but the Russian considers it a comfortable luxury, sometimes remaining there naked on these benches for an hour or two, occasionally, according to inclination, having warm or cold water poured upon the body by the attendant. During the sweating process, the body is well rubbed or gently whipped with leafy branches of the birch tree, to promote perspiration by opening the pores of the skin. All this difference is the effect of habit; and that we might become accustomed to it, and find an advantage from it, as the Russians do, no reasonable person can doubt. Bathing, in its simple form, is a remedy of nature's procuring; and, when it was universally practised in Rome, she had comparatively but little necessity for physicians.

A Russian thinks nothing of rushing, says an eminent writer, from the bath room dissolved in sweat, and jumping into the cold and chilling waters of an adjacent river; or, during the most piercing cold to which his country is liable in winter, to roll himself in the snow; and this without the slightest injury. On the contrary, he derives many advantages from these sudden and abrupt exposures; because he always by them hardens his constitution to all the severities of a climate whose cold and snows seem to paralyze the face of nature. Rheumatisms are seldom known in Russia; such great and sudden transitions from heat to cold, seem to occasicn no detriment to the people there; and, on the other hand, it has been supposed that their great longevity, their healthy and robust constitutions, their exemption from certain mortal diseases, and their cheerful and vivacious tempers, are attributable to their baths and their generally temperate habits of living. When a young man, we boarded in the house with a Presbyterian clergyman whose health had been greatly impaired—some thought to an irreparable degree; but, at last, he resolved to be his own doctor, and completely restored his health. If living, which we believe he is—we know he was a few years since—he must be about seventy-five years old. He effected his cure in this way—by systematic temperate habits of living—one day in the week comparative fasting, and by cold water bathing and other exposures to the cold, and by habitual bodily exercise. We have known him, when the ground was covered with fresh snow, to rise early from his bed and to walk in the snow, barefooted, half a mile, before breakfast. We never knew him injured from it; and he uniformly affirmed he felt the better for it. We have, too, a little experience of our own in confirmation of the same theory. During much of life we have had occasional attacks of rheumatism. Towards twenty years since we were induced to spend a winter at the West, where we often lodged in cabins and log-houses, affording little more shelter from the elements than a large apple tree top, sometimes finding the bed in which we slept, pillows, head, and all covered with snow; yet, although we left home amply furnished with rheumatic nostrums, not an atom of them was needed. We did not suffer a rheumatic pang in the whole season of our absence.

It has been affirmed that the French owe much of their cheerfulness and vivacity of disposition to the warm bath; and, it would be extremely difficult to devise a punishment more to be deprecated by that people, male and female, than to be deprived of it, as it becomes from frequent and habitual use almost a component of their existence. The soft, delicate, and beautiful skins for which French ladies are so much distinguished, are very much owing, says the same writer, to the tepid bathing, being far preferable to all the cosmetics and other preparations sold for the purpose of whitening and beautifying the skin. The habits of persons are very different as to perspiration and sweating—some perspire very much, and others very little—from some no offensive effluvia arises in perspiring, whilst from bodies of others there arises a perfect fetor; and, it is a fact, that of all possible odors, that arising from the perspiration of some human bodies is the most loathsome; and to persons thus liable to exhale from the pores of the skin these fetid emissions, bathing is the best antidote. Hence, bathing may be used to free ourselves, under such circumstances, from being an offence to others, as well as from those foreign accumulations received from the air, and from objects connected with our respective avocations.

Among the Asiatics, baths are in general use. The Turks, by their religion, are obliged to make repeated ablutions daily. Besides these, men and women must bathe in particular circumstances and at certain times. For this purpose there is, in every city, a public bath connected with a mosque; and rich private persons possess private bathing houses, adorned with all the objects of Asiatic luxury. Besides the baths, the Turks have also the dry bath of the ancients. The buildings, which they use for this purpose, are built of stone, and usually contain several rooms, the floors of which are of marble. These rooms are heated by means of pipes, which pass through the walls, and conduct the heated air to every part. After undressing, they wrap themselves up in a cotton coverlet, put on wooden slippers, in order to defend the feet against the heat of the floor, and then enter the bathing room. The hot air soon produces a profuse perspiration; upon which they are washed, wiped dry, combed, and rubbed with a woolen cloth. At last the whole body is covered with soap, or some other application, which improves the skin. After this bath, they rest upon a bed, and drink coffee or lemonade. The Turkish ladies daily bathe in this manner—the men not so frequently.

A peculiar kind of baths are used in the East Indies, of which Antequil gives the following account: "An attendant stretches the bather upon a table, pours over him warm water, and begins afterwards, with admirable skill, to press and bend his whole body. All the limbs are extended, and the joints even made to crack. After he has done with one side he goes on with the other; now kneels upon the bather; now takes hold of the shoulders; now causes the spine to crack, by moving the vertebræ; and then applies gentle blows to the fleshy and muscular parts. After this, he takes a cloth of hair, and rubs the whole body, removes the hard skin from the feet with pumice-stone, anoints the bather with soap and perfumes, and finishes by shaving and cutting the hair. This treatment lasts about three quarters of an hour, and produces the greatest refreshment. An agreeable feeling pervades the whole body, and ends with a sweet slumber of several hours.

The best place for bathing is the sea shore. For various kinds of disease manifesting themselves upon or through the skin the salt water of the ocean exercises a medicinal as well as cleansing power. This is too well known to require corroboration; for persons living at a distance from the ocean are occasionally under the necessity of making long journies thither, in order to derive this benefit from it. Nor are we ignorant of the salubrity inhaled in breathing air that has passed over those strongly impregnated waves of the briny deep. Comparatively a few only live so near the ocean as to make its shores a place of resort when using the bath. Remotely in the interior a tolerable substitute may be provided for the class of invalids needing it, by dissolving in the water to be used for a bath, a quantity of salt sufficient to render it of the same saline consistence as the water from the ocean. This indeed is attended with inconvenience; but not to the extent of making the long journies sometimes rendered necessary where such a substitute cannot be had. Besides a vat filled with prepared salt water, like the ocean water, will remain fresh and pure, far longer than water not so impregnated, so that a new solution is not daily required.

Those living in the neighborhood of rivers, lakes, and ponds, may also be enabled to bathe at pleasure without any fixtures save what are thus provided by nature. Hence an adequate supply from one of these sources of water, on a farm, for bathing as well as for cattle and other purposes, is an important consideration in selecting a farm, or in estimating its value. Those who have experienced its convenience will fix no small price on such an appendage to a rural establishment. However, every farmer cannot have on his premises either river, lake, or pond ; yet, every farmer and his family are in daily or weekly need of the ablutions we are now recommending. Hence, if all have not the natural conveniences, they should provide artificial ones. Every one able to do it, and the cost would not be great, should have a room reserved for bathing purposes. It need not be large ; but, it should contain a bathing tub, fixtures for the shower bath, and if practicable, fixtures for heating water, so as to have the temperature of the water regulated according to pleasure, or the bodily health of the bather. We imagine that twenty-five dollars would provide all such accommodation for the life time of a family. Besides the comfort annually to be derived from such fixtures, the cost of them may be saved many times over in the preservation of health, and in the discharge of doctor's bills.

But, if this cannot be afforded, a common tub, or half an iron bound hogshead, holding four barrels, would make a decent bathing tub, and might last ten years. If kept well painted, it would last double that time. And if this cannot be obtained, with a good piece of sponge and a bowl of water, a person in ten minutes time, on retiring, or when rising in the morning, would enable one to undergo a pretty good cleansing from head to foot. Perhaps there is something better still. Let a person be covered over, and closely wrapped up in an old sheet, having been thoroughly wet in a pail of water ; and, then, after remaining ten minutes or so, be well rubbed with a coarse towel.

This will be found a pretty good bath, and will be attended with no cost, and with but little trouble or consumption of time. Many people of wealth use the sponge every morning for an ablution, preferring it to a bathing tub. There are ways enough in which it may be done, if persons feel the importance of it, and possess the resolution to do it. Were all to do it, much would be added to the aggregate of human life. If all living on farms would habituate themselves to it, many of the offensive and repulsive features of such a situation would be removed or neutralized.

AGRICULTURAL PROSPERITY.

Men are accustomed to discourse on national prosperity and individual prosperity—on mercantile prosperity and manufacturing prosperity. All this is understood, and becomes a part of our current literature. For instance, if the industry of a country is duly stimulated and duly rewarded; if the poor are well fed, well clothed, and well educated; if the people generally are happy, contented, and thrifty; if they are becoming wiser and morally better; if they are, as a whole, increasing in enterprise and wealth; and if the country is exporting more than she imports; if she sells more than she buys; if the balances of capital are in her favor, and if her means of defence and progress are ample, it is said that nation is prosperous. On the other hand, if her industry is not properly encouraged; if her poor are hungry, ragged and ignorant; if the people generally are becoming deficient in energy, means for a living, and the progressive elements of the age; or if her commerce and manufactures are languishing; if she is buying more than she sells; if she is in debt more than she can pay; and if she has little or no means of self-defence, it is said she is not prosperous. It is much the same with individuals: if they are increasing their means of subsistence, and are constantly adding to the accumulation of previous earnings, we say they are prosperous; whereas, if they are spending all or more than they earn—or if they are constantly consuming the little they possess, we say, they are not prosperous; that they are becoming paupers.

In the same way decisions are made upon mercantile and manufacturing prosperity. When are merchants said to prosper? Evidently when their canvass is whitening every ocean and sea; when their ships are causing a hum of business in foreign ports as well as our own; when the exported cargoes command ready and high prices; when the return cargoes yield fair profits; when specie too is flowing in upon us to cancel balances of trade in our favor; and when they can afford to pay substantial wages to our hardy seamen; then, it is affirmed in truth, that merchants are in a prosperous career of enterprise. But let the reverse of all this occur, and bankruptcy, like a frightful ghost, will show its terror-stricken grimace in the marts of trade; ships will be dismantled, stores will be closed, and banks will be declared insolvent. What a contrast! How many family establishments will be broken up? How many clerks will be thrown out of employment? Will merchants then be pronounced the princes of the land—the sole arbiters of a nation's wealth—the only legitimate candidates for ease and affluence? Let a period of commercial revulsion respond to these interrogatories. With such contrasts before us there will be no difficulty in deciding what is mercantile prosperity, and what is mercantile adversity.

Similar in its results would be a truthful exhibition of the manufacturing interests of a country. It is well known that there have been periods when the raw materials could be converted to their appropriate uses, enabling the capitalist to pay liberal wages for every description of labor employed, and then to realize a fair profit on requisite investments. Such periods give a business-like vitality to a community the most delightful to an active mind; sending its galvanic impulses to every branch of productive industry; the poor furnished with occupation and means of subsistence; the farmer provided with a near and a ready market for his surplus productions, and, what is equally beneficial, the factor stimulated and made buoyant with the hope of ample remuneration for well-directed devices. At such periods the curling volumes of smoke and flame may be seen rising like volcanic eruptions—only inspiring joy instead of terror—from the furnaces of every mineral locality; at such periods thousands of foaming waterfalls will be taxed to the utmost extent of their capacity; machinery will fill the broad land with the noise of its ceaseless revolutions; mechanical genius and enterprise will be on tiptoe; and on Sundays may be seen at church factory girls dressed like ladies, and orphan children, recently in rags and half-starved, clad like the children of rich men. This is manufacturing prosperity. This is one of the attributes of a nation's renown and glory! But, let these furnace-fires become extinct; let these foaming waterfalls become valueless; let this buzzing and clattering machinery become inactive, like a palsied animal limb, and what dismay and haggard looks will take the place of life, and joy, and plenty!

Agriculture, too, may be prosperous; not simply in providing competence for those immediately engaged in it, but in giving occupation and means of subsistence to those in other departments of industry. There is, however, a vast dissimilarity in the means of producing prosperity among manufacturers and merchants, and causing it in an agricultural community. So complicated are the elements of trade; so fluctuating are the circumstances on which commerce depends, that the shrewdest merchant is frequently disappointed in his schemes for gaining competence and wealth. Nor does he simply fail to increase the capital already his own; but, on account of the numerous disasters having a relation to his own interests, he frequently loses what he had, and becomes poor. It seems next to impossible for an active merchant, so to entrench himself as not to be liable to the most ruinous reverses. He may be rich one day and a pensioner on the kindness of friends the day after. It is proverbial, that not more than one person of an hundred merchants, becomes a man of enduring wealth. The others, like shipwrecked mariners, are submerged in a boiling whirlpool and are forgotten. It always has been so, and it always will be so. The difference between the past and the future will be this; the future chances for wealth will be even less reliable than they heretofore have been. The nature of society and of commercial business seems to render this unavoidable; and in manufacturing business, contingencies of the same nature, although it may be in less numbers, will always cause an analogous uncertainty in making plans for prospective prosperity.

Agricultural prosperity rests on a different foundation. Is there any security to the merchant that he will be able to sell his goods for more than cost? Has he any pledge that his ships will not sink to the bottom of the ocean? that floods and conflagrations will not sweep away his hoarded treasures? True, he may partially secure himself against loss from shipwreck and the ravages of fire, at the insurance office; but he can have no guarantee that trafficking in merchandize will afford him a remunerative gain; or even that his goods will not remain so long on hand as to be ruined by age and vermin. Or, has the manufacturer security, on investing capital in machinery, that the business to be pursued will uniformly and always be productive of profits? Surely not! Yet, the farmer has the pledge of Omnipotence, that there shall be forever, seed and harvest. Here is his dependence; here lies his confidence! He cleaves to the earth, which, like a fond mother, never forgets or casts away her own children. So long as well fed and consequently in health, her breasts will always be full. On her fair bosom he may always repose, finding rest and sustenance! She may not indeed enable him to become corpulent, and bloated, and by consequence diseased, like those fed on dainties and luxuries, but she will afford him a competent supply of nature's nutriment, so that he will enjoy health, and comparatively long life.

It is asked, in what consists agricultural prosperity? It may be replied generally, in raising good crops. However, to have a correct and full view of the subject, some specifications should be made. Possibly there may be a fair, and even a large crop of corn, wheat and grass, and yet the cost of its production exceed its commercial value. Under such circumstances, no one would be correct in saying, it is an evidence of agricultural prosperity. The merchant, to be prosperous, must make a profit on his goods, over and above paying all incidental charges; so must the manufacturer; and so must the farmer's crops exceed in value the cost of producing them. Otherwise, these men are not increasing in wealth; and, it may be, provided they spend freely, they are annually becoming poorer; and the degree of their prosperity is, moreover, to be estimated by the per centage of their profits. If one farmer can raise wheat at seventy-five cents per bushel, and another can raise it at the cost of a dollar per bushel, when its commercial value is one dollar and ten cents per bushel, it is easily seen that if both may be called prosperous, the former is more thrifty than the latter. So if one person can raise pork at four dollars per hundred, while it costs another six dollars per hundred pounds. So, also, if the dairy of one yields a profit of forty dollars to each cow, while the dairy of another yields a profit of only ten dollars to each cow. Similar specifications may be made in relation to every individual article made a part of rural economy. Accordingly, the enlightened farmer not only makes it a rule that each portion of his labor shall be profitable; but as highly profitable as practicable—in no case less profitable than when applied for corresponding purposes by any other person. This rule, rigidly observed, will ordinarily be attended with the desired success.

No one can be expected to form and carry forward such high views, unless he have what may be termed a good agricultural education. When we say education, it is not to be understood that the knowledge implied must have been obtained from books; it matters not how it was obtained; if it is possessed, that is sufficient; whether obtained from long experience—or from careful observation—or from familiar intercourse with others who possess it—it is the same in substance; or, it may have been obtained in all these several ways, and it is still the same. A knowledge of agriculture obtained at the plough, we call education, as though it were obtained from books. We contend for the reality, and not for some contingent or accidental circumstance. We should like to have agriculture made a branch of Common School education, so that all young men in the country—indeed, girls too—should be as familiar with its fundamental elements, as they are with arithmetic or geography. We should like to have it studied in our High Schools, Academies and Colleges, that individuals may be there qualified to teach it systematically, particularly in giving lectures throughout the country. But, if we cannot have this, we will accept any fair equivalent. If our farmers have a good agricultural education, it is of minor importance how or where they acquired it. If they have it, the means are at command for causing the prosperity of the interests of their vocation. It depends, therefore, on farmers themselves, whether agriculture shall be prosperous—whether or not it shall add to our national or individual wealth. Here, as already remarked, is an exemption from the uncertainties attending many other pursuits. It is for them to say, as a general thing, whether they will become rich, or remain poor; whether their country shall or not derive the full benefit of this great department of labor.

At the foundation of practical agriculture is a knowledge of soils and of vegetable and animal physiology. Animals derive their constituency from vegetables, and vegetables derive theirs from the soil. The soil, therefore, must contain whatever food or vegetable element enters into the formation of plants, or they cannot flourish and acquire their full perfection; and, unless they do this, there will be a corresponding deficiency in the growth of animals. It is just as absurd for a farmer to undertake to raise pork, without knowing the best and cheapest kinds of feed for swine; or to keep a dairy without knowing the best modes of rendering milk rich and abundant, so far as depending on the feeding and management of cows, as it would be to go into the manufacture of clocks and watches, without knowing the materials wanted in the construction of their machinery. What mortifying and ruinous results would attend the labors of the ship carpenter, who attempts to construct a steam-boat, without understanding the several kinds of wood and metal, and the respective quantities of each, wanted in its formation! And, yet this is no more absurd, and will no more be attended with mortifying and ruinous consequences, than for the farmer to undertake to grow a crop of corn, wheat, or clover, without knowing the elementary substances of which they severally consist, unless these substances are already in the soil. The case is so clear, it seems childish to offer arguments upon it. To attempt convincing farmers of this, might appear to be an insult to their understanding, inasmuch as it implies a disreputable deficiency of common sense, to say nothing of agricultural chemistry. And, the fact cannot be disguised, that unless farmers understand this, agriculture cannot be expected to prosper though seed-time and harvest are to continue till the day of doom. Many of our farmers seem to expect results without the use of requisite means; this God never ordained; this a wise man will never anticipate; to attempt thus to secure them is a war upon nature.

Perhaps, next in importance to the farmer, after a study of the soil and vegetable and animal physiology, is the theory and practice of thorough tillage. In this we include ploughing, harrowing, rolling, irrigation, and drainage, because on all these different soils depend, in their adaptation and efficiency, for the best agricultural remuneration. Tillage is based on principles of sound philosophy. The soil is to be made permeable as well as rich; and it should be thus to the depths penetrated by the roots of any particular crop. Hard and clammy soil can no more be assimilated into vegetable fibre than whole kernels or ears of corn, cobs and all, can be kneaded into dough for bread. In the former case as well as the latter, the compact substance must be made fine, so that each particle can be brought into contact with other particles, and perhaps with other substances, for new combinations. Or solid masses of soil can no more be moulded into new substances or compounds, than solid masses of metal, tumbled together in blocks, without being melted, can be formed into alloys. Hence, the necessity for ploughing, harrowing, or otherwise pulverising the soil; and incidentally to the reasons given for it, the temperature of it is changed by being better exposed to the sun, the essence of it is changed by being exposed to the action of the atmosphere, exhaling what is not needed, and inhaling what is, and it becomes a more copious reservoir for the moisture which is to nourish the growing plant. A thin soil can hold but a small quantity of moisture for the nourishment of vegetation. In a few days under a scorching sun, all will be evaporated, and the soil will become as dry as ashes. Then, for the want of moisture, vegetation will languish and die; but if the soil were in a condition to hold moisture to the depth of fifteen inches instead of five inches, this supply of vegetable nourishment would last three times as long. To suppose a thin soil will hold moisture as long as a deep one is as irrational as to suppose water in a wash-tub, eighteen inches in depth, will evaporate under a hot sun as soon as that in a milk-pan only three inches in depth, whereas five-sixths in the former will remain after all in the latter has disappeared. This is another argument for deep or sub-soil ploughing, as well as for good tillage to promote agricultural prosperity.

Irrigation and drainage, where respectively wanted, are essential to the best results in husbandry. Their influence may be rendered forcible by a familiar illustration. Suppose two quarts of water were the requisite quantity for being used in converting a given measure of flour into dough for bread, but one person engaged in doing it puts in but one quart, and another one equally unskilled in the process puts in four quarts, would either of them produce good bread? By no means. The first would have to put in another quart, and the last would be obliged to double the amount of her flour. So it is with moisture in the soil. If it is too dry water must be applied to it. If too wet, the surplusage must be removed, or good crops cannot be expected. Nature is the best of chemists.

All her compounds are in just proportions. It is from her laboratory that human chemists derive all their skill. Of her they take their best lessons. The best chemists are those who succeed best in operating according to her teachings. To imitate her is all they attempt to do in their compends of agricultural science. If they proposed to subvert the laws of nature, or to make improvements upon her domain, well might the farmer denounce their labors as hostile to agricultural prosperity. Such is not the fact, and it will be found, that if agriculture ever becomes in this country universally and greatly prosperous, science and practice must become handmaids in tilling the soil, and in every department of husbandry.

The importance of good tillage is shown in the products of the well cultivated garden. It is well known how much is produced from a small piece of ground, cultivated as a garden. An acre here will commonly yield as much as four or five acres with ordinary field culture. This is well known. We have frequently thought there is as much real profit from our own garden as from our farm ; yet the former is by no means superior nor the latter inferior to farms generally found in the neighborhood. Why is there such a difference ? For no other reason than the difference in the care of the two. Who would think of getting daily supplies from the kitchen garden, if no more attention were bestowed upon it, than is customary in field culture ? In the one the soil is from twelve to twenty-four inches in depth ; in the other, perhaps from four to eight inches. In the one the soil is made fine and permeable; in the other much of it is in lumps of from half a pound to two pounds in weight. In the one the soil is amply furnished with the best fertilising agents ; in the other it is comparatively as destitute of vegetable element as a worn-out animal is of blood and muscular power. It is a matter of course that an acre of one will yield as much as four or five acres of the other. If farmers generally would cultivate one acre where they now cultivate two, giving the same quantity of manure and applying the same amount of labor in one case they have done in the other, and their products will be increased, and their profits doubled. This is the way to make agriculture prosper. This is the way to make farmers think they get paid for their toil. This is the way to make them feel proud and contented, that they are farmers.

The necessity for manures is pretty generally admitted; by many, however, who do not seem fully to comprehend the philosophy of that necessity. The admission is made, because it has been so frequently seen, that without manures there is a partial, if not total failure of the crops; but why it should be so is not always clearly apprehended. If this subject is not understood, one might suppose there is no difficulty in understanding it. Let an attempt be made to simplify it. Let it be supposed a farmer has raised upon his farm forty horned cattle; oxen and cows; the bones of each one, to say nothing of the other portions of the animal structure, will weigh on an average, seventy-five pounds; three thousand pounds in all. The principal part of these bones is lime in some of its varieties. Whence did these forty cattle obtain the lime for these bones, as well as for their teeth? Evidently from the grass they eat and other food, for there is no other way in which they could get it. But how came lime to be in the grass, hay, and corn? They obtained it from the soil. Accordingly it is evident, the owner's farm contains thirty hundred pounds of lime less than though these cattle had not been fed from it. Now if these cattle are to be slaughtered and consumed on the premises, bones and all, there will be in the soil the same quantity of lime there was before; if they be sold to the butcher for slaughter and consumption elsewhere, the soil of this farm will be impoverished to that amount. Should the same process be continued on this farm, of raising cattle and sending them away for market, the lime will all be extracted from the soil, unless it be from time to time again furnished with it. Every gallon of milk also, and every pound of cheese or butter, or every dozen of eggs sold from a farm, unless again supplied with the materials of which they are composed, reduce its capacity for future crops.

Were the farmer's wife daily to draw a quart of vinegar, she must as often as doing it, supply the cask with the same measure filled with cider or some other liquid, or the time may be predicted when the cask will contain no vinegar which can be drawn. It is precisely so with the soil of her husband's farm; if he keep extracting from it the elements of vegetable life, without the restoration to it of their equivalent, he will ultimately render it incapable of yielding crops. The subject not being understood, this has been done in hundreds of thousand cases in our country, thus rendering those localities almost worthless for agricultural purposes. This has deprived agriculture of the means of prosperity. This has brought the business into disgrace. The fault is not in the soil, but in those who have the care of it; not in its original organization, or any allotment of Providence respecting it, but in those who have robbed it of the elements of vegetable sustenance. A good financier is aware that unless he put money into his pocket as well as draw from it, he will soon have none to spend, and will become bankrupt, like the farm of which we are speaking. It is the business of science, and of the scientific farmer, to preserve this equilibrium of the soil; and if preserved no occupation is more desirable than that of the husbandman—none is better remunerated.

The secret of having good crops lies in the manure heap; and they will be good in proportion to its size and its quality. If no manure heap is to be found on the farm, the owner of it understands not his business, and is destined to a career of poverty. He may have thousands of acres of land, but they will be like ships of merchants having no cargoes, or like the houses and stores of capitalists having no tenants, or like grain mills having no corn to be ground; in each case worthless, temporarily at least, because they yield no income. A farm without a manure heap is not very unlike a bank without specie, which will be unable to make dividends because its bills will not circulate. Hence the manure heap is the same to the productiveness of the farm, that the gold mine and the mint is to the bank; on each the auxiliary is worth more than its principal, because the whole vitality and profit of the principal is derived from the auxiliary. A farm without the manure heap is as incompetent to subserve its legitimate purpose, as the grindstone without a crank, or the plough without a plough-share, or the frame work of a harrow without teeth. Who but a stupid fool would chain his oxen or horses to the latter and drag it over his soil to pulverise it, day after day; yet this would be no more stupid than to spend the season in tilling ground as destitute of the power to promote vegetation as a steam engine to operate machinery when there is no fire in its furnace. Or the farm without the manure heap is like the horse having only three legs; the fences and taxes for the one being about equivalent in cost to the feeding of the other; yet both of no available consideration, because neither brings to the owner any income.

Let it be seen how well the farmer understands the philosophy of manuring his soil; let it be seen how well he performs the duties rendered imperative by this philosophy; let it be seen whether he is enriching his soil, as the judicious stock-proprietor causes the accumulation of fat and muscle in hogs and beef cattle, or is exhausting its energy as leeches draw blood from the sick patient, and his future progress, whether prosperous or adverse—whether ultimately in affluence or poverty—may be predicted with the certainty that the astronomer calculates an eclipse. Indeed, his results may not have the character of mathematical accuracy; but they will approximate very near to it—as near to it as anything can, depending in the smallest degree, on aught not of nature's law. It is affirmed, therefore, that the prosperity of agriculture may be made to rest on a basis more sure than that of any other industrial branch of enterprise and labor. To see a skilful and industrious farmer become a pauper, and the inmate of a poor-house, unless induced by disease or drunkenness, or some providential calamity, would be a most extraordinary occurrence—an event never to be expected. Our paupers, both vagrant and genteel, come from other localities than the farm. Our poor-houses are rarely to be tenanted by those who have been habitually devoted to tilling the ground.

There are other auxiliaries to prosperity in agriculture. Next in order, good agricultural implements may be named. It is not here proposed to go into an enumeration of those needed on a farm, or the particular models they should respectively exhibit, in order to answer in the best manner the purposes for which they are designed. We have not space for this in the present essay. Nevertheless, the prosperity of the farmer depends so much on a due appreciation of this subject, we cannot pass over it without saying a few words upon it. Why has the plough been generally substituted in place of the spade for ameliorating the soil? Why has the fanning mill come into use for cleaning grain in place of old borealis, proverbially capricious and coquettish in the dispensation of its aid? Why is the revolving horse-rake taking the office of the hand-rake, and the reaping machine the office of the hand-cradle? It is to save time! It is to do things at less cost! Suppose one man were to do the tillage of his cornfield with a spade, as once practiced in some places, and another one were to do the same labor with one of the best modern ploughs, and what would be the consequence? Plainly this: the crop of the former would cost from double to quadruple of the cost of that where the plough is used. Why is steam employed instead of horse-power for the conveyance of merchandise and all burdens? Because it is so much cheaper and so much more expeditious! Why is machinery taking the place of manual labor in making yarn and cloth, nails and cutlery, waggons and window sashes, chairs and cabinet work, muskets and agricultural implements? For a similar reason. Machinery will do this quicker, cheaper, and better than could be done by hand. Those who operate machinery for this purpose may become rich, while those who attempt the same without machinery, will remain poor and without patronage. The principle here disclosed may be of universal application; it appertains to every description of improvement in the construction of implements and machines designed to facilitate, cheapen, and perfect the processes of agriculture. That farmer can never prosper who neglects to avail himself of such improvement, while his neighbors, who do avail themselves of it, produce their crops at half or two-thirds the cost of them to him. We may, therefore, judge tolerably well of the character of the farmer, and his prospects of success, when we inspect the tools with which his work is to be done.

Good stock is also a powerful auxiliary in advancing the prosperity of the farmer. Why is he accustomed to keep sheep instead of goats, and hens and turkeys instead of domesticated crows and owls? For the obvious reason, that there may be a profit in the former, while none could be expected in the latter. This is sensible; and if the principle were carried out in all species of stock, rural life would present some new features of interest; for, if it is wise to select some species of animals for the farm, because of the remuneration they afford, and to reject others which do not afford remuneration, it is wise also to select choice varieties of the same species, and to neglect other varieties less valuable; especially as the cost of feeding the latter is about equal to that of feeding the former. A yoke of large, well-built, and well-broken oxen will do more labor than two yoke of inferior animals not well trained to work. Hence, if the former can plough as much in one day as the latter can in double the time, and at the same rate for any longer period, it is clear that all is saved between the cost of feeding one yoke and the cost of feeding two. Nor is this all. If the inferior ones are worth seventy five dollars each yoke, and the others are worth one hundred, by keeping the one instead of the other there is a gain of fifty dollars in the investment. Similar is the difference in the economy of keeping good horses and poor ones.

The same principle applies to sheep, cows, and hogs. The expense of keeping good and poor sheep must be about the same; yet the wool of the one may be worth double the wool of the other. With large wool-growers the advantage from the choice breeds swells to a large amount; and in small flocks it is one of the several items on which all wise farmers look for prosperity. In cows the gain is perhaps still more obvious to the superficial observer. The milk of prime well fed cows may not cost the owners more than a cent per quart, while that from inferior breeds, illy selected, and poorly fed, will cost two cents per quart. Nor is the cost of producing pork much less in favor of prime breeds over those which are not so. If there were nothing else besides stock on which farmers are to receive remuneration, it is easy to comprehend why one portion of them is thrifty and prosperous, and is becoming independent, and another portion of them is always complaining, and truly so far as they are concerned, that no money can be made by farming. Merchants and manufacturers, with their close habits of calculation, would quickly decide, and all in the same way, under analagous circumstances, which kinds to select. If the manufacturing of one description of goods would afford double the profit of another kind, would any time be spent in resolving which to produce ? If one description of merchandise could be bought and sold at double the profit of another kind, how long is it imagined the merchant would have to hesitate in which to invest his capital ? When will our farmers become equally wise and decisive ? On the arrival of that time they will all be prosperous and contented.

The farmer who calculates on prosperity will not neglect his fences. Fences are not indeed, like manure and good stock, his productive capital ; they are rather the means for preserving such as is productive. They are analagous to the boxes containing merchandise, or casks containing sugar, or flour, or wine, or molasses. Without these enclosures, the articles for which they are designed, would suffer injury—in some instances would perish. Packing boxes, and sacks, and casks are not then to be deemed irrelevant and a superfluity in the business with which they are connected : nor is the cost of them an excrescence which is to be removed. It is so with the farmer's fences ; without them he would often be unable to preserve what belongs to him. His crops would be in perpetual jeopardy from the ravages of animals running in the highway and upon common lands. Without them everything that had grown would be swept away by the lawless depredation of such hungry marauders. And, where fences are seen in a ruinous condition ; fallen down from violence or decay, the conclusion is unavoidable that the owner of them is poor and will be poorer still, or that he has none of that intelligence or ambition that makes a farmer independent. Strong and well made fences on a farm, and particularly by the road-side, are the best kind of evidence to the transient observer of good husbandry ; and, when near the mansion, all well devised and well painted as well as strong, they are also evidence of good taste in the proprietor, not less than of thrift. Such exhibitions of good taste create a peculiar charm in rural life, no where else to be found ; in simplicity resembling the manners of those who dwell there, and in rustic beauty the fair forms and the delicate hues, that spring up like enchantment amid the undefiled bowers of nature's realm.

Another incident and evidence of agricultural prosperity is the fair condition of the farm cottage or mansion, as the case may be, and the out-buildings connected with it. The early settlement of our country was unfavorable to that degree of taste and comfort which should be inseparable from rural architecture, but which is frequently wanting. The first residences in a new country are usually cabins or other cheap structures, simply affording a defective shelter, and wholly without ornament. These from time to time are removed, or enlarged by successive additions, according to the necessities or the means of the occupants. Of course nothing like uniformity, or architectural skill was seen in them. They as little resembled each other as the different trees in the adjacent forests. Perhaps, however, an exception to this remark should be made for the log cabins, which could not well admit of great diversification in external appearance, or internal arrangements. The effect of this was a general want of taste on the subject ; and, if now and then, an individual was found having the means and the disposition, to erect a house calculated to please the eye, as well as to be in conformity to established principles having reference to domestic comfort, it was difficult to secure the services of one competent to perform the labor. Hence, till within a few years, in most sections of our country, it was no easy thing to find any evidence that scientific architecture ever entered into the imaginations of the people.

There has not only been a neglect in the country to cultivate taste for agriculture, but the whole subject has been generally viewed with indifference, and in some instances, not rare, with disdain. Perhaps this, under the circumstances, is not very unnatural. However, it is easy to convince most persons, even of moderate mental culture, that such views are not only unsound in themselves, but are inconsistent with views entertained by the same persons, on other subjects. A vast many farmers who have burlesqued everything like architectural taste in the construction of enclosures for their wives and children, have erected barns, horse stables, and cow pens, on the most approved plans, both in reference to appearance and utility. Nor are they wanting in due appreciation of all nice points in the formation of ploughs, waggons, harrows, and other agricultural implements. In the latter they are mechanical philosophers of the first grade, and their quick and well balanced perceptions upon all matters of fancy, are equal to their skill in deciding on their respective adaption to the purposes for which they are designed. To us it seems absurd that such regard should be had to the things named, to the fashion of our garments and other personal ornament, for which even savages are not insensible, and to pleasure vehicles of every description, and there should be such an apathy for every feature of beauty or family convenience in the form and style of dwelling houses. To us this is perfectly absurd and inexplicable.

Extravagance and great expenditure in the construction of them, especially where the means of the owners are inconveniently impaired thereby, is always to be discouraged. Our theory is to build cheap houses, as well as expensive ones, upon correct principles and in good taste. It costs little or no more to make a pair of boots or shoes, perhaps not as much, as they are now made by first-rate workmen, than to fashion them in the form of a ram's horn, as worn in the reign of William Rufus; or, as worn in the time of Chaucer, when the points were so long that they had to be chained to the knee, in form resembling a church. It is much so in the construction of houses. Good taste, economy, and adaption to their legitimate purposes are all in harmony in rural architecture; neither is at variance with the others. And, it is a fact which cannot be controverted, because it is sustained by the best experience and the most genuinely good taste, that the display in the country of plain and neat architectural skill adds much to the elegance of rural scenery, much to the rational pleasure of rural life, and is an evidence of agricultural prosperity as well as of a growing refinement in those by whose agency this prosperity is advanced. Human pleasure, save what is falsely so called coming from a gratification of the senses, is the result of harmony between nature and well cultivated intellect, or of a commingling of social affinities of kindred mould, or of the moral tendencies of the soul in their higher efforts to assimilate with the undefiled spirit world. The first of these sources of human pleasure embraces all matters of taste, not the least of which is the rich display of artistic skill in the economy of civilized society, architecture being perhaps the most prominent.

It is difficult to conceive where can be found landscapes of more beauty than are exhibited in rural districts among the farm houses, out buildings, or rustic or well painted fences of an independent and enlightened yeomanry. Here the sojourner may sometimes pursue his winding way for half a day's ride, where every knoll and every greenwood opening presents some new type of cottage devised ingenuity. At one point the native forest still lifts its lofty and wide spreading tops, as if all were the work of creation's dawn; now and then projecting cliffs, with here and there dark ravines and sparkling cascades. Near by stands the New England mansion of the last century, with its sides of sheathing, its rusticated corners, its massive window caps and cornices, and its heavy substantial balustrades. Not far distant is a brace of mammoth barns, a corn crib, a workshop, cattle sheds, and a carriage house, with the entanglements of fences and gates, so as to appear in one group. Around the first bend in the highway comes to view the above in miniature, but rendered even more attractive by its shrubbery and flowers. Here too is the peach and the nectarine, bending under a mass of rich fruit. Adjacent is the full orchard with the best varieties of the apple and the pear in full perfection and ready for harvest. The road side leading either way is lined with transplanted forest trees of vigorous growth and rich foliage, the maple, the birch, the hickory, the ash, and the black walnut, alternately mingling their pendant branches. In passing along through fertile glades and well cultivated hills, sloping in every direction for miles, every house presenting some peculiarity of form or ornament, the traveller may observe the most simple kind and the slightly ornamented farm residence, but each in good taste—the Swiss cottage and the Gothic cottage—then those, having in conjunction or separately, features of the Egyptian, the Roman, the Grecian, and the Italian styles ; but none appearing bleak and desolate for the want of nature's garniture. Far from it ; the mountain side and the nursery have both contributed of their treasures to beautify each chosen spot. By one may be seen in near proximity the venerable elm and the tall pine ; by another the hemlock and the balsam ; by another the gum with its healthful odors, and the delicate mountain ash with its enduring berries ; by another the catalpa, the locust, the alianthus or the linden ; and others too diversified in their lineaments to be even hastily specified unless by a practiced botanist. Such a development of rural beauty and simplicity is surely an evidence of a highly progressing state of agricultural prosperity.

Glances at rural occupations and rural interests are as interesting to the philosopher as to the political economist ; for they not only lead to commercial results, but exhibit nature in an attitude to modify and improve the moral susceptibilities of man. The heart must indeed be cold and hard which receives no new impulses from the teachings in the volume of physical science. There is not only a beauty in nature seen no where else, but there is a magnificence in her, which enlivens faith, and secures a homage commensurate with her broad domain. What is there in the ledgers of merchants and bankers that can fill the soul with admiration like the illimitable prairie of

LIFE ON THE FARM.

wheat, its surface rising and falling as the waves of the sea? Look at the cornfields of the farmer, bowing down under their weight; look at his rolling meadows, as it were sinking under their heavy burden; look at his flocks and herds, large, well-formed, and sleek; look at his orchards, displaying rich fruit of every shade of color ; look at his kitchen-garden daily supplying his table with an abundance of its choice products, and say if it is in the power of a gold mine to cause so high an appreciation ? Is there not a moral suasion in the one not in the other ? And does not this suasion cause an ecstacy and fervor in the soul never induced by that metal, or any conventional evidence of its existence ? We apprehend so ! We apprehend the latter is powerless compared with the former, in such ministrations. The influence of gold on the affections is sordid and degrading. The influence of nature upon them is the reverse—it warms and expands the heart, creating aspirations that rise to the author of nature.

To the farmer nature is an august temple ; the broad horizon its gilded walls ; the trees its firm pillars ; the mountains and hills its high altars, and the sky its lofty dome ! Here he may daily hold communion with her ! Here he may daily burn incense ! In this temple is no difference of cast ! All are of one brotherhood, one faith, one hope ! In other temples there may be heresies, bigotry, and false worship ! But if heresy, or bigotry, or schism, or false worship spring up here, they come from other teachings than those of nature herself ! How the young heart here bounds and rises upward ! The blush of the morning in the season of flowers is not more beautiful and fragrant, than that of youth when inhaling their odorous breath, and before having been defiled by disease or sin. The nearest approximation of our species to the image in which we came from the hand of God, is when in the joy of childhood we thus sympathise in the harmonies and unite with the melodies of the wide creation. Here, too, mankind in its mid-day energy and glory, is confirmed and invigorated according to its first faith and love ! And here, also, tottering old age is occasionally rendered buoyant, and cheerful, and happy, and then lies down in peace, as upon the bosom of a beloved mother, to sleep till the morning of another life ! We may then say, in devout aspiration, of the home of the husbandman, as the inspired bard said of Jerusalem, "Peace be within thy walls, and prosperity within thy palaces !"

A RURAL WEDDING.

Early in the morning of July 23d, 1816, when a resident of the village to which allusion has been repeatedly made in the foregoing pages, a young man of fine personal appearance, but of plain and respectful manners, indicative of the best rural society, applied to us to go about ten miles that forenoon into the country to officiate at a wedding. As we had no acquaintance in the town where it was to be, we felt a little curious, as Yankees usually are in regard to similar unexplained incidents, what should have induced an application of the kind, attended with so much inconvenience on his part as well as of inconvenience to ourself, when persons might doubtless be had, at a less distance to perform the ceremony. He replied that there was no minister of the denomination to which the families most interested belonged in their neighborhood; and although they did not belong to the English Church, as he designated the denomination to which we belonged, yet the parties to be married, of which he was one, of a Sunday morning, three months previous, had been induced to attend the services in which we officiated, at the close of which a couple was married, and they were so much pleased, that they resolved some day to be married according to the form then used in the Church. Hence, it had been arranged, by the parties, that this intention should be consummated, and all their friends, for four or five miles around, had been invited to be present, with a special view to witness such a wedding as the one they had seen.

The fact cannot be disguised that we felt gratified at the personal compliment; and, that the Matrimonial Service of the Episcopal Church had made an impression on these strangers from the country so decidedly favorable. It was natural that we should have considered the facts stated evidence of the good sense and good taste of the couple becoming so prominent in the details of the present reminiscence. We did not hesitate, therefore, to give an affirmative answer to the request made. We and the principal invited guests were expected to be present at a mid-day dinner, soon after which the ceremony was to be performed. The details of this arrangement were not quite in harmony with the usages with which we were most familiar, but there was nothing objectionable on the score of taste, and with the old-fashioned practice of eating dinner at noon, when persons most need it, we have ever been pleased, notwithstanding the great innovation now general in regard to the hour of eating this most important meal. It will be perceived we could have had no time for unnecessary delay; and, as soon as practicable commenced preparation for the mission on which we were to engage. The young gentleman took his leave of us with the assurance that we might be expected soon to follow him, and to reach the house of his intended wife by eleven o'clock or soon after.

Before proceeding in our narrative, it may be interesting to young readers especially, to introduce some particulars connected with the institution of matrimony. That one of each of the sexes was designed exclusively for the other is probable from the uniformity of numbers in the sexes. Had that not been the design of Providence, the equality in the respective numbers of the sexes would not be so apparent as it is—with small variations, so far as statistics are possessed, it has descended from the first human pair to the present time, and it has extended to all places of which we have knowledge. Had a plurality of one sex been intended for the embraces of a single one of the other, it may be presumed they would at first have been created in proportions suited to such an end. Or, if the sexes were to be left to an accidental or promiscous intercourse, the proportional numbers of the two would not be alike at different periods and in different places; but, would be varied by external circumstances, as they might arise. It is well known, that in nature everywhere, each part is adapted to the end for which it exists. This adaptation is perfect, and so far as depending on a law or principle in physical science, is unvariable. Where can exceptions be found? In what part of the material kingdoms is not this principle cognizable? And is it not from such demonstrations that we prove not only one Great First Cause for everything, but the perfection and infinitude of his wisdom.

The number of males born is indeed a little greater than of females. Between different writers there is some slight difference of opinion in regard to this inequality. Mr. King, an English author, calculates that in London there are ten males to thirteen females; while in other cities and towns, villages and hamlets of that country, he says there are one hundred males to ninety-nine females. But Major Graunt, connecting the population of England, in one estimate, supposes there are fourteen males to thirteen females; and from this hypothesis infers that the Christian religion, prohibiting polygamy, is more agreeable to the law of nature than Mahometanism. Mr. Derham, another writer agrees substantially with Graunt, and says, that for an hundred years in his own parish register of Upminster, though the burials of males and females were nearly equal, being 633 males and 623 females in all that time; yet there were baptized 709 males, and but 675 females. In other places the proportion is about the same. Dr. Price, however, has sufficiently shown, that although more males are born than females, there is a considerable difference between the probabilities of life among males and females in favor of the latter; so that the males are more short lived than the females, there being a greater mortality among children as well as in adult age, when males are more liable to die from accident, great exposure, and excesses of living. Keresboom informs us that for a period of one hundred and twenty-five years in Holland, females have lived on an average from three to four years longer than the males. With respect to the difference between the mortality of males and females, it is found to be much less in the country than in cities and populous towns; and hence it is a fair inference, that human life in males is more brittle than in females, only in consequence of adventitious causes, or of some particular debility that takes place in polished and luxurious societies, and especially in great towns. The above writers, from these premises, have drawn the conclusion, already given as our own, that there is ordinarily one man for every woman, and one woman for every man, the surplusage of male births being needed for the waste of life among males, at sea, in war, and other extra hazards, from which females are exempt.

An able French writer takes a more extended view of the subject, and as it is concise and clear, we introduce it with some modifications. He says, in Europe there are always more boys born than girls, in the proportions of 21 to 20, or according to others, of 26 to 25. On the other hand, he adds, as we have already stated, the mortality is greater among the male children, in the proportion of 27 to 26; in consequence of which, about the fifteenth year, the number of the two sexes are brought almost to an equality—there being, however, a small surplus in favor of the male. But this surplus in the number of the men, even though it were three or four times greater, is carried off by wars, by dangerous voyages, and by emigration, to the casualties of which the female sex are much less exposed. Thus the final result of this is, that in our climate, and in the latter periods of life, the women are always more numerous than the men; that is, although there are more boys than girls, there are to be found more old women than old men. This difference is particularly observable at the conclusion of long wars. According to Wargentin, it amounted in France, after what is called the seven years war, to 890,000, in 24 or 25 millions of souls; and in Sweden, after the Northern war, about 127,000 in a population of two millions and a half.

If the plurality of women were in the early instead of the later portion of life, there would be more plausibility in the argument against the system of monogamy, that is, of marriages between one man and one woman. Such marriages only are fitted to ensure domestic happiness, and to maintain pure morals; such marriages only are indicated by the law of nature, and they are sanctioned also by the soundest maxims of political economy. Some travellers have indeed imagined, that in warm climates there are more girls born than boys; and as the male sex is more liable to rapid destruction in such climates than in ours, the surplus of women must there become very great. Hence, Montesquieu unsoundly coucludes that polygamy among those people admits of a very plausible excuse, because the position from which he makes the inference is not founded in fact. The researches of Father Parennin in China, the lists of baptisms kept by the Danish missionaries at Tranquebar, the various censuses taken by the Dutch at Amboyna and Batavia, and the observations made at Bagdad and Bombay by the judicious Niebuhr, have demonstrated that the number of children of both sexes is not more disproportionate in hot than in temperate climates.

In speaking of the institution of matrimony it is curious to consider the different ages at which the sexes under different circumstances arrive at a marriageable state. It has generally been supposed that this state is attained at the earliest age in extremely warm countries. The Barbary women are generally mothers, says Malte Brun, at eleven years of age. Buffon, quoting from Thevenot, says, that in the kingdom of the Deccan, boys are married when ten years old, and girls when eight, and that there are some which bear children at that age; so that they may be grandmothers before they are twenty. But if this were purely the effect of climate, as Buffon imagines, a very singular consequence would follow. The climate which the negroes of Senegal inhabit, is certainly warmer than Barbary, and even that of the peninsula of the Deccan. If, then, it is the influence of climate alone which accelerates the proper period for marriage among the nations of India, and which fixes it at ten or eleven years, the same influence should also fix it at seven or eight among the negroes. But so far is this from being the fact, all the accounts which we have consulted, seem to prove that the physical phenomenon in question depends rather upon the difference of race than that of the climate.

There are other facts still more conclusive in support of Malte Brun's inference. All the Russian and Danish travellers who have given us written accounts of Lapland, and the other countries in the neighborhood of the Frozen Ocean, agree in saying that the women in those regions become marriageable at a very young age. A Frenchman, who has seen much, and seen with observation, assures us that that is true of the Swedes. In Russia the peasants often marry at the age of twelve years. In the Vivarais, a mountainous and cold tract of country, the natives are as soon marriageable as in the other provinces of the South of France. And it is well known that the savages of our own country, living near the equator, are wont to marry no sooner than those living near the pole. With them the men do not marry before they are thirty nor the women before twenty. It appears then that we should consider this physical difference rather as inherent in the particular race than as dependent on climate. The cause is often found in the extreme corruption of manners ; but, let it proceed from whatever cause it may, women who marry and bear children at the early ages named, are liable to become old in all physical attributes at thirty. We know that it so in India.

Perhaps it may be expected, after detailing the above facts, that an opinion will be given as to the proper age for persons in our own country to marry. Were all to be materially influenced in their plans for forming and consummating matrimonial alliances by any rule furnished touching this point, the utterance of a judicious opinion would be more important than it now is. We believe such a question is philosophically of vast importance; that upon a due regard to it, the future vigor of the race, physically and mentally, will greatly depend. Nevertheless, not one in a thousand contemplating matrimony will be likely to give any heed to it. Some persons marry when their business or their personal convenience renders it desirable, not paying the least attention to age—perhaps not realizing that any important consequences can result from it. Others marry altogether from impulse, putting reason wholly aside. No one should deny that the former of these two classes are wise in making marriage to such an extent a matter of business, to be decided like other business questions, on principles of reason and expediency. Still the age of the parties should not be overlooked in treating a subject of such magnitude. Excepting in rare instances, a regard to the latter might not overrule due regard to the business features of the case. Our own persuasion is, that marriage should not take place except where there are strong counter considerations, till the physical developements of the whole system in each of the parties are complete, and till their mental powers have reached, if not maturity, a vigor that verges towards maturity. It is upon such a hypothesis only that we explain the difference of health, of muscular strength, and of intellectual stamina, in different individuals. We do not indeed have credence in the doctrine of innate ideas, but we have it in the doctrine of transmitting from parents to children mental capabilities, as well as in the similar transmission of bodily disease or of its counter attribute—a healthful constitution. And were there space for it, illustrations could be furnished, rendering our hypothesis probable, to say the least. Where have children ever lived to old age whose parents early died of consumption which they inherited from their parents? And where has an intellectually great man arisen from parents whose united ages when he was born, did not exceed thirty or thirty-five years? Possibly such a case may be found; but, for each one such, a thousand born under similar circumstances will be imbeciles.

As marriage is a connexion existing in all ages, and probably in all nations, though with very different degrees of strictness, it constitutes one of the most interesting phenomena for the inquirer into the various manifestations and different developements of the common principles of our nature. In almost all nations the day of marriage is celebrated with religious ceremonies. Nothing is more consonant to the feelings on such occasions than to propitiate, by prayers or otherwise, the blessing of Heaven on such an union; and the intercession of a priest was generally esteemed, in the early ages of the world, most efficacious. With the most ancient inhabitants of the East the bride was obtained by presents made, or services rendered to her parents. The patriarch Jacob served fourteen years for Rachel, double the time stipulated, in consequence of the deception practiced upon him by her father. It is a little amusing to witness amidst the first seven years of service—a pretty strong evidence of the sincerity and strength of his affection indeed—his credulity in not discovering for that period the deception of Laban—a credulity or simplicity that would be likely to find no parallel now-a-days. To the present day the practice prevails among the Circassians, and the poorer Turks and Chinese, of obtaining the bride by presents or services to her parents. Respecting the customs of the ancient Persians, Babylonians, Indians, and other inhabitants of Asia, ancient writers have left us little or no information. It is only known that polygamy was customary with them. In Assyria the following curious custom prevailed. Marriageable girls were put up at auction. Such as were handsome and attractive sold for large sums. The money paid for these was applied in the form of marriage portions to secure husbands for such as had not personal charms sufficient to secure husbands.

From the time of Moses, polygamy was prohibited among the ancient Hebrews; and, if Solomon and others took several wives, they rendered themselves guilty of a violation of the laws, particularly if these wives were foreigners. The Hebrews married as the Jews even now do, very young. On the day of the wedding, the bridegroom, being duly anointed and ornamented according to prescribed forms, proceeded in company with a friend and followed by several companions to the house of the bride; thence, veiled according to custom, and followed by her friends, she was conducted to his own house, or the house of his father. The procession was rendered cheerful and joyous by songs and music, and at later periods by torch lights. Allusion to these lights is made in the parable of the ten virgins. On the arrival at the place of her future home the wedding feast was celebrated, which continued for seven days, unless the bride was a widow, then only three days. The nuptial formality seems to have consisted in pronouncing a blessing over the couple. After the wedding meal, the bridegroom and the bride were conducted to the bridal chamber, where they were saluted with the melody of bridemaids. Among the Jews, if no physical objection is found to exist, males are allowed to marry when thirteen years and one day old, and females one year younger than that. Friday is the appropriate day for those of that nation now to marry; and, it is ordained that on the evening previous to the ceremony, the bride, accompanied by female friends, shall perform regular ablutions.

The early Greek marriage ceremonies, it may be inferred from Homer's account of them, had their origin in those of the Hebrews, for several of them are the same in substance. The Greek bride, veiled from the shoulders, was accompanied from her father's house to that of her husband, with torches, merry songs, and playing upon the lute and the harp. On their arrival, there was dancing and songs, and the bride attended the bath and received appropriate personal decorations. The house was also ornamented with wreaths; and fruits were cast upon the bridegroom and the bride, as a symbol of plenty. As a pledge of their perpetual fidelity, the axle of the carriage in which they rode was burnt, to indicate that she could not return to her former home. At Athens the pair ate a quince, in connexion with other rites, probably in imitation of the pomegranate eaten by Proserpine while walking through the lovely fields of Elysium on occasion of her nuptials with Pluto. At Lacedemon, among the marriage ceremonies, it was customary that the bridegroom should carry off the bride by stealth. Rarely on these primitive marriage occasions was there any lack in the profuse display of flowers, universally the tokens of joy and gladness; and, on some of them the female attendants of the bride, carried a sieve, a spindle, or some other implement of female handicraft, as symbols of domestic life.

The Romans were not deficient in nuptial observances. On the day previous to marriage, the bride made a sacrifice to Juno, or Venus, or both. On the day of wedding, she received bridal ornaments; a wreath of flowers, the tunic of matrons, a woolen zone about the waist, a red veil about the face in token of bashfulness, and shoes of the same color. After the completion of the sacrifices and other religious rites, the couple seated themselves upon the fleece of the victim offered in sacrifice, in allusion to the original dress of man, and the domestic duties of the wife. In the evening, the wife was led home by the bridegroom; she resting in the arms of her mother, or one of the next female relatives; the whole in allusion, as supposed, to the rape of the Sabines. The music of the lyre and the flute accompanied the procession, during which the bridegroom scattered fruit among the people. The bride was lifted, or stepped gently over the threshold of her parents' house, and of that where she entered, this part of the dwelling being held sacred to Vesta, the protectress of virgins. These thresholds were strewed with flowers, and she was preceded by the boy Camillus, carrying in a small box her personal ornaments and the amulets for the future offspring. Arrived at her new residence, she hung woolen bands, as signs of chastity, at the doorposts, and rubbed the posts with the fat of hogs and wolves, to guard against enchantment. Her first step into the house was made on a fleece, the emblem of domestic industry. The keys were then handed over to her, and both she and the bridegroom touched fire and water as signs of chastity and purity.

Of the marriage rites of the ancient Celtic and German tribes, as little is known as of the ancient Asiatic tribes; and, in the little which is recorded, the ancient authors, in some particulars, contradict each other. They are almost unanimous, however, in stating that the ceremony of buying the wife was customary with them; but, it is doubtful whether polygamy existed among them or not. Cæsar says it prevailed among the Britons; others say the same of the inhabitants of Spain. The Germans and Gauls seem to have had generally but one wife; yet, exceptions are known. According to the historian, Adam, of Bremen, polygamy was common with the ancient Saxons. Among the ancient Germans, the marriage of a free person with a slave was punished; and if a slave seduced a free girl, he was beheaded, and she was burnt. Although the suitor paid a price to the father of a girl and she became betrothed, if he delayed the marriage longer than two years, the engagement was dissolved. After marriage, the wife was deemed to be inseparable from her husband, so that divorces were rare, and a violation of matrimonial fidelity was punished by death.

The Mahometans consider matrimony as a mere civil contract; and they practice polygamy. They may have four regularly married wives; in addition to these, concubines which have been purchased, are held permanently; and they are also allowed to hire wives, whose obligation to live with a man lasts only for a limited time. Generally, nevertheless, the Mahometans have but one wife; the wealthier sort have two; the very rich, still more. With the Turks, the marriage is concluded upon between the parents, and at the most, the contract is only confirmed before the Cadi. Generally, the bridegroom has to buy the bride; most commonly they do not see each other before marriage. The bride is conducted on horseback, closely veiled, to the bridegroom. It is a real misfortune for a Turk to be obliged to marry a daughter of the sultan; for he prescribes the present to be made to the bride, and the husband is required to follow her in all things. So many presents are required, the husband is frequently ruined. In Arabia, if a young man is pleased with the appearance of a girl in the street, where the women always appear veiled, he endeavors to get a sight at her face, by procuring admission into a house where she frequently comes, and remaining concealed there by the aid of some kind relatives. If he is pleased, he makes a bargain with the father, and the contract is signed before a shiek. After the usual ceremonies, baths, and entertainments, the Arab awaits at his tent the arrival of the bride, who is conducted thither by female friends. On entering she bows, and receives a gold piece pressed on her forehead. She is then carried by him to the interior of the tent; around which the bride and other women dance all night.

The Laplanders marry very early; but a youth is not entitled to take a wife till he has caught and killed a wild reindeer. His friends first court the father of the object of his choice with presents. If the proposal be accepted, the young man is admitted to the presence of his fair one, and offers her something to eat, which she rejects before company, but accepts in private. He also promises wedding clothes, and makes presents of rings, spoons, cups, and silver coin. The richest also give silver girdles, and silk or cotton neckerchiefs. The parties being thus betrothed, the young man is allowed to visit his mistress from time to time; and as every visit is purchased from the father with some fresh token of obligation, the courtship is sometimes prolonged for two or three years. At last the banns are published in the Church, and the marriage immediately succeeds their publication; but the bridegroom is obliged to serve his father-in-law for four years after marriage. He then takes home his wife and her fortune, which ordinarily consists of a few sheep, a kettle, and some trifling articles; but the dowry of the wealthy consists of from thirty to forty, or even eighty reindeer, besides vessels of silver and other utensils.

In the Russian provinces, when a man has fixed upon a young woman whom he wishes to marry, he repairs to her dwelling, and addresses himself to her mother or nearest female relative, in language to the following effect—"Bring forth your merchandize; we have money to exchange for it!" The young woman is then introduced, and the terms are settled. More commonly, however, the match is made up by the parents, or friends, before the parties have seen each other. The bride is afterwards carefully examined by a number of females; and if they pronounce her to be free from personal defects, and of a good disposition, preparations are made for the wedding. On the day appointed, she is crowned with a garland of wormwood; and when the priest has concluded the ceremony, the clerk, or sexton, throws a handful of hops upon her head, wishing she may prove as fruitful as that plant. She is then led home with abundance of coarse ceremonies; one of which consists in the bride presenting her husband with a whip of her own making, in token of submission, and he fails not to employ it as an instrument of his authority. But the barbarous treatment of wives by their husbands, which formerly extended to the putting them to the torture, or even to death, is now guarded against, either by the laws of the country, or by particular stipulations in the marriage contract. Marriage among the Tartars consists in the purchase of a female from her friends—from twenty to an hundred reindeer is the common price—but when they are not to be had, the husband elect agrees to serve the father a certain time for his daughter. The Tartars admit polygamy; but the first wife is the chief, and the others are her attendants.

Marriage is celebrated among the Japanese with many ridiculous ceremonies, and often with great pomp. The princes receive their brides from the hands of the sovereign; and the marriages of the vassals are regulated by their lords. Among the middle classes in the cities, the business is arranged by the parents. The wives bring no portions, but are purchased of their parents and relations, to whom handsome daughters, or wards, are often a source of great wealth. The bridegroom most commonly sees his bride for the first time, upon her being brought to his house from the temple, where the nuptial ceremony has been performed, and where she is closely veiled from head to foot. On the wedding day, the bride's teeth are blackened with corrosive liquid, and they ever remain so. In some parts of the empire her eyebrows are also shaved off. After marriage, the wives of the rich are mostly confined to their own apartments; those of the other classes visit their relations, and appear in public, but are distinguished by great reserve and modesty. The marriages of the Calmucks are celebrated on horseback. On the appointed day for the nuptials, the bride, mounted on a fleet horse, rides off at full speed. Her lover pursues; and if he overtakes her, she becomes his wife without further ceremony. But if the woman be disinclined towards her pursuer, she will not suffer him to overtake her, and continues her flight at pleasure.

The modern Greeks still retain many of the ancient nuptial ceremonies. On the eve of the wedding day, the bride is led by her female acquaintance in triumph to the bath, attended with music. She proceeds at a solemn pace, profusely adorned, and covered with a red veil. The bridegroom and bride, before their presentation at the altar, have each a crown, or chaplet, placed on their heads, which during the ceremony, are interchanged by the priest. A cup of wine, immediately after the benediction, is given first to the married couple, then to the sponsors, and finally to the witnesses. The bride is accompanied home by her friends, who sedulously prevent her from touching the threshold as she enters the house, which would be reckoned unlucky. The splendid torch of Hymen blazes in the procession, and is carried to the bridal chamber, where it remains till it is all consumed. If by some accident, it be extinguished, the most dismal presages are drawn ; and, to prevent this, unremitting vigilance is used.

Marriage in Abyssinia is generally a simple contract, over which the priest has no control. When a man is derirous of marrying, he applies to the parents, or nearest relatives of the female of his choice, and their consent ends the business ; the girl being rarely consulted on the occasion. The settling of the dowry which she is to bring, is of much more importance, and sometimes attended with serious difficulties. When, however, this is adjusted, the friends of both parties assemble, the marriage is declared, and after a day spent in festivity, the bride is carried to the house of her husband, either upon his own shoulders, or those of his friends. The wife does not change her name ; and her dower is kept apart from her husband's property, to be returned, should his ill treatment force her to abandon him. Among the Caffres polygamy is general, and wives are purchased for cattle. The marriage ceremony consists in the bride's drinking milk presented by the bridegroom in the presence of the kraal, after which a number of cattle are slain, and a festival, accompanied with music and dancing, is kept up as long as they last.

With the Hindoos, any one who marries out of his caste, loses all his privileges belonging to it. They marry their children very early, often in the seventh year. When the marriage is agreed on, gifts are sent, with song and music to the bride. Similar ones are returned to the bridegroom. On the day before the marriage, the bridegroom, adorned with a crown and flowers, proceeds through the city, accompanied with music, and attended by young men of his own occupation, in palanquins, carriages, and on horseback. The bride does the same, on the day of the wedding, attended by her young female acquaintance. In the evening the wedding takes place. A fire is lighted between the couple, a silk cord wound round them, and a kerchief, folded up, is placed between them, after which the Bramin pronounces a certain formula, the purport of which is, that the husband ought to give sufficient support to the wife, and that she ought to be faithful to her marriage vows. In Pegu, the women are bought, and generally only for a certain time. In Siam, the husband may have, besides the legitimate wife, others whose children, however, are not legal, and are sold as slaves. In China, the wife is bought; poor people ask wives from the foundling houses. The young couple do not see each other before the contracts are exchanged. The bride is then conveyed with music and torches to the husband. She is carried in a chair, securely enclosed, the key to which is given, on her arrival, to the bridegroom. Here he sees her for the first time. The bride is then led into the house, where she bows low before the family idol. Entertainments then follow, each sex being separate from the other. After marriage, the wife sees only the husband, and on particular occasions, the father or some other relative, unless express provision is made for more liberty in the contract. The Parsees, or worshippers of fire, consider Matrimony a holy state, conducive to eternal felicity. At the wedding, the priest asks the parties whether they will have each other; if they answer in the affirmative, he joins their hands and strews rice over them.

Although the nuptial celebrations of various countries are affected more or less by the indolent or poetic, the energetic or superstitious temperaments of different nations, it is believed, it will be generally found, that in proportion as women are reverenced, and as civilization becomes advanced marriage festivities are conducted with proportionable increased solemnity and simplicity. The cases already given go to the establishment of this point. Others may be adduced. Let us see how matters stand in the South Sea Islands. There, if the union contemplated is between parties of rank, four large piles of plantain, yams, cocoa-nuts, bread, fruit, fish, cakes, bananas, with a baked pig on top of each, are, early in the morning, arranged in front of the house of the bridegroom, and the spectators assemble around them decked in new dresses, and their bodies anointed with sweet oil. Then the bride, closely veiled in fine matting made from the bark of the mulberry tree, is brought to the same place, and her feet, hands, and face being first anointed with sandal wood and tumeric, she takes her seat, and mock duels with clubs are performed in her presence, followed by boxing and wrestling matches, after which the bride and bridegroom, accompanied by their friends, who sing as they walk, enact a sort of procession before the spectators, who greet them with loud acclamations. The bridegroom then commences a dance with his young men attendants, during which the bride is led into her future habitation; the heaps of provisions are next distributed or scrambled for, succeeded by another boxing match; and the lighting up of the abode of the bridegroom, with singing and dancing in the evening, concludes these somewhat barbarous festivities.

In Spain, the warm climate and temperament of its people are exhibited in the poetical ceremonies attendant on courtship and marriage. When a mutual understanding has taken place between the young people, a night is appointed for the betrothment, and the lover seeks the fair one's abode, which is decorated with festoons and flowers. He is accompanied by torch-bearers, musicians, and attendants, who form a circle round the house, and a senerade is performed of the most flattering kind; and when she had been sufficiently woed, the coy maiden opens a little window, and asks what the gentleman wants? This leads to another rapturous burst of musical tenderness, and at last the lady throws down the garland from her hair, and promises everlasting constancy ; the musicians immediately strike up a triumphant allegro ; the windows are illuminated ; the maiden and her parents come out and conduct the serenaders into the house ; and firing of guns and shouts of joy resound through the calm, delicious air of Valencia. The day of the marriage is celebrated with musical entertainments, horse races, and divers other amusements, and at midnight the bridegroom bears away by main force the bride, who is detained as long as possible by her companions, to the beautiful arbor for their retirement on the terrace upon the roof of the house.

The woer of the Swiss cantons commences his courtship by the more truly romantic offering of a bouquet of flowers, gathered on the brink of a precipice ; and to see his beloved, he is often forced to journey many leagues over the mountains at night, exposed to the risk of being waylaid by jealous rivals. When the object of this nocturnal wooing has been accomplished, the wedding day is fixed, and, preceded by musicians and bridemen, decked in gay ribbons, the younger people walk to church, followed by a woman bearing a basket of flowers. The bride is dressed in a plaited apron, red hose, a floral crown, and a stomacher upon which are inscribed her Christian and surname, and the date of the year, and the chief brideman holds her by the apron. When the religious forms are completed, the spectators obstruct the way of the bridal party, who are obliged to give them wine before they can proceed to the village public house, where the festivities are to be held. Here Swiss dances are succeeded by the appointed person taking off the bride's virgin crown and casting it into the flames, whose crackling indicates that the young couple must not expect to be free from mankind's common portion of ill fortune during their career. Food is also distributed to the poor in an adjoining meadow, and with the simple fervor of religious faith in mountainous countries, the newly married are then conducted to the bridegroom's house, which everybody enters, after first kneeling down and praying for the welfare of the wedded pair.

The Illyrians usually appear well armed, and have their

hats adorned with peacock's feathers, in compliance with ancient prejudices, on nuptial occasions, and, even now, bloody encounters are too common, when rival suitors insist on trials of skill. As their wedding lasts several days, each guest is daily furnished with a small tub of water wherewith to wash himself, and each leaves in the tub some money for the bride, which thus augments her little dowry, of one cow and her wearing apparel. In some districts a ridiculous custom is observed, of the parents depreciating their daughter in set speeches before she is conducted to the house of the bridegroom, who says in return, to the young wife—" Well; I shall find means to bring you to reason: and to begin with you in time, I shall let you feel the weight of my arm." He then pretends to beat her, though his part of the business is not always confined to a mere form. Another curious ceremony at Illyrian weddings is during the wedding dinner, in the midst of which all the company rise up, and the bride is expected to throw over her husband's house a cake, made of hard coarse dough; the higher she can throw this, the happier will the marriage prove; and if the cake falls on the other side without breaking, it is considered a convincing proof that she will make a good house-wife. The firing of pistols is common in these provinces on festive occasions; and sometimes for a week before the wedding a bride is expected to kiss all the men who come to see her, in token of the regard which she shall henceforth feel for the sex of her husband; and, on the day of her marriage, the bridegroom's friends ride forward and present her with a white silk handkerchief, which she returns, and the messenger then gallops back to the rest of their party, among whom the kerchief is divided, and who, ranging themselves in a circle, partake of refreshment, amidst the discharge of fire-arms. On arriving at the bride's abode, the attendant maidens fasten an apple, encircled with flowers, to the top of the standard bearer's lance; and, on reaching the church, the bride is the last to alight, though she has the privilege of assisting her father-in-law to alight.

It would be needless to give further details of the celebration of matrimony. Our object is simply to show the importance that has every where, and in all ages, been attached to it. An institution inferior to no other one, and taking precedence of all others, both in its antiquity and its influence on society, merits this popular demonstration. The details given are mostly illustrative of the sentiments of mankind on the subject, in the subordinate classes. Among the corresponding classes in modern society, under the influence of Christianity, the celebration of marriage is relieved from much of the publicity and offensive ceremony known in early times, and at the present day, where civilization and refined feeling do not give tone to the current of popular usage. Marriage is no less than formerly viewed as leading to the most prominent developments of the human character, and to those primary combinations in society, on which the distinctive features in families and communities, in future times, will mainly depend. Were it not for these new combinations, the human family would degenerate, losing most of its elasticity and buoyancy, as the ocean would become stagnant and putrid were the tides to cease flowing. Matrimony in the one case is analogous, in this respect, to the tides in the other. By these two agents, the realms in which they respectively operate, are preserved in ceaseless freshness ; and to the end of time, without dimunition, will subserve the great purposes of Infinite Wisdom in giving them existence. Were these combinations not to take place—were families, century after century, to maintain, in all cases, like the castes of the Hindoos, an unmixed perpetuity, it requires but little of philosophy, and none of the spirit of prophecy, to see that the most lovely and valuable exhibitions in our species would be wanting.

The celebration of matrimony, in our time, and especially with the masses of society, as intimated, is essentially relieved from unmeaning ceremonies, and from public notoriety inconsistent with good taste, and is rendered impressive by the sanctions of religion connected with it. If there is any occasion in the whole current of man's earthly pilgrimage where an invocation for divine illumination, and for an increased sense of moral responsibility is especially necessary, it is when at the nuptial altar, before God, and in the presence of his people, two individuals, with the sanctity of an oath, pledge themselves to each other *till death do them part,* to perform faithfully all the duties, *in sickness and in health,* of the new relation into which they are thus admitted. How could the young and tender bride go out from the habitation and tear herself forever from the arms and the throbbing bosoms of father and mother ; and, how could they sustain themselves under such an abruption, were it not that the bridegroom thus solemnly and religiously promises to be to her even more than father or mother? If man or woman ever has occasion to supplicate divine wisdom to enable them well to understand their duties, and grace faithfully to perform them, it is when standing at that altar, and hearing the annunciation in the name of the Holy Trinity, that henceforward they are man and wife, and that no one is to put asunder those whom God hath thus joined together! If the angels ever gather up human tears, and with them, on unwasting scrolls, make record of human events ; and if the signet of Heaven is ever so deeply impressed on such record of human events as never—never to be broken, it is on occasions of which we here speak.

At half past nine o'clock of the morning named in our first paragraph, well furnished with a good horse and chaise, we were on the road to C., and the house of Mr. Mann, father of Elizabeth Mann, the young lady so soon to become a bride. Although it was so near the reign of old Sirius, from time immemorial slandered with causing the most oppressive temperature in the whole calender year—a slander of which we were never guilty, for we no more believe the Dog Star has anything to do with the suffocating heat of the latter part of July and August, than the moon has in stealing from the farmer's dinner pot the pork that is slaughtered when that lunar functionary is in its wane; notwithstanding the season of the year was so unpromising we had a delightful ride. The weather was quite cool, there having been a recent fall of rain. Of course there was no dust to annoy one, and as there had been no previous deficiency of rain to occasion drought, vegetation was nearly as fresh as in the month of May. Besides the air was balmy and fresh from the scent of new made hay. Everything, therefore contributed thus far to render the journey agreeable ; and, if one cannot enjoy rural excursions under such circumitances, according to our notions, his taste is not a subject of envy. We did once in early life, journey alone on horse-back, when in ill health, day after day, for months, but never to our recollection, felt oppressed with solitude, although passing through dense forests. There was a language in the trees and the waterfalls that seemed to supersede the necessity for human speech. In silence we could receive instruction, and be inspired with reverence, when meditating on the pathos of their moral suasion. So it was also, as we glided along to the mansion in which we were to officiate at Hymen's altar, on the twenty-third day of July, 1816.

On emerging from a noble piece of woodland of chesnut, oak, maple, and hickory, at about eleven o'clock, we observed in the distant opening, one of the best class of New-England farm houses, what is there called a double or square house, with four large rooms on a floor, besides projections. It was built in the best style of that day, and was surrounded with fruit and ornamental trees, and farm buildings, so as to resemble a little village. As we advanced, we soon discovered a large number of carriages, which satisfied us that this was the terminus of our mission. On reaching it we were met on the lawn by Mr. Timothy Scott, the young gentleman seen a few hours previous at our own house, and who was to be the bridegroom. Speedily we were by him conducted into the house, and introduced to the members of the two families, now to be so closely allied ; all in their manners and dress denoting the influence of rural competence and independence. The candidate for the bridal honors to us appeared particularly interesting, although evidently not rendered easy and flippant from much intercourse with what is called the best society. Still her manners were not so stiff and formal as to be repulsive. Her form was decidedly good, her features regular, her complexion as fair and ruddy as nature could make it, and, especially with splendid black hair and piercing black eyes, that had won victory from the Scott of her choice, and would have won it, had he been as tall and as brave as Winfield Scott, the present senior Major General of the armies of the United States. Such eyes in the farmer's daughter, with the concomitants, fine womanly developements and rosy cheeks, are as sure of a good shot as the best rifle. Let them have fair play, even on the philosophical bachelor, and the game is secured. His philosophy will yield to the milder impulses of his nature. Resistance will be useless.

At the hour of noon the dinner tables were spread and laboring under the weight of the most bountiful provision for a wedding entertainment. Not less than fifty of the relatives of the parties were present to partake of it. All were at ease, cheerful and happy ; all seemed to appreciate the proprieties of such an occasion ; all were joyful in behalf of the fortunate couple, upon whom the future was opening with such splendid scenery ; yet, these fortunate ones and their friends did not manifest boisterous exultation, as if peradventure thorns might not spring up where fragrant flowers were budding in the greatest profusion. If it is ever wise to throw restraint about an over buoyant spirit ; if it is ever the part of wisdom to let our moderation be known, and to rejoice with caution, if not with fear and trembling, it is at a time like this. Such appeared to be the conviction of all present. At the house of a wealthy farmer it would be a vain effort, to convey by words an adequate idea of a sumptuous dinner—its tender and well savored meats, the young porker, the lamb, the mutton, the bacon and the chickens—the fresh and nutritious vegetables, new potatoes, squashes, beets, green corn, lima beans, cauliflower, and tomtatoes—and not less palatable the rich fruit of the horticultural department of the premises, summer apples, pears, and apricots. We scarcely observed an article, not the product of their own premises, or of the immediate neighborhood.

However, it is by no means a part of our present purpose to give a minute account of that excellent dinner. Never in our life have we thought enough of what we eat to spend the time and paper requisite for writing such an account. Probably if we were now to attempt it, the most interesting portions to the gastronomer have been forgotten by us. The most we now remember of it is the general impression then made upon our mind, and that incidentally from its connexion with what succeeded ; for we positively remember more of the incidents of this wedding than of all the others we ever attended prior to the last twenty years, at the commencement of which we relinquished the care of a parish. This dinner was soon finished and the tables cleared. The rooms were then made ready, chairs on one side of the largest parlor being reserved for the bridegroom and bride. Till the hour of two o'clock there was a constant succession of other invited friends and neighbors, not at the dinner, entering and taking their places, wherever vacant chairs could be found, so that finally not an unoccupied niche remained. The whole number could not have been less than one hundred—possibly more. As soon as all had come that were expected ; or as soon as there ceased to be accessions to the party, impatience increased for the performance of the ceremony, and especsally because the rooms were so crowded. A message was sent to the nuptial suit, that all was in readiness ; yet they came not—for fifteen minutes at least, all was silence. Then, another message of the same tenor was sent. They still came not ; but, there was a response ; to wit, that the bridegroom and bride desired first to see the minister. Their request was promptly granted.

We were not a little puzzled to imagine what could be the trouble. However, on entering their room, the revelation was soon made. The bridegroom and bride, with their attendants about them, looking more demure than the day required, met us at the door. We perceived some dilemma in the way, and that they had not courage or tact to make a joke of it, and tell us what they wanted. For a moment they were silent. To relieve them of the embarrassment, we blandly said to them—"Well, my good friends, what is the trouble?" Mr. Scott, with evident discomforture, replied—"We never used one of your Prayer Books, and never saw but one couple married, according to your form of marrying, and are afraid we shall make mistakes in the service. If we do, we shall feel mortified, and our friends will laugh at us! Will you not marry us, as the Baptists and Presbyterians marry persons?" "It is a hard case," said we, "if persons in your present situation cannot be indulged in having their own way, in a matter like this. We should like to gratify you—nothing would give us more pleasure; but, we apprehend, that afterwards you would be sorry, that we did not advise you otherwise! It has been understood by your friends, that they were to witness a Church wedding; it has been your wish and request to be married in that way; we have been invited to be present for that purpose; and, now, if we were to substitute another service, they would be disappointed, and would have the best ground to laugh at you for your timidity. Nor have you the least occasion to fear perplexity in the service; it is perfectly simple; you cannot well make mistakes in it; each part suggests what is to come next; when once commenced, you will glide along in it, as readily and as naturally, as a carriage will descend an inclined plane; and, you will be through it, almost unconsciously, so well pleased, it may be you will like a repetition of it" Hence, we took the Prayer Book, and opening at the marriage service, run it over with them, showing its brevity and simplicity, saying nothing could be easier. They seemed to yield a tacit consent, which led us to suppose all was then right. Accordingly, we returned to the parlor, expecting they would speedily follow, so that our labors would forthwith be at an end.

Nevertheless there was farther delay. The nuptial train were still missing. They came not, and minutes to those in waiting for them began to be tedious like hours. All began to be dissatisfied at the procrastination, each asking some one near by what it could mean; or making sagacious explanations for it; and, the venerable mothers of the bridal couple became so fidgety, they could neither sit still or refrain from the desponding use of their tongues; Mrs. Mann was distressed for fear that Elizabeth was suddenly unwell; and Mrs. Scott improved the alternate moments of silence in wondering why Timothy was causing such delay. Thus another quarter of an hour was spun out, when a second request was made that the minister would hold another parley with Hymen's unfledged disciples. No time was lost in complying with the request, for time began to be unusually precious; there was no telling what new type of perplexity was to exhibit itself; possibly, thought we, our journey hither will be to no purpose. On again entering their room, without an instant of hesitation, and with good natured surprise, we exclaimed—"Why, Mr. Scott, what in nature is the matter now; has the bride any conscientious scruples against being endowed with your worldly goods, as required in the service, when you place the ring on her finger; if she has, perhaps you would be willing to dispense with this portion of it?" "No, no," says the bride as quick as lightning. "I have no idea of that." We perceived we had hit her in the spot, for the mischievous play of her bright eyes was fully manifest; the strong tension of the muscles in her countenance gave place to a bewitching smile; and, in an instant the entire party was in a titter. Not disposed by delay to lose any advantage we had thus gained, we instantly added—"What in the name of goodness, then, can be the matter—do tell—be quick, for the old folks are impatient to have it over!" "The truth is," said Mr. Scott, "Elizabeth don't want to promise—*to serve and obey*—and I had as lieve she would not—we wish you to omit this part of your service."

"Oh, nonsense," we exclaimed, "do not all good wives, having kind husbands, serve and obey them—that is, devote themselves to their interests, when nothing unreasonable is expected, whether they promise to do it or not? We are not aware, that wives who have made this promise are oppressed with improper requisitions, or that they feel the most slightly degraded ; or, on the other hand, that husbands who have received at the nuptial altar such a promise, are accustomed from that cause to become unkind or tyrannical." But, even after this, the bride rejoined, "I had rather *not promise to do it ;*" her manner, however, satisfied us, that she seemed more intent to be thought by her friends to carry her point, than to maintain conscientious scruples. "Well, well," we added, "follow me to the parlor, as cheerful as you have occasion to be; be married without any mutilation of the service ; speak up boldly and honestly in every part of it; and, when it is over, provided either of you shall feel dissatisfied, you shall be married over again, precisely in your own way. Have no fear, it shall be done ! Are you not satisfied now ?" The flag of truce was promptly accepted ; all sorry looks disappeared ; one of the bride-maids, a little more ready in an emergency than the rest of the company, gave impulse to our remarks, saying, "come, Lizzy, you cannot object now—don't wait any longer. I should like two weddings to-day—if you are not pleased with the first, you can have the second—we have plenty of cake and wine !" The others said—"Oh yes, oh yes—now go along !" The silk dresses began to rustle, all fell into due order for an onward movement, and in less than five minutes, they were duly marshalled in the parlor, the friends and guests all rising as they entered.

For a moment all was profoundly still ; we then said in the language of the Marriage Service—"dearly beloved, we are gathered together in the sight of God, and in the face of this company, to join together this man and this woman in holy matrimony, which is commended of Saint Paul to be honorable among all men, and, therefore, is not by any to be entered into lightly or unadvisedly, but reverently, discreetly, advisedly, soberly, and in the fear of God. Into this holy estate these two persons present come now to be joined. If any man can show just cause why they may not lawfully be joined together, let him now speak, or else hereafter forever hold his peace ;"—then turning to the persons to be married, it was added, as part of the same service—"I require and charge you both, as you will answer at the dreadful day of judgment, when the secrets of all hearts shall be disclosed, that, if either of you know any impediment why ye may not be lawfully joined together in matrimony, ye do now confess it; for be ye well assured, that if any persons are joined together otherwise than God's Word doth allow their marriage is not lawful." We never witnessed a more respectful attention. The novelty of the service to the persons present doubtless made it the more impressive. It would be difficult to conceive how language could be better adapted to such an occasion than that here used.

After the customary pause, that these admonitory addresses might be duly considered, we then said to the bridegroom—"Timothy, wilt thou have this woman to be thy wedded wife, to live together after God's ordinance, in the holy estate of matrimony? Wilt thou love her, comfort her, honor, and keep her in sickness and in health; and, forsaking all others, keep thee only unto her, so long as ye both shall live?" Manfully, boldly, and promptly did he answer—"I will!" Upon which we said to the bride—"Elizabeth, wiit thou have this man to be thy wedded husband, to live together after God's ordinance, in the holy estate of matrimony? Wilt thou obey him, and serve him, in sickness and in health; and, forsaking all others, keep thee only unto him, so long, as ye both shall live?" Gladly did we witness her own response, equally loud and distinct—"I will." True, some of the girls, on hearing those obnoxious words—serve and obey—shook their heads, as if that were a hard condition of matrimony. The old folks appeared to think well of it.

The hands of the bridegroom and bride being now duly united—and never had we witnessed a more sincere and hearty grasp—each in turn, then repeated audibly the language of the above interrogatories with a pledge of ceaseless fidelity. Whereupon the bridegroom placing a ring on the fourth finger of the bride's left hand, repeated after us, in an emphatic tone,

—"With this ring I thee wed, and with my wordly goods I thee endow, in the name of the Father, and of the Son, and of the Holy Ghost. Amen." The girls, at this, assuredly nodded and smiled, as it were, saying, we would promise to serve and obey, too, if we could get this hold of Timothy Scott's purse-strings. Thus ended, with the usual prayer and benediction, the formal services of one of the most memorable rural weddings at which we ever officiated.

Restraint was now thrown aside, while a profusion of cake and wine seemed to form the crowning epoch of the day. We are wholly unable to say, how long it is since so palatable an appendage was decreed for the marriage ceremony. The longest memory of the present generation goeth not to the contrary, we believe; and, this we do know, that more than eighteen hundred years ago, so important was the juice of the grape esteemed, on one of these occasions at least, a deficiency of it was supplied by the performance of a miracle. This may be considered tantamount to all other authority for such an appropriation of it. The use of wine, like many of the good things given to us, often becomes an offence from its abuse. How far this is ground for being discarded at weddings, as well as on other occasions, we pretend not now to decide. Whatever is found injurious to health and morals should evidently be rejected. Perhaps fewer objections can be made to this rule of conduct than to most others.

Time passes away swiftly where all are in good humor, and consequently happy. This is the great secret for which all are desirous. So it was here. The old family clock, which had, for the quarter of century, here been a faithful sentinel, striking five, admonished us that there were but two hours to sunset, and that preliminaries should be adjusted for our homeward course. These preliminaries at a wedding, where the company is large, consume not a little time. Accordingly, we entered upon them. As soon as this was perceived, a huge piece of the wedding cake, the customary perquisite of the officiating functionary's better half—provided he has one—was duly made ready. We were not wanting in respectful attention to the parents of the bridegroom and bride, extending to them those congratulations, at such a season always expected. Rarely is the future on marriage occasions more bright and cheering than it was to these excellent individuals. We failed not to impress this on them; and we affectionately endeavored to encourage in them a thankful heart. Apparently they sincerely reciprocated every expression of kind regard for them. We felt like rejoicing with them. Others too in this numerous company had rendered themselves so agreeable as to be entitled to something more than a wholesale, hasty, and laconic recognition in bidding them adieu. Not a few of them were formally taken by the hand, where the grasp was mutually ardent, as we withdrew from them.

Nor did the wedded couple fail to receive more than ordinary attention. The incident that had led to a delay in performing the duties of our own mission seemed to open the way, not only for an unusual sympathy for them, but to correct some misapprehensions, presumed to exist with others present, in relation to the obligations enjoined in the service just had in requisition. We thought it wise not to let the opportunity to pass unimproved. Intimating this, the whole company became ready to listen to our remarks. Hence, we said to the bridegroom and bride—"You cannot be unaware, it is a matter of course, that, on our departure, we express an affectionate solicitude for your future happiness. Be assured, it is no common place formality with us for yourselves. As little as it might by some be imagined, considered that our personal acquaintance commenced in the morning and perhaps will terminate for ever in the evening of the same day—being only a few hours—we cherish an inexpressible desire to contribute personally in causing you to be happy; to show, by our act, that we are sincere in wishing you so. The matrimonial union is often viewed, in prospect, as a rich and beautiful garden, filled with fragrant flowers and luscious and nutritious fruit; it may indeed be made such, when its relations and responsibilities are duly appreciated and regarded. On the other hand, it is not considered, that it may ever be like waste ground cursed with a rampant growth of noxious weeds; yet, it may be so, if those to whom its care is entrusted, misapprehend and neglect these relations and responsibilities. It depends, then, much on yourselves, individually and unitedly, whether this union, into which you have just entered, be the one or the other of these states; whether it shall be blissful like ancient Eden before man's defection, or like a polluted soil, rank with poison, subsequent to that defection."

"Prominent among the responsibilities alluded to is the one that caused delay in your recognition of it, a few hours since. The promise, on the part of the bride, to serve and obey, has been to day and by yourselves, as well by others on other similar occasions, a subject of misapprehension; sometimes of discord. If the subject is properly examined, all must think alike about it; no difference of opinion can exist in relation to it; the responsibility implied in this part of the marriage contract is analagous to what exists, in perhaps every association of right, interest, and duty; without it there could be no such harmonious and efficient association. Its very essence supposes a relinquishment of some natural rights as an equivalent for others to be secured; an obligation to perform certain new duties as a substitute for an exemption from the performance of others devolving on those where no such association exist. This is peculiarly so in the marriage state. All know it to be so in the body politic, in schools, in families, in business confederacies, and indeed in all confederated rights and duties. Thus in the case under consideration, the husband not only binds himself to abstain from all others—as it were, to forsake father and mother, brothers and sisters, and even friends—to love and kindly treat his wife till death; but to protect and support her, in sickness and in health, in poverty and disgrace—to do which he is under no natural obligation. On the other hand, and as an equivalent for this, she enters into a similar general stipulation of love and conduct towards him; and as an offset for the protection and support he gives her, she makes a promise to serve and obey him. A very little attention to this matter will convince any one, that the marriage contract looks somewhat like a fair business transaction ; no inequality about it. The wife simply gives her services, her time, her ability to labor, to the husband, for his guarantee to afford her the means of living. Those who do not think this a fair bargain had better live singly. If the female think this incompatible with her ideas of personal independance and dignity, the alternative is within her own power; she can take care of herself and be under obligation to no one.

"St. Paul elucidates the subject, so that it might appear no objection can be made to this feature in the Church Matrimonial service, which was thought by yourselves to be unnecessary, and by the bride in particular, as improper. He said to the one party, "Husbands love your wives;" and to the other party, "Wives obey your husbands in the Lord." Now, what is implied in this injunction to the wife? Simply this, that she is to obey him in those things which are agreeable to the Lord; and if he were to command her to do what is not agreeable to the Lord—to lie, to cheat, to defraud, to commit murder, or to defile herself, she would not be obliged by her marriage contract to do it; that is, she is under a higher obligation to her God, her fellow creatures, so far as their own rights are concerned, to her common country, and to the preservation of her own conscience, than she is to her husband. This seems to be the whole of that shadow which so disquieted the imagination of the bridal friend now being addressed. It is imagined you are both now reconciled to it. However, a few words more may be added to what has already been said. The subject may then be still more satisfactory. Such is our belief; and also that you will both consider this one of the most beautiful and convenient features in the marriage covenant.

"It does not follow that the wife is inferior to the husband in talents, or moral worth, because in the marriage covenant she is required to serve and obey. In all confederated bodies there must be one and only one individual head, so far as authority is concerned. Otherwise, all might be anarchy and confusion. Suppose, on shipboard, two persons were in joint and equal authority; one orders the helmsman to steer the ship this way, the other orders him to steer it that way—one orders the sailors to hoist the sails, and the other orders them to lower the sails—what would be the consequence? Disorder certainly; and if in a storm, most likely the ship would founder. So it would be in family government, were there two persons in joint and equal authority. Accordingly, on shipboard of the two first, one is master, and the other is styled mate; and it may sometimes happen that the latter has more skill as a navigator, and is a better man than the former. Still, it is essential to good order that the latter should serve and obey the former. The same may be true of the general of an army; he may not be equal in military talents to many of his subordinates; still, it is his duty to command, and it is their duty to obey. The president of the United States may be surrounded with cabinet officers, altogether superior to him in every attribute of statesmanship, still he is the constituted head of the government, and as such has a claim to their services. As men, they may be more highly esteemed than he is. No doubt, in multitudes of cases, the wife is far more competent to manage their joint interests and labors than the husband is; and, is so estimated and esteemed by all who know them; yet were this made the ground for superseding authority, the evil to the masses of the community would be great indeed.

"Let this subject be thus understood, and there will be no contention between you as to personal dignity and authority; each will most rejoice in the honor and welfare of the other. A spirit of selfishness will never be felt by either. There will be between you no jealousy or discord. Your onward journey in life will be like the current of a gradually descending river, having no obstructions and no disasters; and, at last, as the pure waters of the latter will be swallowed up in one vast ocean, so will you be gently carried along till merged in the great and unfathomable abyss of time; so, living in this world, according to the language of the service by which you are united in wedlock, that in the world to come, you will have life everlasting! —God, bless you, my dear young friends—bless you forever!"

On concluding the above remarks, and giving the married pair an affectionate squeeze of the hand, we bid them adieu, leaving them in the care and under the guidance of a merciful Providence; nor, have we seen either of them, or heard a single word respecting them, from that day to this. Whether the world has ever smiled or frowned on them we know not; whether, they have lived under its bright and perpetual sunshine, or have been oppressed by its cold and blighting shadows we have never been informed. Indeed, we know not, whether either or both of them is in the land of the living, or have descended into the grave. They may still live; and if so, be surrounded and blessed by a virtuous and honored offspring—children and grand-children; for it is thirty-six years since they entered the married state. Or, alas, they both long since may have slept in dust, leaving none to continue their blood and name, and of course now forgotten in the place where we saw them. This we know, that according to the statistics on the subject, those entering wedlock, live on an average only from twenty to twenty-one years. Hence, the mortality among them prior to this period is a full equivalent in numbers and in an aggregate of years to the survivors, beyond that limit. Accordingly, of those married in the year 1816, none could be expected to live on beyond 1837, except, to make an offset for those that died earlier, at any intermediate periods; and, could a record of all married in 1816 be now produced, and compared with the records of mortality up to the year 1852, it would be found that not more than one in five or six of the original number is still interested in the busy scenes—the glory or the shame —of the living world.

It is startling to sober calculation, to be told that all the incidents of nuptial life, is, on an average, crowded into the narrow space of twenty or twenty-one years; that its joys and its sorrows are then to be buried in one deep and broad grave;

and, that the houses erected for this life, and the furniture with which they were rendered comely, convenient, or magnificent, and the equipage connected with them, at the end of this period, are to be demolished, cast away, worn out, or remodelled to suit other tastes, is truly parallizing to the genius of pride and ambition. Such, however, is the fact! What chasms have been made in human society in the twenty-one years last gone by! Let each one tell what they are in the circle of his own observation, and he will be able to imagine what other ones will occur in the coming twenty-one years. Such is the record of human life! Such is the limit to human enterprise and human grandeur!

A FUNERAL IN THE COUNTRY.

The extinction of human life causes a kind of paralysis in all within a sphere to be affected by it; and, a funeral, especially in the country, is more powerful in calling forth human sympathy than any other event. This is natural; we might expect it; and it is confirmed by experience and observation. How could it be otherwise! For, the strong love of life with which man is endowed; the tender affection that binds him to his kindred and to objects around him; the agency that he may have had in the business of society; and especially the changes in his own family, in the body politic, and in the sphere of the productive elements of the community, render his demise, particularly if he has been distinguished in his generation, an occurrence of overpowering influence. For a season, at least, there will be a general and a solemn pause, both in labor and pleasure; for a season, at least, the deep fountains of the soul will be completely broken up; and, for a season, the machinery, mental and physical, of which he was a part, seems to stop its revolutions, if not to roll its ponderous wheels backward. This is a component of the philosophy into which everything spiritual or physical of the great kingdoms of the world is resolved. It is not strange, therefore, that from the time of the first recorded annals of human mortality, funeral solemnities should have been characterized by demonstrations of individual and public grief. A full history of such solemnities would form a deparment in our literature of the most absorbing interest.

With funeral solemnities in our own families we are all familiar. And, who has not witnessed them, surrounded by conventional pomp and magnificence, in the high places of our own beloved realm? Who has not read of them in other realms—in ancient as well as in modern times? The noble obsequies of Marcellus by a patriotic people, has been a theme, in all succeeding ages, of admiration and eulogy. But even these do not compare with those of the venerable founder of the Jewish nation, the splendor of which is without a parallel in history. His remains were followed to the place of their interment, nearly two hundred miles, in a distant country, not only by his son Joseph and his brethren and all his father's household, but by all the servants of Pharaoh, the elders of his house, and all the elders of Egypt, conducting the most solemn lamentations, being an itinerant national multitude, resembling in its progress a mighty river or flood. If to this any counterpart can be found, it is in those national demonstrations of honor and homage which have been shown to some of the venerable fathers of our own republic, when their remains have been followed from city to city, and from state to state, by the most illustrious of their compeers and by crowds of deeply stricken citizens. And, what pomp and splendor were exhibited when the remains of a renowned Gallic chieftain were brought from the sea girt isle where for years they had slumbered, to be deposited in the land he once called his own.

And in less imposing exhibitions of the kind there has often been witnessed a solemnity and a moral sublimity found under no other circumstances. That heart must indeed be hard as marble and those affections cold as polar ice which receive no impression and evince no emotion, when amidst the darkness of night and the lonely howlings of the ocean, the cold and lifeless body of the mariner, or of the female hectic, on a voyage for health, is, in the language of the Church Burial Service, committed to the deep. Not less impressive is the burial by torch light, of the slain of an army, under a flag of truce or voluntary suspension of dire conflict, the death-like silence being interrupted only by the clear voice of the chaplain, the funeral dirge of the band, or the well timed minute discharge of the cannon. That of the British General, Fraser, as described by Madame Riedesel, prior to the surrender of Burgoyne, is full of thrilling interest.

Our present purpose, however, is to depict the mute eloquence of funeral solemnities in rural life, where there is no pomp—no extraneous circumstances to impress the mind. The first funeral we ever attended, was when at the age of eight years; and the whole scene and the impressions we received are as fresh and vivid, as though it were but yesterday, though more than half a century since. It was of a lady, a little past the meridian of life and belonging to the best class of society, in a new country with a spare population. Our own home was three miles distant, but that was no obstacle to a pedestrian attendance. Seemingly, the members of every family in the town, young and old, male or female, were there. It was in the busy season of May but the farmers had all left their work, All joined to mingle their sympathies with those of the afflicted, and to share in the moral instruction furnished by the event. Such was the respect then paid in a new settlement to a deceased fellow mortal, although to most present she had been personally but little known ; and such was the readiness of all to receive the instruction furnished in a scene there so unusal. Such an awakened sensibility to the grief of others, and to the admonitions of religion to be themselves also ready, presents a feature in society, if old fashioned, truly lovely. It gives character to those who possess it, infinitely better than all the affected indifference and levity in relation to such subjects, now a days of so rare occurrence. Now, with one sex it is too often looked upon, as unmannerly, at least, to give such heed to these admonitions of Providence; and even in the other sex it too often appears that there is more regard to the popular conventional exhibitions of what is practiced in fashionable life, than to the indulgence of genuine grief, or the religious culture of their own hearts.

The population of the whole town, as it were, we have already remarked, was there; not only filling the entire house, which was large, but more were collected about the house, than were in it. The lower parts of the windows were all blocked up by human forms—so were the door-ways and the avenues to them. A profound stillness pervaded the whole mass. Not a whisper was heard. Every look was downcast, and the pulsations of every bosom denoted sorrow. The funeral services consisted of prayers and a sermon, and singing Dr. Watts' Hymn,—"Hark, from the tombs, a doleful sound!" producing in us a thrilling awe that will never be forgotten, and from the appearance of the crowd each one felt himself a brother, or of kindred consanguinity.

To this succeeded the removal of the corpse, on the shoulders of the pall-bearers, to the grave some distance off, in a corner of the orchard. First followed the afflicted husband, a portly, gentlemanly looking farmer, pensive and sad, but without visible emotion—probably the more oppressed, because his sorrow had no vent in tears. Near to him in the train, were the two eldest children, a son and a daughter, advanced nearly to adult stature; then a still younger sister, leading by the hand, a little brother of our own age—say eight years. The more distant relatives and the attendants upon the occasion voluntarily joined in a dense and long procession. With what sadness it moved forward! So absorbed were all in deep thought, that nature seemed as if hushed to unwonted stillness and responsive emotion; and, it required little effort of the imagination to see the heavens spreading over the whole group a mantle befitting the occasion. It was a clear, bright day with Spring's genial influences, so that within and without all was harmony.

At length, the grave was reached, and the coffin was lowered into it! To us, the scene being an entirely new one, as well as to the stricken family, this was seemingly more than could be sustained. The little fellow named sobbed aloud, trembled like an aspen leaf, and apparently, had it not been for his sister, who held him by the hand, would have fallen or leaped into the grave upon his pulseless mother! We wept too, and almost lost the locomotive power to withdraw from the spot and return to our home. Such was the impression on us that we ever afterwards cherished for this boy a warm affection. The affection became reciprocal, and then led to personal friendship and great intimacy that continued fifty-two or fifty-three years, when he* followed his mother to the bourne from whence no one returns. Were we near his final resting place, we should not fail to make a pilgrimage to it, both to indulge in those natural impulses which bind kindred hearts in one bond of affectionate union, and to revive and invigorate those moral impressions received by us on the occasion described. This narrative of the incident here given has no connection with our main design. It is given simply to corroborate by our own experience the influence of a funeral in the country. We believe that incident did materially and permanently change the tone of our social character.

Our intended memoir is yet to be rehearsed. The prominent facts are not unlike those in hundreds of country funerals, denoting the respect that is manifested in a rural population for the memory of a deceased member of it, as well as friendship for surviving relations. Among all the endearing amiabilities of which we are capable, we know of no one besides so precious. The idea of seeing a whole community make a deliberate pause when one of their number become sick or dies—or suspend their own labors or pleasures in order to administer to the needy or the afflicted, shows that our common brotherhood is duly recognized by them, and that there is in man a spirit of communion and fellowship, as if we were all the children of the same Heavenly Father. Even the tolling of the church bells, at a country funeral makes an impression on a passing traveller, or

* Jonathan Clark, of Gilmanton, N. H.

the people generally, of more deep seriousness than an ordinary dissertation or sermon on human mortality.

Sometime about the middle of October, 1839, we had occasion to make a journey through one of the best agricultural districts in New England, and to stop a few days at a place, for convenience, we shall call Beemantown. We arrived at the neat village hotel, just as the harvest moon was presenting to view her full orbed disc, in its wonted beauty at that season of the year. The surrounding country denoted a thrift and good taste that usually attend well educated labor applied to agriculture. In the village were several stores, a Gothic Church, a capacious academical edifice, and sundry mechanical establishments; all was neatness and simplicity; and nothing was to be seen or heard indicating idleness, poverty, or a lack of good morals. The loveliness of an evening during the season of the Harvest Moon, especially in rural situations, is too well known to require description, for it has long been the subject of poetic delineation and eulogy. After supper, we took a short walk, to witness the good order and tranquility with which we were then surrounded; and never had we before experienced from a serene sky, a mellowed atmosphere, and an unbroken quietude, such a mental charm—delightfully in contrast with the emotions arising from the bustle and excitement of the city. We felt as one might be supposed to have felt in Paradise, or in a land of undefiled spiritual existencies. Of course we returned to our lodging place well prepared for sleep. Soon were we lost in gentle slumbers. Never to us was sleep more sweet. The little fatigue of the day's journey, and the indescribable effect of meteorological and local influences caused our slumbers to be like those of a healthful infant, without sighing or convulsive throbs, or any change of features.

Thus we probably slept two hours. Then the loud striking of the village clock announced the hour of midnight, and we awoke to behold the silvery light of the moon, rendering every object about us as distinct as if in the light of day. When na ture is thus wrapped in silence, one suddenly aroused in this manner from his slumbers imagines himself a unit in the broad expanse; yet his solitude causes no shrinking from existence—no terror from imagined danger; but his breath, like fragant and gently undulating incense, rises to the great spirit that shelters and upholds him. When in a kind of waking revery the death toned sound of the village church bell was heard; it fell upon us, as might be supposed, like the voice of the archangel, when announcing the end of time, will fall upon the living world; and before its waning cadence was entirely gone, or its echo from the distant hills came back to us, a second one, if possible, still more impressively fell upon us—then a third, and a fourth, till we counted fifty-two, at intervals of about twenty seconds between each, then a pause of a few minutes, which was succeeded by still another one, if possible yet louder than those before—when all again was dead silence, and we seemed, as it were, lost and alone, although it might be that hundreds in the village felt as we did.

Had we not known the usage of former times in the country, thus to toll the church bell, commencing as near as possible upon the last breath whenever one dies—the number of strokes being the number of years in the age of the deceased—afterwards a single stroke as in this case, signifying a male, and if two strokes a female—the above to us would have been a mystery; but, knowing it, we were apprised, that a fellow mortal, a brother of the age of fifty-two years, had, at this still hour of midnight, ceased to live—bidding a last adieu to the pains and sorrows, and disappointments, as well as to friends and relatives, and the once budding hopes and joys in life's panorama. The reader need not be told, that our thoughts became pensive and sad; and for the remainder of the night we knew not sleep. Had it not been so, we might rightly have been judged void of those amiable sensibilities which belong to our nature. Far be it from us to be ashamed of them. If the result of weakness we rejoice in it. And it is a matter of course, that on the return of morning, we enquired for whom had been made those demonstrations of reverence and respect. The following narrative is the reply to our enquiry.

In the year 1802, a youth, whose name we shall call Charles Beeman, left the homestead of his father and took a clerkship in the city of New-York. The father was of the third generation who had tenanted the same mansion and cultivated the same farm, thus rendered dear to the family by a thousand cherished associations.. The house was indeed ancient, but from its age, its stateliness, and the lofty elms in front of it, was truly venerable. The farm, too, was large and productive; and the income of it, under a prudent management, had placed the proprietor in dignified independence and comfort. The family consisted of the parents and three children, two sons and a daughter. Charles was the eldest of the three, and unfortunately had imbibed the idea that the mercantile life was less severely taxed with toil, and was more respectable, and, what was more in his imagination, was the only way to affluence. Accordingly, he resolved to forsake the residence of his ancestors, and to seek his fortune in the city. The resolution might have denoted what is usually looked upon as superior talents and enterprise; but, the sequel will show, that the determination of the younger brother, who remained with his parents, was attended with far better results.

The memoir of Charles Beeman, with no essential variations, is that of thousands of young men, who, under similar circumstances, leave the country and repair to the city. From his career thousands may learn wisdom. He thought agriculture a dull and irksome occupation; and that it was beneath the dignity of one of his imagined talents to spend his life in laying up thirty or forty thousand dollars as his fathers had done, when he might become a merchant and obtain five times that amount. He looked at the rich apparel, and furniture, and equipage of merchants; and at the display of merchandise, and cash, and stocks in the city, and was completely bewildered with the fascinating picture. Of course away he went, and spent six years, as a clerk in one of the most respectable houses in that commercial emporium. The daughter of Squire Beeman, as the father of Charles was generally denominated, soon married a young man of property, talents and character, and lives only a few miles distant, and James, the other son, with less brilliancy but more soundness than his brother, remained on the homestead, married, and became one of the most respectable farmers of his State. He was elected several times to the Senate of his State; and might have been sent to Congress, but preferred remaining at home. He and his father so labored and managed, that when Charles was ready to engage in business on his own account, he received as his portion of the estate ten thousand dollars. At length the parents of James died, and he of the fourth generation with a family of children of which a prince might be proud, was left in sole possession of the Beeman homestead. As might be inferred he was independent as though he had possessed a million of dollars; and, it would be difficult to imagine anything not within his reach, that would have added to his enjoyment or his reputation.

Charles Beeman commenced his career as a merchant under the most favorable auspices. His character was pure; his talents were quite respectable; and he had a cash capital of ten thousand dollars with unlimited facilities, from his mercantile acquaintance. For twenty years his course was most prosperous. Everything seemingly on which he placed his hand turned to gold. He became the president of a large banking establishment. On the land and on the water he gave employment to hundreds of persons. He married, and as usual under such circumstances, had a splendid family establishment. His charities, too, were on the most liberal scale. His annual expenses could not have been less than eight or ten thousand dollars. His children, three sons and two daughters, as a matter of course, grew up with the most expensive habits for dress and amusements, without acquiring the habit of doing anything, or having the least reference to the means for a living. The consequence was two of the sons became dissipated and diseased, and early sunk into the grave. The other son was amiable and not immoral, but without efficiency for business. The two daughters married young merchants, who after a few years were unfortunate, and involved their father-in-law in heavy responsibilities. By the aid of friends both were provided with foreign agencies. The broken hearted daughters were in a few years relieved by death from poverty and mortification. Their days were few. Their sun rose in beauty and brilliancy, but was soon overcast, and went down in the darkness of night. Their early hopes of happiness budded in great profusion, but a blighting mildew caused them all to wither and die without fruit. Their fond mother soon joined them in the world of spirits.

The affairs of Charles Beeman were hastening to a perilous crisis. As usual, misfortunes come not singly. His own family expenses had been enormous. Sundry ordinary business losses, which at former periods would scarcely have been thought of; but now, in connection with an accumulation of disasters, became insupportable. There was no other alternative. Bankruptcy was the unavoidable result. His carriage and horses were first sold; then his furniture and house, and he took lodgings in a hotel. His business being completely broken up, he retired from his office in the bank. The only alleviating circumstance in all this revulsion was that his feeble surviving son was furnished with the office of porter and messenger in the same institution on a salary just sufficient to give him a decent living. What a desolation for one who a few years previous had estimated his wealth by hundreds of thousands of dollars, and had prided himself, at least with beautiful daughters who with their mother were the centre of attraction in every sphere of fashionable life. If there is anything to break down the spirit of a man, it is this. To such an one the light of day or the blackness of midnight are much alike. An unwavering paralysis settles down upon the soul. His energies are completely prostrated!

For a few years the friends of Charles Beeman attempted to inspire him with vigor, to engage in mercantile brokerage, but it was to no purpose. He seemed to feel no motive for making the effort. No one depended upon him now for subsistence, and to him the world had completely ceased to offer any charms. His faint spirit seemed to yearn after some untried and unknown panacea for a curative for his malady. The opportunity was soon discovered. Disease, which like a ravenous beast, lying in ambush to discover the weakest position of the destined victim, soon seized upon our life wearied, broken down merchant, It would be difficult to tell whether a consciousness of the fact gave him joy or sorrow. He had seemingly but little remaining susceptibility for either. The world to him had become a dismal blank ; and, as soon as rendered morally certain that his days were nearly numbered, he acceded to the pressing invitation of his brother James, to return to Beemantown, and become one of his family at the old Homestead. He did so, and a few trunks contained his wardrobe and every vestage of his once large estate. He was there treated with the utmost kindness : Mrs. James Beeman and her lovely daughters were like angels of mercy ; but they were never able to raise a smile on his pale emaciated features, and in three months, ten days, and nine hours, the solemn tones of the church bell, at the still hour of midnight, as we before stated, announced to the people of the village and township, that he had gone to the land of his fathers. There is one leaf more in the memoir of Charles Beeman.

Becoming greatly interested in the narrative, of which the above is a mere abstract, we resolved to attend his funeral, which was to be the second day afterwards. During the interval there was a chastened sensation which we had never before witnessed, for the death of an individual in private life. It however, was a sensation of that calm and unobtrusive character, which is manifested only in a general and deep tranquility of spirit, and in the slow measured pulsations of labor and business. It was a sensation springing from the deep recesses of the heart, and receiving no type from popular impulse or conventional power.

The day of funeral arrived. It was one of peculiar loveliness, so well known at that season of the year. There was not a cloud to be seen, and seemingly not a breath of wind was felt. Nature was in harmony with the mellowed and subdued passions visible on every countenance. For some time previous to the hour appointed for the funeral obsequies, the people from the remote parts of the town began to assemble. There was a general suspension of labor and of business. Little groups were constantly arriving, and the village green was soon dotted over in every part. The church was filled to overflowing. Its ponderous bell announced the hour of two o'clock in a loud sepulchral cadence. The bell a mile north of the village, in a few seconds gave a similar response ; then, at a corresponding interval of time, echoing among the surrounding hills, was heard the other church bell of the town ; thus the three in regular alternations, uttered their mournful thrilling notes, till the corpse and the procession of mourning relatives and friends, with slow and solemn steps, measured the distance from the Beeman Homestead to the House of Prayer. Such a spectacle must be seen in order to be fully appreciated. One might as well impart to the marble statue, or the canvass, the varying hues and the breathing of the living human form, as to give sufficient delineations of it with ink and paper.

There was not indeed in the procession a deeply afflicted widow, bowed down in sorrow for the loved one of her youth. There was not indeed a cluster of orphan children, convulsed with anguish for the loss of their only guide and support, a fond father. They were in mercy saved from this heart-rending hour, in being called away before him. But still, there was no want of sincere mourners; all present knew the sad history of his life ; most of them had been faimliar with his visage ; all were impressed deeply, and saw in that history a memento of the vanity of the world's brightest jewels. There were no boisterous lamentation ; there were no passionate outpourings of stricken hearts ; but the wide spread undisturbed silence ; the downcast demeanor throughout this sad living mass, with here and there a silent tear, made demonstration that in the human bosom there is a chord that vibrates to another's woe. Here was an irresistible admonition not to make haste in striving to become rich ; nor to despise those frugal bounties which reward the cultivators of the soil ; and especially not with unguarded pace to press amid temptations and perils, the consummation of which, not unoften, is in the chill darkness of the grave !

We have read many essays and heard many homilies on the vanity of this world's glory—the folly of pride and pompous exhibitions in social life—and, especially, the canker which fixes itself upon the heart of those who despise the charms of rural pleasure for the delusive splendors of the city; but the history of Charles Beeman made an impression upon us without a parallel. These few days in his native village made the country seem more lovely to us, and the city fraught with more perils and mental agonies than ever before imagined. Oh, that we had a pen to delineate all the incidents, and a pencil to place on canvas all the lights and shadows, blended into each other in his memoir, from the cradle to the grave ! Well would it be if hundreds of others known to us, could have been present to witness what we witnessed, and to have received that impressive culture we received, and which we apprehend will never be effaced from our recollection ! And, if the unnumbered young men in our country, now impatient of the toils and the moderate but sure gains to be expected upon their native hills and valleys had witnessed it, there would not be such a perpetual rush to embark in adventures, and to grasp at objects presenting them to the inexperienced visual powers, in magnitude surpassing the reality, as did the men first seen by St. Paul, after the scales had fallen from his eyes—"like trees walking !"

DOMESTICATION OF THE OX FAMILY.

From the earliest periods the ox has been regarded in the light of property; nor has its intrinsic value been less appreciated after a lapse of ages. Who does not know how intimately the well-being of a nation and of individuals is connected with the various uses of this animal; and how closely the interests of commerce and agriculture are bound together. It is proverbial that he who makes two stalks of corn, or two spires of grass, grow where only one grew before is a public benefaction; and by parity of reasoning, he who improves the breeds of domestic cattle, feeds two on the land which before had supported only one; and he who devises superior modes with regard to the extension of their utility, also, serves the interests of the community. On the labor of the male and the milk of the female, and on the flesh of each there is a dependance, greater than generally realized, unless by persons familiar with the statistics of these branches of rural economy. All the ox tribe are, as well known, gregrarious in their habits; and no quarter of the world, except Australia and South America, is destitute of its indigenous species, existing in a state of freedom. They roam over hills or plains, or tenant the glades of the forest. The immense herds of wild cattle which now roam the pampas of South America, are the descendants of those domesticated ones which were originally introduced by the Spaniards; and, it may be added, that although in North America the Bison, and Musk ox, are indigenous species, the domestic ox, now of such value, is in like manner an importation; while the indigenous species are disappearing before the advance of colonization.

That the ox has been domesticated, and in the service of man from a very remote period, is as certain as any fact recorded in history; for we learn from the book of Genesis, that cattle were kept by the early descendants of Adam. Preserved by Noah from the waters of the deluge, the original breed of our present oxen must have been in the neighborhood of Mount Ararat; and from thence, dispersing over the face of the globe altering by climate, by food, and by cultivation, originated the various breeds of modern ages. And that the value of the ox tribe has been, in all ages, and all climates, highly prized, we have the ample evidence from history. The natives of Egypt, India, and of Hindostan, seem alike to have placed the cow among their deities; and judging by her usefulness to all classes, no animal could perhaps have been selected, whose value to mankind is greater. In nearly all parts of the earth cattle are employed for their labor, for their milk, and for their food. In Southern Africa, they are as much the associates of the Caffres as the horse is of the Arab. They share his toils and assist him in attending his herds; they are even trained to battle, in which they become fierce and courageous In central Africa the proudest ebony beauties are to be seen on their backs. They have drawn the plough in all ages; in Spain they still trample out the corn; in India, raise the water from the deepest wells to irrigate the thirsty soil.

Intelligence, says Mr. Martin, appears to be more limited in the ox than in the horse. The brain is comparatively smaller in the former than in the latter; and the ratio of intelligence is probably in about the same proportion. But we must not regard the ox as remarkably stupid. The working ox knows his driver, and readily obeys his word of command, displaying, at the same time, considerable docility and willingness. The cow not only knows, but often evinces decided affection towards the person by whom she is regularly milked and fed, and not unfrequently refuses the attentions of another. Cows, pastured in the fields, draw towards the accustomed spot, at the usual milking time, and, by their lowing, seem to give notice of their readiness. In Switzerland, the herds, feeding on mountains, are called home by the sound of the Alp-horn. The Alp-horn is merely a wooden tube, of simple construction; and its deep, mellow, prolonged note, heard at a distance floating over the upland pastures, and frequently echoed in succession by crags and rocks, makes a pleasing impression on the traveller. On many of the higher Alps this horn is sounded regularly at sunrise and at sunset.

At whatever part of the day it is blown, to collect the cattle, the cows are seen, as soon as its note can reach their ears, scampering away, for their home, often at a full gallop. In one of the volumes of the " Menageries," the writer says, " A correspondent informs us that he once witnessed the evident dismay of a cow in not being able to obey the summons, She was a very pretty creature, and was lying ruminating, in a little dell, by herself. Presently the Alp-horn was heard ; and the peasant, who sounded it, was seen opposite on a green hill at no great distance. The poor creature instantly arose, but could not proceed, for she was lame and sorely wounded, most probably from some fall she had met with. After dragging herself along for the distance of some yards, she stood still and lowed. The horn sounded again, and again making a great effort, she went on a few yards. She then laid herself down and lowed most piteously, fixing her eyes, all the time, on her companions, who were running from every direction. The poor animal seemed to be suffering, not merely from her hurts. but also from that pain a punctual person always feels on breaking an engagement, or failing in the performance of a duty. The traveller hastened to inform the peasants what had happened, and returned with them to see them assist the grieving absentee. By the time they had reached the dell, she had made a little more progress, but was again prostrate on the ground, and lowing in a melancholy manner. When the peasants came near to her she rose up, and ceased her complaints at once ; and though, with all the aid they could give her, she was obliged to lie down again, she did not repeat them ; and at last she reached the dairy, where she lowed in a very different tone.

In Norway, a wooden trumpet, about five feet long, made of two hollow pieces of birch wood, bound together by slips of willow throughout its whole length, is used to call the cattle feeding on extensive hilly pasture grounds. Its notes may be heard at a great distance ; but, instead of being mellow, they are extremely harsh and discordant. In Terceira, one of the Azores, where cattle are very abundant, the oxen are remarkable for docility and intelligence. They are very fine and large, with horns of prodigious size, and are so gentle and familiar, that when among a thousand collected together, the owner shall call one by its own name, for they all, like our dogs, have each its own proper name, the ox will not fail to come to him. Every person, too, familiar with our cattle, knows that when called by some general name with which they are usually designated, will instantly run at the sound; and in the same way individuals will run in answer to particular names by which they are called.

But of all instances of intelligence in the ox, those of the backeleys, or war oxen, among the Hottentots, while that people were in a state of pastoral independence, are the most extraordinary. The ox was the sharer of its master's toils and wars, his assistant on the plain, and the guardian of his flock; it seemed indeed to lose its ordinary character in his service, and rise into a higher state of being. "The Hottentots," says Kelben, "have oxen which they use with success in battle. They call them backeleys; the word backeley, in their language, signifying war. Every army is always provided with a large troop of these oxen, which permit themselves to be governed without trouble, and which their leader lets loose at the appointed moment. The instant they are set free, they throw themselves with impetuosity on the opposing army; they strike with their horns, they kick with their heels, they overthrow, they rip up and trample beneath their feet, with frightful rapidity, all that opposes; so that, if they be not promptly driven back, they plunge with fury into the midst of the ranks, throw them into disorder and confusion, and thus prepare for their masters an easy victory. The manner in which these animals are trained, does great honor to the talent of these people."

In another place the same writer says, "These backeleys are, moreover, of great use to their owners as guardians of their flocks. When out, in the pasture lands, at the least sign of their conductors, they will hasten to bring back the cattle which are straying at a distance, and keep them herded together. They rush on strangers with fury; whence they are of great service against robbers who may attempt to plunder the flock. Each kraal has at least half a dozen of these backeleys, which are chosen from amongst the most spirited oxen. On the death of one, or when, in consequence of old age it becomes unserviceable, in which case its owner kills it, another ox is selected from the herd to succeed it. The choice is referred to one of the old men of the kraal, who is thought to be the most capable of discerning that which will most easily receive instruction. They associate this noviciate ox with one of long experience; and they teach him to follow his companion, either by beating, or some other method. During the night they are tied together by the horns, and they are thus also kept the greater part of the day, until the learner has become a good guardian of the flock. These guardians know all the inhabitants of the kraal, men, women, and children; and testify towards them the same respect that a dog displays to all those who live in the house of his master. There is, therefore, no inhabitants of the kraal who may not, with all safety, approach the flocks; the backeleys never do them the least injury. But, if a stranger, and particularly a white man, should offer to take the same liberty, without being accompanied by some Hottentot, he would be in great danger—these guardians of the flock which usually feed around it, would come upon him full gallop; and then, unless he be within hearing of the shepherds, or have fire-arms, or good legs, or unless a tree be near in which to climb, he is sure to be killed. It would be useless for him to have recourse to sticks and stones; a backeley has no fear for such feeble weapons." Vaillant confirms this statement of Kolben; yet it does not appear, as far as we can learn, that any of the tribes of Southern Africa, rear backeley oxen in the present day.

The only importance that can be attached to facts like these is this, if cattle have been thus trained by the Hottentots, it shows their docility and intelligence, and that they are susceptible of similar instruction again and elsewhere. There is nothing of food or climate in the south of Africa not existing in other countries to induce such capability. If the Africans can thus teach their oxen, Europeans and Americans can do it also. Nor are we without forcible illustrations of this. It is well known that individuals occasionally find the fullest evidence of great tracticability in working oxen. They are taught to plough without a driver; and, they seem frequently to understand as well as a driver the extent of their power to draw large burdens, as well as the best mode of applying this power. There is little or no doubt that by a proper course of training oxen, particularly when steers, their value for labor would be greatly increased. This is a subject farmers have too much neglected. Their attention cannot be too frequently called to it. A well trained yoke of oxen are as much superior for labor to a yoke not so trained, as a first rate laboring man is better than another that does not understand his business.

The cow, too, as well as the ox has similar intelligence. It is not indeed exerted for the same purposes; but this shows the versatility of it, or its susceptibility of being directed to any object necessary. The cow very quickly learns, when suckling a calf, the particular seasons when this office is to be exercised; and if there is any delinquency, on the part of the keeper, in putting the dam with the offspring, she manifests the greatest impatience. So likewise when the flow of milk is very abundant, and the udder becomes painfully distended; she seems to be fully aware of the relief afforded her by being milked, and even seeking opportunity for it, and afterwards manifesting a kind of affection to the one that thus habitually relieves her. In the chapter on the artifices of animals several tricks of the cow were introduced; but, there was one unintentionally omitted, which is so curious, a place is here given to it, although more appropriate in that chapter than this. There have been instances known of her selecting a thicket in which to drop her calf, so excluded and so dense, as to escape human observation. After remaining with her nursling a due length of time, by some means or other she teaches it to lie down and remain perfectly quiet, so as not to be noticed by any living creature. To this spot she steals away twice or thrice a day, to yield to it the contents of her udder; but so adroitly, that with the most vigilant pursuer, she at last eludes his eye, and disappears from his view. This is sometimes continued for weeks with perfect success, the younger as well as the elder understanding how to keep the secret; and in one case, which fell under our own observation, was obtained by a person concealing himself for the last half of the day in the known vicinity of her retreat. And, he stated, that when at the approach of sundown, she drew near, with the utmost caution, every few yards stopping to see if she was perceived; and, at last, supposing she was not discovered, again stopped and gently lowed. This was the signal for recognition. The calf, with the celerity of a deer, darted from the place of his concealment, took his evening repast, and then receiving a species of a kiss from the mother, took to his nest. When this was done, she departed as stealthily as she came. However, as much amused as the proprietor was with this specimen of bovine cunning, another place was provided for them.

Among the more useful specimens of intelligence in the cow, the following is given. Three or four years ago, we had occasion to visit the premises of a very large stock owner about six miles west of our own residence. The proprietor was a man of wealth, and had a very large farm, the pasture of which consisted of many hundred acres, in which all his young cattle roamed at pleasure from spring to fall. The cows were turned to it each morning, but voluntarily returned to the barn-yard at the close of the day. To raise stock, and to deal in it, was an important part of his business. In addition to those raised, he was always ready to purchase good animals at low rates; and equally ready to sell oxen, cows, steers, heifers, or calves, when purchasers were willing to pay his price. Indeed, a selling price was attached to nearly every animal belonging to him, as systematically as the price is fixed to all merchandize in a store. We reached his house when the sun was about two hours in height. Our object was to purchase a cow. After spending a little time in general conversation, we started for the pasture in company with his son, who appeared familiar with all the details of the business. Soon after entering it, we discovered through the opening of an adjacent wood lot, a venerable looking old cow, red, with white face, and a shrill bell fastened to her neck, moderately coming towards us. Her gait and demeanor indicated the dignity and the self-complacency usually attendant on the exercise of authority, although there was nothing else about her that attracted our notice. She had evidently attained an age too much advanced for our taste. Following her in the same direction, was one succeeding another, in a kind of Indian file, sometimes, however, irregularly, two abreast, in all the number of twenty-five or thirty.

As they all seemed to occupy the hard trodden path, which doubtless they had traveled an hundred times, and the leader evinced no consciousness that others, even her owner had a better right to it, we readily turned aside, commenting upon the appearance of one after another in the train, as happening to be opposite to them. On reaching the rear point in the train, the cows having pretty much come to a stand, either to feed or to be admitted to the barn-yard, we inquired the properties and the price of such as seemed to attract particular notice, receiving the most prompt answers, as if the merit of each was perfectly understood. At last, on our way back to the gate, we said, not because we wanted her, but to make conversation— "Well, young man, what is the price of the old cow with a bell?" "She is not for sale," replied he. "That is probably because she is too old, to bring a good price, is it not," we rejoined; "and if a purchaser wanted her, you would fix a price, would you not?" "No," he answered, and that, "his father would not take an hundred dollars for her." "Perhaps," we added, "she is of a choice breed, and you retain her for the calves." "Not so," was his response—"she is a native breed; pretty fair, as a milker, but not extraordinary; we hold on to her, because she is worth more in taking care of the others, than any hired man on the farm. Her authority with the whole herd is complete; wherever she goes, they will follow after;

and, if any one strays away, or falls too far behind the others, she will quick discover it, retracing her own steps till the delinquent is found. The means used by her in exercising this kind of jurisdiction, and making every individual one willing thus to follow her, day after day, may be somewhat peculiar, and not understood by us, yet they understand it as well as an army understands the orders of its commanding general. She is as uniform, each day, cloudy or sunshine, as though she had one of the time-pieces, to be seen advancing, with little variation, an hour before sun down. If she is two miles off, at the most distant point in the pasture, her homeward course is commenced enough sooner, so as to be here at this hour. And what makes her supervision the more valuable, she is as familiar with every spring of water, and with all the best food in the entire pasture, as any man could be ; being particular to appearance each morning when conducting the herd in pursuit of forage, to go to those parts that are the best."

With this account of the old bell cow, we did not wonder that she was not for sale, and that so high a value was placed on her. On hearing this account of her, we viewed her with peculiar pleasure, as furnishing new evidence of the satisfactory results to be expected, when judicious efforts are made to improve the brute creation. By them the pleasures of rural life are greatly advanced. They help to render pleasurable, what might otherwise be repulsive in the labors of the husbandman. Where the cow, the ox, the horse, or even the hog, seems, in consequence of assiduous and kind treatment shown to it, to rise above its species, we necessarily become attached to farm animals beyond what we should be, if none of this docility and intelligence were evinced by them. Administering, therefore, to their wants, and to the modifications in their nature, of which they are susceptible, furnishes one of the legitimate branches of labor, as well as of enterprise and pleasure in the life of the agriculturalist. All who have had experience in it, knows well the influence it has on the affections.

The ox family has generally been described as consisting of

two great classes; one having hunches on the back, called aurocks; and the other without hunches, and called bisons, or buffaloes. The best writers apply the name of buffalo to the wild ox of the eastern continent, and bison to that of the American continent, although these names are sometimes applied without reference to such limitations. To each one of these two general divisions belong a great number of varieties, occasioned by difference of climate and food, and other external influences. The bison and buffalo are said to be more obstinate or less tractable than the domestic ox; obeying with greater reluctance, and having a temper more coarse and brutal. This, however, may be wholly the result of a more limited and imperfect education or training than the ox has received during the successive generations of his domesticated existence. Possibly, if the one were to be as long under the immediate jurisdiction of man as the other has been, these differences between the two would be less apparent than they now are, provided they were found to exist at all.

The breed of the aurochs is generally the habitant of the cold and temperate zones. It is not very much dispersed towards the southern countries. On the contrary, the hunched ox dwells mostly in southern countries; throughout the extended territory of India; the islands of the South Seas; in all Africa, from Mount Atlas to the Cape of Good Hope, no other oxen but these are found; and it even appears that this breed has many advantages over others in hot countries. The hunched have the hair much softer and more glossy than our oxen; they are always swifter and more proper to supply the place of a horse; and they perhaps vary more than our oxen in the color of the hair, and the figure of the horns. The hunch does not depend on the conformation of the spine, nor on the bones of the shoulder; it is nothing but an excrescence, a kind of wen, a piece of tender flesh, as good to be eaten as the tongue of an ox. The wens, or hunches, of some oxen weigh from forty to fifty pounds; others have them much smaller; and, their horns are sometimes of prodigious size. One of them has been referred to in the French cabinet of Natural History, three feet and a half in length, and seven inches in diameter at the base. Many travellers affirm that they have seen them of capacity sufficient to contain from fifteen to twenty pints of water. The bison, or wild hunched ox, is generally stronger and much larger than the tame ox of India, yet, sometimes, it is found smaller; its size doubtless depending much on its food. At Malabar, at Abyssinia, at Madagascar, where the meadows are naturally spacious and fertile, they are frequently of prodigious size; yet, in Africa and Arabia Petrea, where the land is dry, they are of diminished stature.

The hunched ox differs from the common ox in the color of its hide; his body is thicker and shorter; his legs are longer and proportionably much less; the horns not so round, black, and partly compressed, with a tuft of hair frizzled over the forehead; the hide is likewise thicker and harder; the flesh is black and hard; and the milk of the hunched cow is not so good as that of our cow, although she yields a great quantity. In hot countries of the eastern continent most of the cheese is made of this milk. It is computed that two buffaloes harnessed or chained to a wagon, will draw as much, when properly trained, as four strong horses. As they carry their heads and their tails naturally downwards, they employ the whole force of their body in drawing. The form and thickness of the buffalo alone are sufficient to indicate that he is a native of the hottest countries. The largest quadrupeds belong to the torrid zone of the eastern continent; and the buffalo, for his size, might be classed with the elephant, the rhinoceros, and the hippopotamus. The camel is more elevated but slenderer, and is also an inhabitant of the southern countries of Africa and Asia; nevertheless, the buffaloes live and multiply in Italy, in France, and in the other temperate provinces. It appears that these animals are gentler and less brutal in their native country than in others; and the hotter the climate is, the more tractable is their nature. In Egypt they are more so than in Italy; and in India they are more so than in Egypt. Those of Italy have also more hair than those of Egypt, and those of Egypt more than those of India. Their coat is never entirely covered, because they are natives of hot countries; and in general, large animals of this climate have either no hair, or else very little.

There are a great number of wild buffaloes in the countries of Africa and India, which are watered with many rivers, and furnished with large meadows. They go in droves, and make great havoc in cultivated lands; but they never attack the human species, and will not run at them, unless they are wounded, when they are very dangerous; for they make directly at their enemy, throw him down and trample him to death under their feet; nevertheless, they are greatly terrified at the sight of fire, and are displeased and sometimes become furious at the sight of red color. The buffalo, like all other animals of southern climates, is fond of bathing, and even of remaining in the water; he swims very well, and boldly traverses the most rapid floods. As his legs are longer than those of the ox, he runs also quicker upon land. The negroes in Guinea, and the Indians in Malabar, where the wild buffaloes are very numerous, often hunt them. They neither pursue them nor attack them openly, but, climbing up the trees, or hiding themselves in the woods, they wait for them and kill them, the buffaloes not being able, without much trouble, to penetrate these forests, on account of the thickness of their bodies, and the impediment of their horns which are apt to entangle them in the branches of the trees. These people are fond of the flesh of the buffalo, and gain great profit by vending their hides and their horns, which are harder and better than those of the ox.

The American bison was once dispersed, in countless herds, over immense regions of our country, although it was rarely found on the Atlantic coast. Lawson, an early historian of the country, thought it a fact worthy of being recorded, that two were killed in one season on Cape Fear River. As early as the first discovery of Canada, it was unknown in that country. Theodat, whose history of Canada was published in 1636,

AMERICAN BISON.

merely says that he was informed that bulls existed in the remote western countries. At a period not very remote they were found in most of our western states and territories; but, as the white population advanced onward in that direction these animals disappeared, pressing in vast numbers to the back parts of Louisiana and the prairies watered by the Arkansas, La Plata, Missouri, and its tributaries. But from many of these retreats they have since disappeared from a similar cause. As far as I have been able to learn, says the editor of the Boston edition of Buffon's Natural History, the limestone and sandstone formations, lying between the great Rock Mountain ridge and the lower eastern chain of primitive rocks, are the only districts in the fur countries that are now greatly frequented by the bison. In these comparatively level tracts, there is much prairie land, on which they find good grass in summer; and also many marshes overgrown with bulrushes, which supply them with winter food. Salt springs and salt lakes also abound on the confines of the limestone, and there are several well known salt licks, where bisons are sure to be found in all seasons of the year. Their migrations to the westward were formerly limited by the Rocky Mountain range; but of late years they have found out a passage across the mountains, and their numbers to the westward are said to be annually increasing. Further to the southward, in New Mexico and California, the bison became numerous on both sides of the Rocky Mountain chain.

Bisons wander constantly from place to place, either from being disturbed by hunters or in quest of food. They are much attracted by the soft tender grass which springs up after a fire has spread over the prairie. In winter they scrape away the snow with their feet to reach the grass. The bulls and the cows live in separate herds for the greater part of the year; but at all seasons, one or two bulls generally accompany a large herd of cows. The bison is in general a shy animal, and takes to flight instantly on winding an enemy, which the acuteness of its sense of smell enables it to do from a great distance. They are less wary when they are assembled together in numbers, and will often blindly follow their leaders, regardless of, or trampling down the hunters posted in their way. It is dangerous for the hunter to show himself after having wounded one, for it will pursue him, and although its gait may be heavy and awkward, it will have no difficuly in overtaking the fleetest runner. The same writer says, when I resided at the Carlton House an incident of this kind occurred. Mr. McDonald, one of the clerks of the Hudson Bay Company on descending the river in a boat, and one evening, having pitched his tent for the night, went out in the dusk of the evening for game. It had become nearly dark when he fired at a bison bull, which was galloping over a small eminence, and as he was hastening forward to see if the shot had taken effect, the wounded beast made a rush at him. He had the presence of mind to seize the animal by the long hair on his forehead as it struck him on the side with its horn; and being a remarkably tall and powerful man, a struggle ensued, which continued until his wrist was severely sprained, and his arm was rendered powerless; he then fell, and after receiving two or three blows, became senseless. Shortly after he was found by his companions bathed in blood, being gored in several places; and the bison was couched beside him apparently waiting to renew the attack had he shown any signs of life. Mr. McDonald recovered from the immediate effects of the wounds he had received, but died a few months afterwards.

Many other instances might be mentioned of the tenaciousness with which this animal pursues his revenge; and we have been told of a hunter having been detained for many hours in a tree by an old bull which had taken its post below to watch him. When it contends with a dog, it strikes violently with its fore feet, and in that way proves more than a match for a prime bull dog. The favorite Indian method of killing the bison is by riding up to the fattest of the herd, horseback, and shooting it with an arrow. When a large party of hunters are engaged the spectacle is very imposing, and the young men have many opportunities of displaying their skill and agility. The horses appear to enjoy the sport as much as their riders, and are very active in eluding the shock of the animal, should it turn on its pursuer. The most generally practiced plan, however, of shooting the bison, is by crawling towards them from the leeward; and in favorable situations great numbers are taken in pounds. When the bison runs, it leans very much first to one side for a short space of time, and then to the other, and so alternately. The flesh of a bison, in good condition, is very juicy and well flavored, much resembling that of well fed beef. The tongue is considered a delicacy, and may be cured so as to surpass in flavor the tongue of the common cow. The hump of flesh is also tender and of fine flavor. The fine wool which clothes the bison, renders its skin, when properly dressed, an excellent blanket; and they are valued so highly that a good one sells from five to ten dollars, in Canada especially, where they are used as wrappers by those who travel over the snow in carioles. The wool, or fine hair, has also sometimes been manufactured into cloth.

Godman, in quoting from Long's expedition to the Rocky Mountains, gives some interesting facts of the habits of this animal, particularly of his sense of smelling. The exploring party, he remarks, were riding through a dreary and uninteresting country, which at that time was enlivened by vast numbers of bisons, who were moving, in countless thousands, in every direction. As the wind was blowing fresh from the south, the scent of the party was wafted directly across the river Platte, and through a distance of eight miles, every step of its progress was distinctly marked by the terror and consternation it produced among the bisons. The instant their atmosphere was infected by the tainted gale, they ran as violently as if closely pursued by mounted hunters, and instead of fleeing from danger, they turned their heads towards the wind, eager to escape this terrifying odor. They dashed obliquely forward towards the party, and plunging into the river, swam, waded, and ran with headlong violence, in several instances breaking through the expedition's line of march, which was immediately along the left branch of the Platte. One of the party, perceiving from the direction taken by the bull who led the extended column, that he would emerge from the low river bottom at a point where the precipitous bank was deeply worn by much travelling, urged his horse rapidly forward, that he might reach this station, in order to gain a nearer view of these interesting animals. He had but just reached the spot, when the formidable leader, bounding up the steep, gained the summit of the bank with his fore feet, and in this position, suddenly halted from his full career, and fiercely glared at the horse which stood full in his path. The horse was panic struck by this sudden apparition, trembled violently from fear, and would have wheeled and taken to flight had not his rider exerted his utmost strength to restrain him; he recoiled, however, a few feet, and sunk down upon his hams. The bison halted for a moment, but urged forward by the irresistible pressure of the moving column behind, he rushed onward by the half sitting horse. The herd then came swiftly on, crowding up the narrow defile. The party had now reached the spot, and extended along a considerable distance; the bisons ran in a confused manner, in various directions, to gain the distant bluffs, and numbers were compelled to pass through the line of march. This scene, added to the plunging and roaring of those who were crossing the river, produced a grand effect, that was heightened by the fire opened on them by the hunters.

The herds of bisons wander over the country in search of food, usually led by a bull most remarkable for strength and fierceness. While feeding, they are often scattered over a great extent of country, but when they move in mass, they form a dense and almost impenetrable column, which, once in motion, is scarcely to be impeded. Their line of march is seldom interrupted even by considerable rivers, across which they swim without fear or hesitation, nearly in the order they traverse the plains. When flying before their pursuers it would be vain for the foremost to halt, or attempt to obstruct the progress of the main body, as the throng in the rear is still rushing onward, the leaders must advance, although destruction awaits the movement. The Indians take advantage of this circumstance to destroy great quantities of this favorite game, and certainly no mode could be resorted to more effectually destructive, nor could a more terrible devastation be produced than that of forcing a numerous herd of these large animals, to leap together from the brink of a dreadful precipice, upon a rocky and broken surface, a hundred feet below.

When the Indians determine to destroy bisons in this way, one of the swiftest footed and most active young men is selected, who is disguised in a bison skin, having the head, ears, and horns adjusted to his own head, so as to make the deception very complete, and thus accoutred, he stations himself between the bison herd and some of the precipices, that often extend for several miles along the rivers. The Indians surround the herd as nearly as possible, when, at a given signal, they show themselves, and rush forward with loud shouting and tumultuous noise. The animals being alarmed, and seeing no way open but in the direction of the disguised Indian, run towards him, and he, taking to flight, dashes on to the precipice, where he suddenly secures himself in some previously ascertained crevice. The foremost of the herd arrives at the brink—there is no possibility of retreat, no chance of escape ; the foremost may for an instant shrink with terror, but the crowd behind, who are terrified by the approaching hunters, rush forward with increasing impetuosity, and the aggregated force hurls them successively into the gulf, where certain death awaits them.

A better and more common way of killing bisons is that of attacking them on horseback. The Indians, mounted and well armed with bows and arrows, encircle the herd, and gradually drive them into a situation favorable to the employment of the horse. They then ride in and single out one, generally a female, and follow her as closely as possible, wound her with arrows until the mortal blow is given, when they go in pursuit of others until their quivers are exhausted. Should a wounded bison attack the hunter, he escapes by the agility of his horse, which is usually well trained for the purpose. In some parts of the country the hunter is exposed to a considerable danger of falling, in consequence of the numerous holes made in the plains by the badger. When the hunting is ended and a sufficiency of game is killed, the squaws come up from the rear to skin and dress the meat, a business in which they have acquired a great degree of dexterity, as they can, with inferior instruments, butcher a bison with far more celerity and precision, than the white hunters.

When the ice is breaking up on the rivers in the spring of the year, the dry grass of the surrounding plains is set on fire, and the bisons are tempted to cross the river in search of the young grass that immediately suceeeds the burning of the old. In the attempt to cross, the bison is often insulated on a large cake of ice that floats down the river. The savages select the most favorable points for attack, and as the bison approaches, the Indians leap with wonderful agility over the frozen ice, to attack him ; and as the animal is necessarily unsteady, and his footing very insecure on the ice, he soon receives his death wound, and is drawn triumphantly to the shore. Sometimes a pound is made by the Indians, a fence being raised in a circular form, of about one hundred yards in diameter. The entrance is banked up so high as to prevent the animals from escaping after once having entered. Various devices are practiced, first to beguile, and then to force them into it. On these occasions, the mounted hunters display the greatest degree of dexterity, as they are obliged to manœuvre around the herd in the plains to urge them into the path that has been prepared for them; and so soon as this is well accomplished, the Indians raise loud shouts, and pressing closely on them, terrifying them so much, they rush, heedlessly towards the snare prepared for them. No sooner are they thus enclosed, than they are quickly dispatched by guns and arrows.

Allusion has been made to the great numbers of these animals which were formerly known to live together. They have been seen in herds of three, four, and five thousand, blackening the plains as far as the eye could penetrate. Some travelers have asserted, that they have seen as many as eight or ten thousand in the same herd, but from the very nature of the case, any opinion based upon a hasty glance of such a moving multitude, must necessarily be subject to great modifications. It would be difficult to imagine any data or process of calculation, by which it could be decided, with apparent certainty, whether there were five or ten thousand in such a herd. In the neighborhood of one of these herds, at night, it has been found nearly impossible for persons, not accustomed to the noise, to sleep; for it has been said, the incessant lowing and roaring of the bulls, very much resembled distant thunder. Although frequent battles take place between the bulls as among domestic cattle, the habits of the bison are peaceful and inoffensive, seldom or never offering to attack man or other animals, unless first attacked or excited.

It is well known that the Indians are accustomed to set fire to the long grass of the prairies, which spreads with fearful rapidity on all sides. The wild horses, bisons, elks, deer, and other animals, conscious of the danger, fly from the approaching flame in the greatest terror. Sometimes they are encircled in the flames, and burnt to death, and sometimes are slain in attempting to make their escape by the Indians, who lie in wait for them. Such scenes present a terrific wildness, far surpassing any effort of the imagination. At the same moment may be seen thousands of acres in one vivid sheet of rising flame and smoke, rolling and curling like the waves of a sea in conflagration. The whole driven forward from one extremity to the other. At one point may be seen a company of emigrants terror-stricken, and struggling with inexpressible ardor to make escape ; at another, a few hunters, with lightning speed, are seen forcing their way from fierce destruction to a place of safety ; and if, with less thrilling interest to the spectator, with no less of frightful grandeur, while the crackling of burning grass and rushes sends up its deafening noise, the ground beneath seems to tremble and to rumble as in agony from an earthquake, as the thousands of these frantic creatures emerge from the smoke and falling cinders, bellowing, and foaming, and snorting, as if the world itself were quickly to become enveloped in the devouring elements.

The object of the present article is to give the reader some of the extremes of peculiarity in the ox family, to satisfy the reader how much change in animals of the same species can be effected by external circumstances. Thus, if the buffalo of tropical regions, and the musk ox of the polar regions, are indebted to these circumstances for all the dissimilarity between them ; and if the difference from each in the several varieties of the domesticated ox is produced in the same way, it shows how it is in our power to vary, and consequently, improve the existing breeds of these domesticated quadrupeds, to almost any indeffinite degree. Some writers have been disposed to assign to the musk ox a genius between that of the ox family and the sheep ; but popular usage has not recognized this distinction, as the animal has invariably, so far as we know, been represented under the name here given it. The form of it will be seen from

MUSK OX.

the accompanying cut ; and the hair with which it is covered resembles that of the bison, and so much resembling coarse wool that it might, if found in sufficient quantities, be manufactured into cloth.

The musk ox inhabits the barren lands of America, lying to the north of the sixtieth degree of north latitude. Only a few indications have been discovered further south than that parallel. Although not latterly found there alive, their sculls and horns have occasionally been discovered near the Great Slave Lake ; and hence the presumption is that it might once have been scattered over the whole country lying between that great sheet of water and the Polar sea. They have been said to range over the islands which lie to the north of the American continent as far as Melville Island, in latitude seventy-five ; but they do not, like the Reindeer, extend to Greenland, Spitsbergen, or Lapland. And, it has been represented by the Indians, that west of the Rocky Mountains which skirt the Mackenzie river, is an extensive tract of barren country, which is also inhabited by the musk ox and reindeer. If so, it must be known to the Russian traders ; but it is probable, that owing to the greater mildness of the climate to the westward of the Rocky Mountains, the musk ox which seeks a cold barren district, where grass is replaced by lichens, does not range so far to the southward on the Pacific coast, as it does on the shores of Hudson's Bay. It is not known that this animal has ever crossed over to the eastern continent, and, evidently, it does not exist in Siberia.

The districts inhabited by the musk ox are the lands of the Esquimaux, and neither the northern Indians, nor the Crees, have an original name for it, both terming it bison, with an additional epithet. The country frequented by the musk ox is mostly rocky and destitute of wood, except on the banks of large rivers, which are generally more or lesss thickly clothed with spruce trees. Their food is grass at one season, and lichens at the other ; and the contents of its paunch are eaten by the natives with great avidity. When the animal is fat, its flesh is well tasted, and resembles that of the caribou, or reindeer, but has a coarser grain. The flesh of the bulls is high flavored, and both bulls and cows, when lean, smell strongly of musk ; their flesh at the same time being very dark and tough, and certainly far inferior to that of any other ruminating animal existing in North America. The carcass of the musk ox weighs, exclusive of the offal, about three hundred weight.

Notwithstanding the shortness of the legs of the musk ox, it runs fast, and it climbs hills and rocks with grest ease. One pursued on the banks of the copper-mine, scaled a lofty sand cliff, having so great a declivity that the pursuers were obliged to crawl on their hands and knees to follow it. Its foot marks are very similar to those of the caribou, but are rather longer and narrower. These oxen assemble in herds of from twenty to thirty, and the females bring forth one calf about the end of May, or the beginning of June. Hearne, from the circumstance of two bulls not being seen, supposed that they kill each in their contests for the cows. If the hunters keep themselves concealed when they fire upon a herd of musk oxen, the poor animals apparently mistake the noise for thunder, and crowd nearer and nearer together as their companions fall around them; but should they discover their enemies by sight, or by their sense of smell, which is very acute, the whole herd seek for safety by instant flight. The bulls, however, are very irascible, and, particularly when wounded, will often attack the hunter and endanger his life, unless he possesses both activity and presence of mind. The Esquimaux, who are accustomed to the pursuit of this animal, sometimes turn its inimitable disposition to good account; for an expert hunter, having provoked a bull to attack him, wheels round it more quietly than it can turn, and by repeated stabs in the belly, puts an end to its life.

THE INDIAN OX.

In Wright's Natural History may be found an interesting account of the Indian ox, or zebu, as it is sometimes called.

From that account the following abstract is made. He says there can be little doubt that it is merely a variety of the common ox, although it is difficult to ascertain the causes by which the distinctive characters of the two races have been, in the process of time, gradually produced. But whatever the causes may have been, their effects rapidly disappear by the intermixture of the breeds, and are entirely lost at the end of a few generations. The intermixture, and its results, would alone furnish a sufficient proof of identity of origin; which, consequently, scarcely requires the confirmation to be derived from the perfect agreement of their internal structure, and of all the more essential particulars of their external conformation. These, however, are not wanting—not only is their anatomical structure the same, but the form of their heads, which affords the only certain means of distinguishing the actual species of this genus from each other, presents no difference whatever. In both, the forehead is flat, or more properly, slightly depressed; nearly square in its outlines, its height being equal to its breadth; and bounded above by a prominent line, forming an angular protuberance, passing directly across the skull between the basis of the horns. The only circumstances, in fact, in which the two animals differ, consist in a fatty hump on the shoulders of the zebu, and in the somewhat more slender and delicate make of its legs.

Numerous breeds of this humped variety of the ox family are spread, more or less extensively, over the whole of Southern Asia, the islands of the Indian Archipelago, and the eastern coast of Africa, from Abyssinia to the cape of Good Hope. In all these countries, the zebu supplies the place of the ox, both as a beast of burthen, and as an article of food and domestic economy. In some parts of India it executes the duties of the horse also, being either saddled and ridden, or harnessed to a carriage, and performing, in this manner, journeys of considerable length, with tolerable celerity. Some of the older writers speak of fifty or sixty miles a day as its usual rate of traveling; but the more moderate computation of recent authors does not exceed twenty or thirty miles. Its beef is considered by no means despicable, although far from equalling that of the well fatted common ox. As might naturally be expected from its perfect domestication and wide diffusion, the zebu is subject to as great a variety of colors as those which affect that with which we are familiar. Its most common hue is that of a light ashy gray, passing into a cream color, or milk white ; but it is not unfrequently marked with various shades of red or brown, and occasionally it becomes perfectly black. Its hump is sometimes elevated in a remarkable degree, and usually retains its upright position ; but sometimes it becomes half pendulous, and hangs partly over towards one side. Instances are cited, in which it had attained the enormous weight of fifty pounds. A distinct breed is spoken of, as common in Surat, which is furnished with a second hump. Among the other breeds, there are some which are entirely destitute of horns, and others which have only the semblance of them, the external covering being unsupported by bony processes, and being consequently flexible and pendulous. One is described by the author, belonging to the London Zoological Society, larger than our common oxen, of a slaty gray color on the body and head, and cream colored legs and dewlap, the latter long and pendulous.

In the Galla country there is a race of zebu cattle, generally of white color, high on the limbs, with a small hump, but on the contrary, with horns of great bulk and length, and sweeping upwards. In Barnou there is a very large white race, with immense horns, which first bend downwards and then turn upwards with half a spiral revolution. According to Clapperton, the corneous external coat of these horns is very soft, distinctly fibrous, and at the base not much thicker than a human nail. The bony core is very cellular, and so light that a pair together scarcely weigh more than four pounds. The following are the dimensions of one of these horns ; length measured on the curve, three feet seven inches ; circumference at the base, two feet ; length in a straight line from base to tip, one foot five inches and a half. From the earliest period this zebu race has been an object of veneration in India ; carved delineations of it are in their oldest temples. Some of these delineations must have been made at least two thousand years before the Christian era; and they are found, as well as similar delineations of the western race of cattle, on the monuments of Egypt, where Apis was worshipped in the form of a bull, of a black color, and where snow white bulls were offered in sacrifice. Rossellini figures a Brahmin or Zebu bull, with a rope round its neck, from which a flower is suspended; probably as if just to be offered in sacrifice.

The large zebu, or Brahmin bull, says Martin, is certainly a noble animal, and much more active than any of our breeds. Olearius describes the procession of an Indian prince, who was drawn in a carriage harnessed to two white oxen, which were as lively and active as horses. Bishop Heber observes that Thakoors, the nobility of the Rajpoots, generally travel in covered wagons drawn by white oxen, whose horns they gild. By some tribes these cattle are bred on a most extensive scale; and, according to Col. Sykes, an army rarely moves in the field without 15,000 or 20,000 bullocks to carry the grain.

It is a remarkable fact, that although the ox is decidedly herbiverous, yet in some countries it is fed, during part of the year at least, on a proportion of animal food. In Norway, for example, as stated in the Farmer's Library, the herds and flocks are driven to the mountains and are there depastured; but, during the long winter they are housed and fed partially on the hay grown within the intermediate precincts of the farm, and brought from the hills, and more plentifully on a kind of food which, to our farmers must appear very strange, if not disgusting, but which it is said the cattle relish very much. This food consists of a thick gelatinous soup, made by boiling the heads of fish, and mixing horse dung with the broth. The boat of the farmer in Norway, supplies not only himself and family with the staple portion of his winter subsistence, but his cows also. That the odor of animal substances ordinarily used for human food is not offensive to the ox family, is evident from what we have often witnessed. It is no uncommon thing that the fat water in which meat has been boiled for the family, and is supplied with vegetable fragments from the culinary department, so as to make a rich broth, is given to the cows. They become fond of it; and it contributes greatly to the increase of their milk. And, it is affirmed by Peall, to be a common practice in some parts of India, to mix animal substances with the grain given to feeble horses, and to boil the mixture in a sort of paste, which soon brings them into good condition and restores their vigor. Likewise, Anderson relates in his history of Ireland, that in some parts of that country, the inhabitants feed their horses with dried fish, when the cold is very intense, and that these animals are extremely vigorous, although small.

Allusion has been made to the value of the ox tribe to the human family; and, consequently, to the presumption that it was domesticated at a very early period of the world. An animal, evidently designed to contribute so much to the wants of mankind, could not for a long time have been overlooked or neglected. Human instinct, or reason, or both, have generally led to an acquaintance with objects, whether in the animal and vegetable kingdoms, susceptible of becoming the staples in sustaining human life, and in the amelioration of human toil. The face of the earth, being smitten by a curse in consequence of the introduction of sin, is, comparatively, infertile, save in the production of noxious weeds, unless reclaimed from its wildness. It is essentially so with the best of our animals. The ox tribe, till tamed and improved, is generally terrific in its appearance and habits; and, if in the least degree contributing to the legitimate end of its existence, has certainly no high claim to consideration. Its value is the result of its domestication, and is proportioned to the care and skill exercised in being thus brought under human supervision. Hence, it is doubted, if the world has ever recognised more decided benefactors than among the ranks of those who have so greatly improved the character of the ox and the cow. It is not easy to make a comparison between the commercial value of such an agency, and an agency in scientific discoveries and mechanical inventions, now revolutionizing the business and the social order of the world. This latter agency may lead to wealth and magnificence; but, the latter sends forth its ministrations of mercy to the poor and destitute, in supplying food and labor upon every farm and hamlet where these animals are kept.

More, it is believed, has been accomplished in Great Britain, than in any other country, towards the beneficial results of which we are speaking. In that country there is a passion for rearing farm animals of great excellence. The business is one of patriotism, of great pecuniary emolument, and even of personal reputation. The nobility smile upon it; and not a few of them are disposed to engage in it. Men of enterprise and capital seek it as affording opportunity for profitable investment. Nor have crowned heads refrained from giving it patronage.

The nation—landlords and tenants—merchants and laborers—princes and peasants seem to feel a common pride on the subject. It is presumed more capital has been invested, and more physiological science has been brought into requisition in that honored country, to develope the capabilities and to improve the breeds of farm animals, than in all the rest of the world together. British cattle have acquired as much of a distinctive character as British manufactures. The one as well as the other becomes a matter of deep interest with the traveller from foreign countries. The one as well as the other is an article of remunerative export. And, in our own country it is diffusing itself under auspices that will, in the end, greatly augment the profits in this branch of rural economy. Supposing there are in Great Britain 25,000,000 of horned cattle, and that each one of them is worth the small sum of ten dollars more than it would be had the old breeds been continued, it is apparent that the few stock breeders have increased the mercantile value of these animals there to the amount of two hundred and fifty millions of dollars above what it would have been had these improvements not been made! And this increased value of these animals pertains to every successive generation of them—say once in every five years! And, supposing there are six millions of cows in that country; and that each one is annually worth twenty dollars more than the old breeds would have been, there is at present an annual profit from British cows of one hundred and twenty millions of dollars above what there would be, had Bakewell and other stock breeders never exerted their efforts to improve these useful animals!

It is but little realized by the mass of our farmers how much they might augment the profits of agriculture if they were to make the same effort as made in Great Britain to improve the different farm animals which they rear. The fact is not to be disguised, that most of these animals are but a necessary evil. Farmers cannot live without them; still they are not enriched by them; in many cases they are impoverished by them. How many cattle are constantly being reared, that could not, on having attained their full stature, be sold for enough to pay the expenses of rearing them! How many cows are kept that do not yield milk sufficient to pay the cost of their feed! How many oxen, whose labor is not equal to what they eat! If farmers were, like merchants, to keep a regular account of what they pay out and what they receive each year, and at the end of each year estimate the value of everything on hand, they would soon perceive where they suffer loss, or find profit. Thus, let stock be charged with all they eat—grass, hay, and other feed, the same as if it were sold to a neighbor; then, in return, let the oxen be credited for the labor they perform and the cows for the milk they give—all at a fair cash price—and it will be readily understood whether there is gain or loss in keeping them. If there is loss, it is occasioned by mismanagement of some sort or other—either in defect in the breed of the animals, or in the manner of feeding. Good animals, well fed, yield a profit. This has been demonstrated. Merchants trafic in goods that yield profit; and preference is given to those which pay the best profits. Manufacturers would be very stupid were they to invest their capital in the production of articles which cannot be sold for enough to pay for the stock and labor used in the manufacture. Farmers should, in all cases, pursue the same policy. There is no more reason, or necessity, for them to spend time and money in tilling the ground without a fair remunerative advance upon the outlay, than there is for merchants to do it in buying and selling goods. Such a principle applied to farm animals, horned cattle, perhaps, in particular, will be attended with the best results.

It does not come within the scope of this work, either to give an elaborate essay on the particular merits of the different breeds of the bovine genus, or to define the processes of improving, rearing, using, and fattening them. For this, we refer the reader to Allen's popular work on Domestic Animals, and to the more extended one of Youatt & Martin on Cattle, edited and improved by Stevens. These works should be in the house of every agriculturist. As familiar as he may suppose himself with such subjects, he may be assured he will learn enough that he did not know before from these books, to pay the cost of them many times over. Our aim is to give hints and isolated facts that will awaken attention to this branch of farm economy; and, by thus creating an interest in it, induce a more careful and general attention to it. Among the breeds imported from Great Britain to this country, and becoming favorites with the best American farmers, are the Durhams, or Short Horns; the Devons; the Herefords; and the Ayrshires; each of these having its admirers; and either kept as a distinct variety, or crossed with our native cattle, or crossed with each other, will do much towards the end we here recommend. It is better that these different breeds should be adopted, than that any particular one should be made the basis for an improved race with us of this valuable animal, inasmuch as they naturally lead to a competition among stock amateurs not otherwise to be expected —a competition laudable in itself, and tending to a perfection in the science of stock breeding, not to be realized without it. The success attending the efforts thus made has generally been satisfactory, and, in some instances, far beyond what was anticipated, even by the most sanguine; and, if it could, in all its details be spread, in tracts, or through the periodical press, before the public liable to be interested in such matters, the business would receive an impulse far more general and intense than hitherto witnessed.

DURHAM PREMIUM COW,

At the Fair of the New-York State Agricultural Society, held at Rochester, in 1851.

The accompanying cut of the Durham cow will give the reader a good idea of the general appearance of the breed to which she belongs. The following description of it, with the sketches of its history, are mostly copied from the treatise on Domestic Animals before named. The Short Horns are assigned a high antiquity by the oldest breeders in the counties of Durham and Yorkshire, England, the place of their origin, and for a long time almost of their exclusive breeding. They are decidedly the most showy among the cattle species. They are of all colors, between a full deep red, and a pure creamy white; but generally have both intermixed in larger or smaller patches, or intimately blended in a beautiful roan. Black, brown, or brindled, are colors not recognized among the pure bred Durhams. Their form is well spread, symmetrical and imposing, and capable of sustaining a large weight of valuable carcass. The horn was originally branching and turned upward, but now frequently has a downward tendency, with the tips pointing towards each other. They are light and comparatively short; clear, highly polished, and waxy. The head and neck are finely formed, and there is great breadth and depth of chest, giving short and well spread fore legs. The crops are good; back and loin broad and flat; ribs projecting; deep flank; and tail well set up, strong at the roots, and tapering. They have a thick covering of soft hair, and are mellow or loose to the touch. They mature early and rapidly for the quantity of food consumed, yielding largely of good beef, with little offal. As a breed, they are excellent milkers, though some families of them in this respect surpass others.

It is claimed that importations of the Durham cattle have been made into this country prior to 1783. They are the reputed ancestors of many choice animals existing in Virginia, in the latter part of the last century, and were known as the milk breed; and some of these, with others, termed beef breed, were taken into Kentucky by Mr. Patton as early as 1797, and their descendants, a valuable race of animals, were much disseminated in the west, and known as the Patton stock. In 1791, and afterwards, it is known that the short horned cattle were imported by Mr. Heaton, of Westchester county, N. Y. These, for many years, were bred pure, and the progeny was widely scattered. Other importations were made into New-York, in 1816, by Mr. Cox; by Mr. Bullock in 1822; by the Hon. S. Van Rensselaer in 1823; and soon afterwards, by Charles Henry Hall. Between the years 1817 and 1825, small importations were made into Massachusetts, by Coolidge, Williams and others; into Connecticut by Mr. Hall and others; into Pennsylvania by Mr. Powell; and into Ohio and some other States, by various individuals early in the present century. Since the first importations, large accessions from the best English herds have been frequently made; so, that at the present time, the Short Horns in the United States have become an extensive breed; a portion of it in purity, although in many cases mixed.

During the speculation time between 1835 and 1840, they

commanded high prices, frequently from five hundred to a thousand dollars. The succeeding years of financial embarrasment, reduced their market price below their intrinsic value; but the tide is again turning, and they are now in demand, but still at prices far below their merits. They have, from the first, been favorites in the rich corn valleys of the west, their early maturity and great weight giving a preference over other breeds. On light lands and scanty pastures, they will probably never be largely introduced. All heavy animals require full forage within limited compass, so as to fill their stomachs at once, and quietly compose themselves. The weight reached by the Short Horns in England, as given by Mr. Berry, have been enormous. Two oxen, six years old, weighed nett, eighteen hundred and twenty pounds each. A heifer of three years old, fed on grass and hay alone, weighed twelve hundred and sixty pounds. A four year old steer, fed on hay and turnips only, dressed eighteen hundred and ninety pounds. A cow reached the prodigious weight of seventeen hundred and seventy-eight pounds. A heifer, running with her dam, and on pasture alone, weighed, at seven months, four hundred and seventy-six pounds. An ox, seven years old, weighed twenty-three hundred and sixty-two pounds. From their comparative small number in this country, most of them have been retained for breeders; few, as yet, have been fattened, and such only as were decidedly inferior. Such animals, extensively introduced by crossing this breed upon our former stocks, have given evidence of great and decided improvement; and the Short Horns, and their grade descendants are destined, at no distant day, to occupy a large portion of the richest feeding grounds in the United States.

The accompanying cut of the Hereford represents an animal of great beauty and extraordinary weight, exhibited at the Shrewsbury Cattle Show, England, October, 1851. The likeness is copied from the London Illustrated Sun; and at the time of exhibition the ox was only three years and ten months old. Marshall says, of this breed of cattle, the following are the usual characteristics:—The countenance pleasant, cheerful, open; the

HEREFORD OX.

forehead broad; eye full and lively; horns bright, taper and spreading; head small; neck long and tapering; chest deep and full; bosom broad and projecting; shoulder-bone thin, flat, no way protuberant in bone, but full and mellow in flesh; loin broad; hips standing wide and level with the spine; quarters long and wide at the neck; rump even with the general level of the back, not drooping, nor standing high above the quarters; ribs broad, standing close and flat upon the outer surface, forming a smooth even barrel, the hindmost large and full of length; thigh clean, and regularly tapering; flank large; flesh every where mellow, soft, yielding pleasantly to the touch, especially on the chine and shoulder, and the ribs; coat neatly haired, bright, and silky. The color of the Hereford cattle is usually a darkish red—sometimes a dark yellow—a few brindled; generally having white faces, bellies and throats, and now and then white on the back. The old Herefords were brown, or red brown, with not a white spot about them. It is only within fifty or sixty years that it has been the fashion to breed white faces. Were it not for the white face, and somewhat larger head and thicker neck, it would not, at all times, be easy to distinguish between a heavy Devon and light Hereford.

Hereford oxen fatten speedily at an early age, and it is more advantageous to the farmer, and perhaps to the country, that they should be taken to the market early, than to be kept long employed as beasts of draught. The Hereford cow is comparatively an inferior animal. Not only is she inferior as a milker, but even her form has been sacrificed by her breeder. Herefordshire is more a rearing than a feeding county, and, therefore, the farmer looks mostly to the shape and value of his young stock; and, in the choice of his cow, he does not value her, or select her, or breed from her, according to her milking qualities, or the price which the grazier would give for her, but in proportion as she possesses that general form which experience has taught him will render her likely to produce a good ox. Hence the Hereford cow is comparatively small and delicate, and some would call her ill-made. She is light in flesh when in common condition, and beyond that, while she is breeding, she is not suffered to proceed; but when she is actually put up for fattening, she spreads out, and accumulates fat at a most extraordinary rate. A middling cow of this breed will weigh from seven to ten hundred pounds; and the Duke of Bedford had one, says Mr. Youatt, which weighed fourteen hundred pounds.

The county of Ayrshire, in Scotland, has a peculiarly fine breed of cattle. The climate of this county is moist, but mild; and the soil, with its produce, is calculated to render it the finest dairy locality in Scotland, and equal, perhaps, to any in Great Britain. There is a great deal of permanent pasture on the sides and tops of the hills; but the greater part of the arable land is pasture, and crop alternately. The pasture ground is occupied by the beautiful dairy stock, a very small portion being reserved for the fattening of cows too old to milk. The origin of the Ayrshire cow is at the present day a matter of dispute; all that is certainly known is, that a century ago there was no such breed, even in Scotland. Did the Ayrshire cattle arise entirely from a careful selection of the best native breed? If they did, says Mr. Youatt, it is a circumstance unparalleled in the history of agriculture. The native breed may be ameliorated by careful selection; its value may be incalculably increased; some good qualities, some of its best qualities, may be for the first time developed; but yet there will be some resemblance to the original stock; and the more the animal is examined, the more clearly the characteristic points of the ancestor may be discerned, although each one of them is improved. Their great value is as milkers, and perhaps that principally in their own native territory, to the feed and climate of which they seem to be constitutionally adapted. Nevertheless, in rich lands, the oxen fatten with considerable facility, and even the cows accumulate flesh; but then they cease to be distinguished for a large yield of milk.

AYRSHIRE COW.

The improved Ayrshire cow of the present day has a small head, but rather long and narrow at the muzzle, though the space between the roots of the horns is considerable; the horns are small and crooked; the eye clear and lively; the neck long and slender; the shoulders are thin, and the fore quarters generally high; the back is straight, and broad behind, especially across the hips, which are roomy; and the tail is long and thin.

The carcass is deep, the udder capacious and square, the milk veins large and prominent; the limbs are small and short, but well knit; the thighs are thin; and the skin is rather thin, but loose and soft, and covered with soft hair. The general figure, though small, is well proportioned. The color is varied with mingled white and sandy red. It has been estimated that there are in the county of Ayrshire upwards of sixty thousand head of cattle, of which more than half are dairy cows. It has been calculated that one of these cows, of the best grade, will yield, for two or three months after calving, five gallons of milk daily; for the next three months, three gallons daily; and a gallon and a half for the following three months. This milk is supposed to return about two hundred and fifty pounds of butter annually, or five hundred pounds of cheese. Mr. Youatt thinks the foregoing account rather exaggerated, although he gives the statistics of a dairy with which he was acquainted, which fall but little below those here given. Mr. Aiton, an ardent advocate of this breed of cattle, says that there are thousands of the best Scotch dairy cows, when they are in their best condition and well fed, yield at the rate of one thousand gallons of milk in one year. The most valuable qualities which a dairy cow can possess are, that she yields much milk; and that after she has done this for several years, she shall be as valuable for beef as any other breed of cows known.

The Ayrshire cow in this country has some devoted advocates. The Massachusetts Agricultural Society has especially encouraged their dissemination. Mr. Webster has them on his farm in Marshfield, and some of the most wealthy and enterprising citizens of his, as well as of other states, have made trial of them. The editor of the Albany Cultivator, whose opinions are always to be respected, thinks favorably of the animal, as adapted to the wants of the American farmer. Mr. Colman remarked, in one of his agricultural reports, that the Ayrshire cows were generally deemed the best dairy stock in Great Britain; and then added, that although his own experience had been too limited to speak emphatically of a statement so strong, their reputation was such, he was not at liberty to subtract from it. And he mentions that he visited a dairy farmer in Ayrshire who had thirty-five cows of this beautiful breed; the best of them, in the best of the season, gave fifty-four pounds of milk per day, which, allowing a pint to weigh a pound, would be twenty-seven quarts a day. The average yield of this dairy was forty pounds, or twenty quarts per day. However, he met in that county with a North Devon cow which, for several weeks in succession, produced twenty-one pounds of butter per week. He also stated, that the North Devon cows of Mr. Bloomfield produced, on an average, upwards of two hundred pounds of butter each, for several years in succession. E. Prentice, Esq., of Mount Hope, a few years since had a small cow of the Ayrshire breed which frequently gave over twenty quarts of milk a day, when fed on grass only.

The following remarks of Mr. Coleman on the improvement of cattle in the United States, are sensible, and well adapted to the consideration of American farmers. He says, which we all know, that our stock is of a mixed and very miscellaneous character. Comparatively few attempts have been made in a systematic manner, and upon an extended scale, for its improvement. Where they have been made, they have frequently failed for want of perseverance, very often for want of encouragement, and have sometimes been met by the sneers of ignorance or the derision of envy. The immense improvements, which have been made in Great Britain strike the observer with grateful astonishment. Few subjects more concern the interest of the American husbandman than the improvement of our stock. Much may, undoubtedly, be done by the selection of the best from our own breeds, and by breeding only from the best; but our stock is so crossed, and mixed up, and amalgamated, that it would be a difficult process to unravel the web, and go back to any original breed. We should certainly avail ourselves of the breeds existing in the highest state of improvement.

The writer further remarked: It is plain that in a selection of breeds regard should be had to the locality where they are to be placed. The improved Short Horns, the Yorkshire, and the Hereford are the best adapted to the rich and deep pastures of the Middle and Western States ; the Ayrshire and the North Devon seem especially suited to New England ; while the West Highland cattle would evidently be fitted to the cold and least productive parts of the country. With us the success of farming must mainly depend on such a conduct of the farm as shall not exhaust its productive powers ; or rather, it shall, from its own resources, furnish the means, not only of recruiting its strength, but of actually increasing its capabilities of production. There is no more obvious way of doing this, than by consuming the produce of the farm, mainly in feeding animals, through whom the riches of its vegetation may be returned in a form to furnish other and better crops. The late Mr. Phinney, of Massachussets expresses himself equally strongly in favor of these animals. There is no better authority. He says the Scotch farmers in and about Ayrshire have been striving for more than half a century to produce a stock in which regard was had almost solely to their value for the dairy ; and there is not perhaps in Europe, nor in the world, a race of cows that will secrete more and better milk from a given quantity of food, that are so hardy, so easily kept, that will fatten so kindly when off their milk, as the Ayrshire.

Mr. Cushing of Watertown, Mass., one of the wealthiest and most intelligent ruralists of the country, has for many years kept cattle of the Ayrshire breed on his farm. He has stated that in the first fourteen days of June, 1848, four of his cows of this breed made over eighty pounds of butter, besides supplying a family of fourteen persons with milk and cream. One of the four made eleven and three quarters of a pound in seven days. In the latter part of September in that year, three of those four cows made thirty pounds of butter in seven days ; or ten pound each in the week

The north of Devon has long been celebrated for a breed of cattle beautiful in the highest degree, and in activity at work and aptitude to fatten unrivalled. The native country of the

DEVON COW.

Devons lies along the Bristol Channel, the breed becoming more mixed in either direction, in receding from certain limits. The Devonshire farmer confines them within a narrow district, and will scarcely allow them to be found with purity beyond his native county. From Portlock to Biddeford, and a little to the north and the south, is, in his opinion, the peculiar and only residence of the true Devon. From the earliest records the breed has here remained the same; or, if not quite as perfect as at the present moment, yet, altered in no essential point until within the last thirty years. The soil and the climate of this district are particularly favorable to the production of these fine animals; and although there was no attempt to improve the breed till the last century, they retained from these natural advantages, and perhaps from other circumstances not known to us, in an extraordinary degree, their distinctive character, and a perfection not to have been expected. Still, in that county as well as in other parts of Great Britian, a spirit of emulation arose on the subject, so these cattle have been materially improved; and it has been affirmed on good authority, that now they sustain so high a grade of excellence they would probably suffer from intermixture with any other breed.

There are few things more remarkable about the Devon cattle than the comparative smallness of the cow. The bull is a great deal less than the ox, and the cow smaller than the bull. The cows, however, although small, possess that roundness and projection of the two or three last ribs which makes them actually more roomy than a careless examination would indicate. The head of the ox is small, very singularly so, relatively to its bulk; yet it has a striking breadth of forehead; and it has a long and thin neck, admirably adapting it for the collar, or as more common for the yoke. Where the ground is not too heavy, the Devons are unrivalled at the plough. They have a quickness of action which no other breed can equal, and very few horses exceed. They have a docility and goodness of temper, and stoutness and honesty of work, to which many horses cannot pretend. The profit derived from the use of these oxen in their native district arises from the activity to which they are trained, seldom reached elsewhere. During the harvest time, and in catching weather, they are sometimes trotted along with the empty waggons, at the rate of six miles an hour, a degree of speed which no other ox but the Devon has been able to stand. The importance of speed in working oxen cannot be estimated too highly. The frequent objection to them among farmers, in ploughing and on the road, is, that they are too slow in their motions. This objection, especially with the Devons, is easily prevented by suitable early training. If oxen can be taught to plough as much, and to travel as far on the road with burdens, as horses, it might be supposed that the latter will generally, for these purposes, be superseded by the former.

The Rev. Dr. Abiel Abbot, of Beverly, Massachusetts, before his death, when in ill health, spent a winter in Cuba, whence he wrote home letters, and which were afterwards published under the title of "Letters from Cuba." In one of these letters he says, that in managing the oxen of that island, the yoke is made fast to the horns, near the roots, behind, so that it does not play backward and forward, and gives to the ox a similar and better chance of backing, as the teamster's phrase is. I have been astonished, said the Doctor, at the power of those oxen in holding back. There is a short hill in one of the streets of the city at an angle of forty-five degrees. Standing at the foot of it, I saw a cart approach at the top, with three hogsheads of molasses, and the driver sitting on the forward cask. The driver did not so much as leave his perch; the oxen went straight and fearlessly over the pitch of the hill, and it seemed as if they must be crushed to death. The animals squatted like dogs, and rather slid than walked to the bottom of the hill. Have we any animals that could have done it? And if they could, have we any docile enough to have done it with the driver in the cart? Thus superior is this mode of yoking in holding back the load in difficult places. And it gives them a more decisive advantage in drawing. A fillet of canvass is laid on the front, below the horns; and over this fillet the cords pass, and the animals press against the most vulnerable part of the frame; his head, his neck, his whole frame is exerted in the very manner in which he exerts his mighty strength in combat. It is the natural way, therefore, of availing yourself of this powerful and patient animal to the best advantage.

The advantage of oxen in farm labor depends much on their discipline. If they are of the right form and spirit, they may be trained to walk as fast as horses, and will do as much at the plough, excepting, perhaps, in the very hottest weather. The late Governor, Isaac Hill, of New Hampshire, who devoted the last ten years of his active life to agriculture, confirms this opinion on the capabilities of the ox. He was accustomed to do all his ploughing with them. We once witnessed how well it was performed. First in the train was a single yoke of oxen without a driver, and with one of the Worcester Eagle ploughs, at quite a rapid step. Then followed, close after, four oxen with a sub-soil plough, scarcely a word being uttered to them, or any urgency whatever applied; and the work was well and quickly done. And there are some oxen that will even stand the heat in the field as well as horses. A few years since, the first premium for ploughing at the fair of the State Agricultural Society of New York was given to a man who used a middling-sized pair of oxen. They did their work quicker and better, it was affirmed, than any other team, although there were several pair of large horses. It was a very warm day, but the oxen were less worried, and were evidently able to perform more in a day, than the horses. Numerous instances of similar testimony might be adduced in favor of selecting the right kind of oxen for the yoke, and then in furnishing them with the proper kind of discipline.

Some four or five years ago, we believe it, there was a report made from the Essex County Agricultural Society, Massachusetts, on the exhibition of ploughing with single teams. J. W. Proctor, Esq., since President of the Society, was chairman of the committee that made the report. On that occasion there were matches with two yoke of oxen as well as with one yoke, and also with a match of horses. The quantity of ground was the same, one-fourth of an acre, in the three matches, but there was but little difference in the time occupied in doing the work—though one of the single teams of oxen ploughed their land some minutes quicker than any of the horse teams. Mr. Proctor's remarks in the report are deserving of particular attention, as showing the capability of oxen in ploughing, and also for a suggestion in reference to the sub-soil plough. From these experiments, he says, we learn that an acre of land may be ploughed by a single pair of cattle and one man in four hours, and probably nearly two acres in a single day. When we take into view the expense of operating a team of this description, compared with those usually employed in this business, it will be quite well for our farmers to consider whether most of their work cannot be done with one pair of cattle, and if two pair are to be used, would it not be better to cut the first furrow of less depth, and apply the power of the second pair to a sub-soil plough to follow directly after? If we do not entirely mistake the signs of the times, our modes of preparing land for culture will ere long be essentially modified by the use of the sub-soil plough. In the county of Worcester, where the management of land and teams is understood as well perhaps as in any part of the country, the premiums in ploughing are confined to one pair of cattle without a driver.

So important do we deem it for the interests of agriculture that more attention be given to working oxen, we adduce the testimony of Edward Stabler, Esq., of Maryland, giving his experience on the subject. He says that he began to use oxen on his farm instead of horses in the year 1822. The first labor was to break up land at midsummer, for a field of wheat. For a day or two the oxen suffered greatly with the heat, in the middle of the day, but by rising early, and resting two or three hours at noon, and feeding on dry food, he was able to plough nearly as much with a yoke of oxen as with a pair of horses, and the work was quite as well done. The horses consumed about one bushel of grain per day, and the oxen none. He found the result, after a thorough trial, so much in favor of oxen, that he has ever since continued their use. For many years there was not a furrow ploughed on his farm except by oxen. He observes that oxen, properly broken, quite as readily, if not more so, take to and keep the furrow, as horses. His rule is to keep two yoke of oxen on the farm to one pair of horses; and, he well remarks, that to judge of the capabilities of the ox by the badly-used, houseless, overtaxed, and half-fed animals we sometimes see in the yoke, is doing him great injustice. Treat the horse in the same unfeeling manner, and where would be his high metal and noble spirit? He would speedily arrive at premature old age, valueless to the owner, and cast off to feed the carrion crows. That the ox can better stand this harsh usage is certainly no valid or sufficient reason that he should be subjected to it. Use him with equal care and humanity, and he will just as certainly, and with more profit, repay it to the owner.

It cannot be doubted that most persons engaged in disseminating improved breeds of cattle in our country, are governed by motives similar to those which actuate men doing other business. They expect to make a profit from it. No one can deny that it is right they should do it. The investment of capital is required. Time is wanted. Yet there is risk in the experiment. Numerous contingences may arise to occasion loss. It is right, therefore, that counter contingences should present themselves, enabling the adventurers to secure a profit proportioned, at least, to the hazards, as in all branches of enterprise. Some of the finest North Devon cattle have been imported into the country, and there are now, in different sections, several established breeders of this stock. The Messrs Hurlbut, of Winchester, Connecticut, are reputed to be the most extensive ones. Ambrose Stevens, of New York, a gentleman of enlarged views and of high attainments as a scholar, has, in his cattle operations, a department devoted to the Devons, said to be next in magnitude to that of the Hurlbuts. Mr. Patterson, of Baltimore, and Mr. Washbon, of Butternutts, N. Y., are also extensively engaged in the same laudable efforts. The breed has succeeded well here, and has proved better adapted to our hilly and unfertile sections of country than larger breeds. They are hardy, generally healthy, thrifty, and active. The best families are fine beef cattle, a very weighty fact, laying their flesh on the valuable parts, and giving that which is of fine grain and well marbled It has been supposed that the common cattle of New England, now called natives, were originally North Devons. The more probable presumption is, that they were South Devons, a variety differing considerably from their northern neighbors, being larger-boned and coarser.

The best farms for making butter are those that lie fair to the sun, where the feed is sweet and of the best quality. Butter made from good, sweet feed will be of good color, and of superior quality to that made from feed from pasturing that lies on the north side of the hill, where the sun shines very little. The land is cold and wet, and the feed is sour and of poor quality, and the butter made from it will be light colored and of inferior quality to that made from good, rich, sweet feed. Dairymen should have plenty of good clear water, where the cows can have free access to it at all times. When the cows are obliged to wade in the mud for water, and drink when there is a scanty supply, and drop their excrements in it, they are obliged to drink an impure mixture, that greatly affects the butter.

Cows should not be allowed to lie in close yards in very warm weather; they should be returned to the pasture, or some convenient place where they have a good place to lie, and fresh air. When cows lie in wet and muddy yards, there will be more or

KENDALL'S CHEESE-PRESS.

less dirt fall from the cows into the milk, while milking, which gives the butter a very unpleasant flavor. All kinds of feed that are of a strong nature, such as turnip and onion tops, and vegetables that have a strong flavor, ought to be avoided, for it is injurious to the flavor of the butter. In the spring every dairyman should feed his cows with a little Indian meal and water every day for two or three weeks before they come into milking, and from that time until they can get a good supply of grass. This not only improves the condition of the cows, but greatly increases the quantity of the butter, and improves its quality.

Dairymen should never undertake to keep more cows than they have plenty of feed for. Twenty cows, well fed, will yield much greater profit than forty poorly kept Every farmer should be very particular to select such cows as give the richest milk, and that which will make good yellow butter. Every one says that it is no more expense to keep good cows than it is poor ones. To have good cows and plenty of good feed; pure water, comfortable barns in winter, where they can be kept dry and warm, and good clean places for them in summer, is the first step towards carrying on the dairy business successfully.

During the year 1850 it appears there were produced in the United States one hundred and thirteen millions of pounds of cheese; and this enormous product was nearly all required to meet the demand for home consumption—the total export amounting to less than nine millions of pounds. By far the largest part of the whole comes from the States of New York and Ohio—the former producing over forty-nine millions and the latter over twenty-one millions of pounds. Massachusetts, Rhode Island, Connecticut, Pennsylvania, Michigan, and Illinois follow next amongst the largest producers. Of the other States, none produce a million of pounds. The quantity produced in the Southern States is very small, in proportion to their population and territorial extent.

Considering the magnitude of the dairy interests in this country—there being one cow to every three or four individuals of our entire population—we close this article, with a few hints more on the milking properties of that useful animal. We cannot possibly do it better than in the language of Mr. Robert Gray, near Fredericton, New Brunswick. The St. John Agricultural Society requested him to state what cows he considered best. "From my own experience in this matter," he replied, "I give a decided preference to Ayrshire cows for the dairy. I believe they will yield a greater quantity of milk, in proportion to the food they consume, than any other breed. Besides this, they are docile and hardy, and will thrive on pasture, and with a description of keep where such breeds as the short-horns would starve. They also possess more than average feeding qualities of their own, and crossed with the short-horn, or Durham Bull, the produce is an animal remarkable for early maturity and a disposition to fatten. If proof were wanted of the excellence of the breed, it would be found in the circumstance that they are carried to almost every quarter of the globe. Large droves are every year taken to England, and during the last ten years considerable numbers have been taken to the Cape, the Isle of France, to Sweden, Denmark, Belgium, and the United States. A knowledge of the points in prime cattle, the peculiarities of the different breeds, and the best modes of rearing, feeding, and fattening them should be a part of every farmer's education".

INDEX.

www.ingramcontent.com/pod-product-compliance
Lightning Source LLC
LaVergne TN
LVHW020928110826
845150LV00004B/798

* 9 7 8 1 4 2 5 5 5 3 3 7 1 *